Animal Biotechnology 1

Heiner Niemann • Christine Wrenzycki
Editors

Animal Biotechnology 1

Reproductive Biotechnologies

 Springer

Editors
Heiner Niemann
Institute of Farm Animal Genetics
Friedrich-Loeffler-Institut (FLI)
Mariensee
Germany

Christine Wrenzycki
Faculty of Veterinary Medicine
Justus-Liebig-University Giessen
Giessen
Germany

ISBN 978-3-030-06407-5   ISBN 978-3-319-92327-7   (eBook)
https://doi.org/10.1007/978-3-319-92327-7

Preface

The domestication of farm animals was a seminal advance that laid the foundation stone for agriculture as it is known today. Compelling evidence is now available that the domestication process started about 10–15,000 years ago at various locations in the world. Choice of breeding stock was initially made by visual selection for specific phenotypes and/or traits, science-based selection only emerging in the sixteenth to nineteenth century with the advance in statistical and genetic knowledge. Progress in selection and propagation of superior genotypes by conventional breeding practices was glacially slow and remained so until the introduction of assisted reproductive technologies (ART), most notably artificial insemination (AI), in the first half of the twentieth century. Artificial insemination remains the most widely used of these technologies and has and continues to play a central role in the dissemination of valuable male genetics around the globe. A means of increasing the rate of propagation of female genomes was only achieved relatively recently with the development of multiple ovulation (MO) and embryo transfer technology (ET) in the immediate post world-war II period. The full potential of MOET is still yet to be realized as it plays a key enabling role in the development of the next generation of technologies, including in vitro production of embryos, somatic cloning and precise genetic modification. Advances in DNA methodology in this century have been truly remarkable. The result is genomic maps now being available for all the major farm animals together with tools that allow precise genome editing at specific genomic loci at even the single base pairs. When combined with ART, this integration of molecular and reproductive technologies has resulted in the development of an impressive range of innovative breeding concepts aimed at improving genetic gain through precise editing of the genome and its rapid dissemination made possible through a dramatically shortened generation interval. In addition to the enormous potential of these advances in agriculture they also open up the prospect of generating new animal products, for example the provision of new models of disease for the health sciences or recombinant pharmaceutical proteins and even regenerative tissue or functional xenografts for medicine. Arguably, the only limit to the scope of animal biotechnology is the human imagination.

However, experience has revealed that translation of these developments into product is not straight-forward; their transformative potential raising many expected and unexpected ethical and legal questions that have already sparked a heated public debate, much of it ill-informed.

This book is designed to provide the reader with the information needed to fully appreciate what is being achieved through the exciting advances in research and application of animal biotechnology with specific focus on the key developments in reproductive and molecular biology that underpin these advances. The book also seeks to address the major issues of concern raised by the public in relation to the social impact of these new methodologies together with the many legal and ethical aspects emerging from this. Gaining a broader public understanding and acceptance of animal biotechnology is seen by the authors as critical to the full realization of the potential of the remarkable scientific advances to address the challenges to food security raised by the ever-accelerating growth in human demand within the production constraints imposed by the diminishing availability of arable land and climate change.

The editors trust that a better appreciation of these technologies and their potential, when applied responsibly, to combat the looming agricultural challenges faced by mankind, will enhance rational debate on these issues.

The editors are extremely grateful to **Susanne Tonks** who provided major assistance in preparation of this book.

Mariensee, Germany   Heiner Niemann
Giessen, Germany   Christine Wrenzycki

Contents

The Evolution of Farm Animal Biotechnology

1

Heiner Niemann and Bob Seamark

Abstract

The domestication of farm animals starting 12,000–15,000 years ago in the Middle East was a seminal achievement in human development that laid the foundation of agriculture as it is known today. Initially, domesticated animals were selected according to phenotype and/or specific traits adapted to a local climate and production system. The science-based breeding systems used today originated with the introduction of statistical methods in the sixteenth century that made possible a quantitative approach to selective breeding for specific targeted traits. Now, with the availability of accurate and reliable DNA analysis, this quantitative approach has been extended to DNA-based breeding concepts that allow a more cost-effective but still quantitative determination of a genomic breeding value (GBV) for individual animals.

The impact of these developments was dramatically enhanced with the introduction of reproductive technologies extending the genetic influence of superior individual animals. The first of these was artificial insemination (AI) that started to be developed in the late nineteenth century. Industry uptake of AI was initially slow but increased dramatically following the development of semen extenders, the reduction of venereal disease risk by inclusion of antibiotics, and most significantly the development of effective freezing and cryostorage procedures in the mid-twentieth century. AI is now used in most livestock breeding enterprises, most notably by the dairy industry where more than 90% of dairy cattle are produced through AI in countries with modern breeding structures.

H. Niemann (✉)
Institute of Farm Animal Genetics (FLI), Neustadt-Mariensee, Germany
e-mail: niemann@tzv.fal.de; Niemann.Heiner@mh-hannover.de

B. Seamark (✉)
Department of Medical Biochemistry, Flinders University, Bedford Park, SA, Australia
e-mail: bob.seamark@flinders.edu.au

© Springer International Publishing AG, part of Springer Nature 2018
H. Niemann, C. Wrenzycki (eds.), *Animal Biotechnology 1*,
https://doi.org/10.1007/978-3-319-92327-7_1

Embryo transfer (ET), a technology that for the first time allowed exploitation of the female genetic pool, was made possible through the major advances in the biological sciences in the later part of the twentieth century. Advances in understanding of the reproductive cycle and its hormonal control, the availability of purified gonadotropins, and improved cell and embryo culture procedures all played significant roles. ET is now being increasingly implemented in top end breeding endeavors, particularly in the top 1–2% of a given cattle population. But its real impact is yet to come as ET is the key enabler in the introduction of the next generation of enhanced breeding technologies. ET has already played a key role in advances such as in vitro production of embryos, sexing, cloning, and transgenesis. With the birth of "Dolly," the cloned sheep, in 1996, a century-old dogma in biology, which inferred that a differentiated cell cannot be reprogrammed into a pluripotent stage, was abolished. Today, through recent developments in molecular cell biology, available protocols are efficient enough to allow commercial application of somatic cloning in the major farm animal species. This will not only further enhance the rate of genetic gain in herds and flocks but through the recent advent of precise genome editing tools allow the production of novel germlines for agricultural and biomedical purposes through the capacity to genetically modify farm animals with targeted modifications with high efficiency. This paves the way for the introduction of the precision breeding concepts needed to respond to future challenges in animal breeding, stemming from matching the demands of ongoing hyperbolic human population growth to the limited availability of arable land and environmental constraints.

1.1   Introduction

The great variety of phenotypes presently seen in domesticated animals is the product of human-directed breeding over many centuries. Compelling evidence of domestication of livestock more than 10,000 years ago is provided by archeological findings showing that milk and dairy products were then already part of a normal human diet. Up to the last century, selection of breeding stock for specific phenotypes or production traits was made by simple observation with science-based quantitation and breeding for specific genotypes only introduced following the development and introduction of statistical methodologies in the late nineteenth century. The accurate and reliable prediction of genetic traits made possible from this introduction revolutionized breeding practices and, together with advances in DNA technology, ultimately led to the quantitative molecular genetic selection procedures used today. The next major advance was the development over the past 50 years, of a growing array of reproductive biotechnologies, most notably artificial insemination (AI) and embryo transfer (ET). The full impact emerging from linking molecular genetics and reproductive technology is yet to be realized. Already one outcome has been that it is now not only possible to precisely and reliably analyze genomes but in an equally precise and reliable way engineer the genome to both enhance desired production traits and introduce novel production traits. Another impressive outcome of this alliance is the development of reliable cloning procedures that utilize somatic cells as the genome source, a major achievement that opens new horizons of possibilities that assures an exciting future for animal breeding enterprises. This chapter covers the cornerstones of the history of animal breeding, from its genesis many thousand years ago to today, with focus on the biotechnological advances that are and will be increasingly employed by livestock breeding enterprises to address the hyperbolical increasing human demand for conventional and novel animal products. Important milestones of this evolutionary process of animal breeding are provided in Table 1.1.

1.2 Evolution of Farm Animal Breeding

1.2.1 From Domestication to Systematic Breeding Concepts for Farm Animals

Domestication of animals was the foundation stone of agriculture as it is known today (Diamond 2002) and a key advance in human development. Classical studies on the historic pathways of domestication, primarily based on archeological evidence, are now being overwritten by a growing body of information provided by DNA studies. Analysis of mitochondrial DNA has been particularly useful in this regard as it is maternally inherited and has various properties, including the lack of recombination, high mutation rates, and the presence of multiple copies (Bradford et al. 2003; McHugh and Bradley 2001). Conjointly these disciplines provide compelling evidence that domestication started around 10.000–15.000 BC, predominantly in the Middle East (Connolly et al. 2011). Archeological findings there and on the British Isles revealed that approx. 14.000–17.000 years ago, humans already kept farm animals and that milk and dairy products were essential parts of their nutrition (Beja-Pereira et al. 2006; Larson et al. 2007). DNA studies of the two main bovine species, taurine and zebu cattle, indicate separate domestications starting ~8000 years BC in Southwestern Asia and the Indus valley (Zeder et al. 2006). The progenitor species was the aurochs (*Bos primigenius*), a tall and well-fortified animal with very long horns, the latter a feature still reflected in most current cattle breeds (Schafberg and Swalve 2015). Domestication of pigs took place independently at predominantly two locations, in East Anatolia and China (Groenen 2016), sheep and goats were domesticated in West and East Asia, and horses stem from the Eurasian steppes (Wang et al. 2014).

The rich variety of geno- and phenotypes in farm animals now extant is the product of man-made breeding over the intervening centuries. Using the technical options that were available in the respective time periods, humans have selected and generated populations of animals matching particular needs and purposes suited to specific climate and production systems. The result is the abundance of great phenotypic and genetic variation now found in domesticated animals including, for example, the more than 3000 cattle and 1300 pig breeds.

Table 1.1 Important milestones in the evolution of livestock breeding and animal biotechnology

~300.000 BC	The first humans emerge in East Africa
~12.000–15.000 BC	Begin of domestication of farm animals
~8000 BC	Separation of taurine and zebu cattle
Sixteenth century AD	Emergence of statistical concepts used for farm animal breeding
1677	Discovery of sperm cells
1780	First successful insemination (dog)
1866	First publication of Mendel's laws
1891	First successful ET (rabbit)
1934	First successful ET in sheep
>1940	Emergence of quantitative genetic concepts to accelerate genetic gain in livestock
1949	First successful ET in goat
>1970	Widespread field application of AI in farm animals
1971	First successful freezing/thawing of mammalian embryos (mouse)
1973	First successful freezing of bovine embryos
1980	First successful production of monozygotic twins by embryo splitting (sheep)
1982	First calf after transfer of in vitro produced embryos
1985	First transgenic farm animals (rabbits, sheep, and pigs) via microinjection
1985	First successful vitrification of mouse embryos
1985	First successful IVF in pig
1986	First successful embryonic cloning in sheep
1989	Birth of the first offspring (rabbits) after use of sex-sorted semen
>1990	Increasing use of QTLs in farm animals
1996	First successful cloning with somatic cells ("Dolly")
1998	First transgenic animal (sheep) after use of SCNT with transfected donor cells ("Polly")
>2000	Growing importance of MAS concepts
2001	Concept of genomic breeding value published; Publication of the human genome
2004	Chicken genome published
2006	Genome of dog and bee published
2009	Genome of domestic cattle and horse published
>2010	Growing implementation of GBV in important cattle breeds
2011	First pigs with a biallelic knockdown induced by the use of gene editing (ZFNs)
2012	Pig genome published
2013	First genetically modified pigs after use of CRISPR/Cas
2014	Sheep genome published
2017	Goat genome published

Abbreviations: *ET* embryo transfer, *AI* artificial insemination, *IVF* in vitro fertilization, *QTL* quantitative trait loci, *MAS* marker-assisted breeding, *SCNT* somatic cell nuclear transfer, *GBV* genomic breeding value, *ZFNs* zinc finger nucleases, *CRISPR/Cas* clustered regularly interspaced short palindromic repeats, *BC* Before Christ

This rich diversity of phenotypes has been a major attractor for evolutionary biologists and geneticists, including Charles Darwin who used the limited data then available as a key component in his theory of evolutionary biology in 1859 (Wang et al. 2014). Their endeavors, together with the wealth of new information stemming from the recent developments that allow detailed, cost-effective studies of individual animal genomes, have led to the accumulation of massive and complex datasets (Gerbault et al. 2014), requiring new modeling approaches to be developed that incorporate the latest statistical, population, and molecular genetics methodologies. The result of the interrogation of the data is an increasingly detailed understanding of domestication processes for all the major livestock species (Gerbault et al. 2014).

A major qualitative step in the evolution of systematic livestock breeding was made in the late nineteenth century with the introduction of statistical methodologies to systemic breeding practices. The initial application of statistical methods to animal breeding and genetics is mainly credited to Francis Galton (1822–1911) and Karl Pearson (1857–1936) who both worked before Mendel's law was rediscovered. One of their key findings was that on average, descendants from tall parents were smaller than their parents, while progeny from shorter parents was taller. This statistical regression of offspring on parent formed the basis of the more general heritability concept (Gianola and Rosa 2015). Subsequent development and application of this and other statistical concepts was critical for the scientifically based animal selection programs emerging in the twentieth century (Rothschild and Plastow 2014). Most animal breeding issues have a quantitative dimension that can be addressed via the application of one or more of the plethora of powerful statistical methodologies developed during the last four to five decades (Gianola and Rosa 2015). Application of these methodologies has allowed the recognition, introduction, and guided expansion of specific production traits to occur at an unprecedented rate. The emerging challenge for the livestock industry is to realize the potential of these advances to specific animal selection programs while maintaining sufficient genetic diversity for future innovations (Groeneveld et al. 2010).

1.2.2 Evolution of Scientifically Based Breeding Concepts

The twin foundations of the science-based breeding programs used in all modern livestock industries are quantitative genetics and reproductive biotechnology. From early on, there were two approaches to applying genetics to animal breeding (Blasco and Toro 2014). The first approach started with the rediscovery of Mendel's law and sought to identify inheritable chemical or molecular markers that could be used in genetic studies. Initial success came from the discovery of enzymatic polymorphisms, through the introduction of electrophoretic technologies in the 1960s that could be related to blood groups and/or coat color. While these studies revealed the potential of using the approach to following genetic variability among animals, it, disappointingly, only led to the identification of a few genetic variants that could be used to guide breeding strategies. The second approach can be traced back to Francis

Galton (1822–1911), a Victorian scientific polymath, who used a statistical approach in his studies of the expression of phenotypes among related animals. Both approaches aimed at promotion of genetic change in economically important productive traits (Blasco and Toro 2014) and subsequently became increasingly intermingled and eventually converged to exploit the genomic maps made available with improved DNA sequencing methods.

The genetic value of an animal is commonly described by its breeding value reflecting the major heritable traits being targeted for improvement in a specific breeding program. Developments in statistics and genomics have led to increasingly more accurate breeding values, thereby improving the rate of gain. In dairy cattle, selection was initially targeted at important milk parameters, such as milk yield, milk protein, and fat contents, other physiological factors being of minor importance or even neglected. Today breeding values recognize the importance of maintaining robust health in the herd or flock and include heritable physiological factors, such as longevity, claw, and udder health with the relative weighting for milk parameters significantly reduced. These breeding values are now recognized globally, thus facilitating the global exchange of valuable genetics.

1.2.3 Advent of DNA-Based Breeding Concepts

The rapid implementation of selection strategies based on DNA analysis became possible through what can only be described as truly impressive advances in DNA-analytical technology achieved since the initial attempts in the late 1960s, made with the simple tools then available (Shendure et al. 2017). Remarkable advances in multiple technologies have been made since that time, particularly in the last two decades. Procedures used to laboriously sequence a few kilo bases of DNA have now evolved to a stage where DNA studies commonly interrogate information derived from massive parallel sequencing of millions and myriads of DNA stretches. Significantly, this advance has been accompanied by a progressive and dramatic reduction in DNA sequencing costs to a point where being able to sequence whole genomes of individual humans and animals for a few hundred € or $US or even less (Shendure et al. 2017). A major driver for these developments have been human health issues, and the challenge of development and application of this capacity together with the growing recognition of the potential of the technology to individualizing medical treatment has, not unexpectedly, resulted in a rapidly expanding medical biotechnology industry. The livestock industry can expect similar major developments following the recent availability of sequences for all the major livestock genomes, including cattle, pigs, sheep, horses, poultry, goat, dogs, and cats (see chapter of Blasco and Pena in Volume II of this book). The first nearly complete draft of the human genome sequence was published in 2001, the outcome of >10 years of intensive work, involving many laboratories and a massive expenditure of money (Venter et al. 2001). The pace of development of cost-effective, reliable, and rapid sequencing procedures since that time is a major factor in establishing and

Table 1.2 Size of genomes of farm animals

	Number of chromosomes	Size of the genome (Gb)	Number of coding genes
Cattle	60	2,86	~22.000
Pig	38	2,76	~22.000–24.000
Sheep	54	2,71	~21.000
Poultry	34	1,2	~18.400
Horse	64	2,4–2,7	~20.000

1 Gb = 10^{-9} bp

refining the ever-growing library of complete animal genome sequences that now includes all major livestock species.

Detailed analysis of this valuable database has shown that animal genomes share a number of important features, most notably the finding that the total number of protein coding genes is only ~21.000–23.000 and that only a small proportion of it, usually 4–6% of the genome, is actively transcribed into proteins, the remaining major part of the genome being made up by repetitive sequences and epigenetic and retroviral elements, presumed, until very recently, to be uninvolved in the regulation of coding genes (Table 1.2). This viewpoint is being increasingly challenged by the finding that gene expression of an individual is being continually altered without any change in the genome's sequence. Recent research has identified some of these now called epigenetic processes, including methylation of DNA, alterations in the histone molecules that hold together DNA superstructures via methylation or acetylation or other biochemical modifications, and various RNA and Dicer protein-dependent processes that inhibit gene expression. In combination, the sum total of all these epigenetic marks in an individual is known as the epigenome.

Clearly, in the light of a growing appreciation of epigenomics and other unanticipated gene regulatory phenomena, our understanding of the significance of these noncoding elements needs analysis and revision. This is currently being undertaken through international collaboration, most notably through a project called ENCODE (Encyclopedia of DNA elements) (Kellis et al. 2014). Future refinement of breeding concepts will be increasingly dependent on the outputs of initiatives such as ENCODE to fully understand gene regulation and the role of both coding and non-coding DNA sequences in the expression of individual traits and their propagation in a given population. This is important to cope with anticipated and the unexpected challenges to future breeding enterprises. Developments in this field are of particular interest to livestock breeders as it is known that the lifetime health and productivity of animals derived by some reproductive technologies may be associated with alterations of the epigenome.

A major advance in the application of DNA analysis to animal breeding was made with the identification and introduction of QTLs (quantitative trait loci). Implemented in the mid-1990s in the dairy industry, it has since led to the discovery of a number of important QTLs in the various farm animal species. An important finding from use of QTLs was the identification of causal mutations for specific traits (Blasco and Toro 2014). The QTL strategy was succeeded by the concept of marker-assisted selection (MAS). This is essentially a three-step process that includes the detection of several QTLs, followed by identification of the gene which causes the respective mutation and finally the increase of the frequency of the favorable allele by selection or by introgression (Blasco and Toro 2014). Early and prominent examples of the use of MAS are the halothane gene in pigs and the Booroola gene in sheep (Dekkers 2004).

MAS systems have now evolved further to what is called genomic selection (Meuwissen et al. 2001). This system was made possible through both identification of a dense set of informative markers that are, ideally, more or less evenly distributed across the genome and on cost-effective genotyping procedures. Genomic selection requires large testing populations and accurate phenotypic characterization (Meuwissen et al. 2001). The insights into gene sequences and their location on the chromosomes revealed through the broad-scale use of genomic selection ensure a constantly improving understanding of the genetic architecture of farm animals and many opportunities for the identification of the molecular identifiers of economically important traits.

The major technological advance already accelerating genomic projects in the major domestic species are chip arrays with several hundred thousand SNPs (single-nucleotide polymorphisms). Chips now available commercially target 750.000 SNPs for cattle, 56.000 SNPs for sheep, and 60.000 SNPs for pigs (Blasco and Toro 2014). Genomic selection by this means has a number of significant advantages over previous programs, most significantly when used to predict the breeding value in the born calves and even in early embryos. Already embryo analysis by this means has been shown to have greater accuracy in predicting breeding value than the classical pedigree index, with the additional benefit of it avoiding the costs and time-consuming maintenance of waiting bulls. Uptake of this approach to livestock selection by the cattle industry is well advanced, and the genomic breeding value (GBV) is increasingly being implemented into the breeding programs of major dairy and dual-purpose breeds, such as Holstein-Friesian and Simmental.

1.3   Evolution of Reproductive Biotechnology

1.3.1   History of Artificial Insemination (AI)

Artificial insemination (AI) was the first and remains the most widely used of the growing armory of reproductive technologies available to the livestock breeder. As a consequence, there is already a library of comprehensive reviews of the origins and history of AI and its impact on the animal breeding enterprises (e.g., Foote 1996; Vishwanath 2003; Ombelet and van Robays 2015; Orland 2017). Only the key advances in this still evolving technology are thus summarized below; for more detailed and informative accounts and references, see the reviews cited above.

The significance of semen in reproduction has been appreciated by most if not all cultures, since the earliest of times, with stories of attempt at AI part of the mythology of several cultures. It is generally accepted that the scientifically based AI traces back to the seventeenth century when development of the compound microscope allowed the discovery and description of mammalian sperm cells from humans and dogs by Antoni van Leeuwenhoek and his assistant Johannes Hamm in 1678 in the Netherlands (Ombelet and van Robays 2015; Orland 2017). However, it was more than 100 years before the first documented success with AI was recorded: in the 1780ties in the human by the eminent scientist surgeon John Hunter, and by Lazzaro Spallanzani, an Italian physiologist, in a dog. Full appreciation of the potential value of AI to animal breeding only became evident in the late nineteenth century when it was made a specific subject of research (Orland 2017). Interestingly, it was Spallanzani, who also made the observation that human sperm became immotile when it accidently came in contact with snow, a seminal observation foreshadowing the use, 200 years later, of cryopreservation to store both sperm and ovum.

A major stimulus to this renewed interest in AI was the report in 1897 by Walter Heape, a British zoologist and embryologist based in Cambridge, of success in AI with rabbits, dogs, and horses. Significantly, his success laid the foundation, in 1932, of the Animal Research Station in Huntington Road in Cambridge, a facility that was to play a lead role in the development of not only AI but many of the other key reproductive technologies now in wide-scale use (Polge 2007). Important milestones in the subsequent history of AI include the development of dilution media to extend the use of single ejaculates and allow long-term storage through cryopreservation of sperm, the addition of antibiotics to semen samples to control bacterial contamination, and the development of freezing and cooling protocols compatible with high survival rates of sperm cells (Table 1.3).

The rate of adoption of AI by animal breeders varied from country to country, impeded in part by religious, moral, and social concerns about interference with the natural order of things. Russia led the way following the pioneering work by Ivanovich Ivanov, a biologist who, by 1907, had extended the use of AI to sheep and a range of other domesticated animals, including foxes and poultry. Japan and Denmark were also early AI adopters and innovators with Edward Sorensen together with Gylling Holm establishing the first cooperative AI-based breeding program in

Table 1.3 Important milestones in the history of artificial insemination (AI)

Year	Discoverer	Main finding
1677	Antoni von Leeuwenhoek	First picture of sperm cells
1780	Lazzaro Spallanzani	First insemination (in a dog)
1790	John Hunter	First vaginal insemination in human
1900	Ilya Ivanov	Development of semen extenders
1939	Gregory Pincus	First conception (rabbit) by AI
1949	Christopher Polge et al.	Discovery of cryoprotective functions of glycerol
1950	Robert Foote and R. Bratton	Addition of antibiotics to semen extenders
1953	Jerome Shumann	First pregnancy after AI with frozen sperm (human)
1978	Robert Edwards and P. Steptoe	First IVF baby (Baby Louise)
Since 1970s		Broad application of AI in farm animals, mostly cattle and pigs

Modified from Ombelet and van Robays (2015)

dairy herds in Denmark in 1936. The clear success of this program proved to be the stimulus needed to encourage the introduction and broad-scale uptake of AI in the USA and throughout the western world (Foote 1996; Vishwanath 2003). The stimulation of demand for animal products triggered by World War II and its aftermath dramatically increased the use of AI, particularly with dairy cows, where it was applied not only to improve the genetics of a given herd but also to gain control over Brucellosis and other prevailing venereal diseases. The accompanying investment in research led to a continuing series of important innovations that have evolved to the plethora of breeding technology options available today. Significant developments in AI resulting from this investment include not only reliable and robust technology for the collection, storage, and insemination of semen but equally importantly accompanying refinements in animal husbandry allowing estrus detection and regulation and standardized measures of fertility assessment. As a consequence, AI remains the primary method of choice for animal breeders around the globe seeking to improve the genetic quality of their stock through the realization of the genetic potential of valuable sires within a given population (Vishwanath 2003). For general breeding purposes, on average, 200–300 insemination doses can now be produced from a single bull ejaculate and stored frozen indefinitely; for a boar ejaculate, usually 10–20 insemination doses can be produced with semen freezing possible, but still at low efficiency and in small ruminants, one ejaculate can be extended to serve 10–30 ewes and successfully cryopreserved.

Today, AI is employed in more than 90% of all sexually mature female dairy cattle in countries with well-advanced breeding programs. The use of AI is also increasing in pig production enterprises with now more than 50–60% of sows served by AI on a global scale. The adoption of AI for use with low unit cost animals such as sheep and goats is less widespread but is still employed in the breeding of greater than 3.3 million sheep and 0.5 million goats annually with further growth anticipated following major refinements in estrus synchronization and insemination techniques and the need for flexibility in genotype of flocks to match fluxes in market demand for meat and fiber. AI is also now widely practiced in the poultry industry with the extremes of genotype found in extensively modified species such as the turkey making it obligatory. The clear benefits of AI have been such that robust and reliable AI procedures are now being available for most domesticated non-livestock species and increasingly for individual breeds of wild animals as a primary means of preserving threatened genotypes (Comizzoli et al. 2000; Comizzoli and Holt 2014).

It is long known that the sperm determines the sex of the potential offspring: when a Y-chromosome-bearing sperm fertilizes the oocyte, the resulting XY-constellation leads to a male phenotype; the XX chromosome set results in a female phenotype. In ancient time, the Greek philosopher Democritus from Abdera (~450 BC) suggested that the right testis produced only males, whereas the left testis produced only females. Subsequently, the lack of understanding of the basic biological principle mentioned above has prompted numerous methodological approaches to be tested in their ability to achieve separation of X- and Y-chromosome-bearing sperm. However, only the recent application of advanced flow cytometric systems, based on the small differences in DNA contents (3–6% depending on species, with the Y-chromosome being smaller than the X-chromosome) between X- and Y-chromosome-bearing sperm, allows effective and reliable separation of living X- and Y-chromosome-bearing sperm for AI. A major breakthrough was reported in 1989, when fertilization with sex-separated semen was achieved with surgical insemination in the rabbit and several pups were born showing the desired sex (Johnson et al. 1989). Later improvements of sexing protocols provided sex-sorted semen in large enough quantities for use in bovine IVF (Cran et al. 1993). Nowadays, flow cytometry has been advanced to a stage that frozen/thawed sexed semen can be routinely supplied for bovine AI (Garner and Seidel Jr 2008) and is now being offered commercially by different companies around the globe. Thus the use of sexed semen in AI has rapidly emerged as an important new tool to enhance efficiency of dairy production.

1.3.2   History of Animal Embryo Transfer

A detailed history of embryo transfer (ET) can be found in the excellent publication from Betteridge (2003). Efforts to establish embryo transfer technology were made as early as the nineteenth century with a Canadian-English evolutionary biologist, George John Romanes (1848–1894), credited with the first, albeit unsuccessful, attempts. The first transfer of embryos resulting in live born offspring was achieved in rabbit by Walter Heape in 1890. Interestingly, Heape did his experiments at his home in Prestwich, near Manchester, using the rabbit breeds Angora and Belgian hare as embryo donors and recipients. This small-scale project typifies work in the biological sciences being carried out at the time. However, technological advances achieved in this way could still attract worldwide recognition through the intense network of interconnections established between biological scientists in the UK and elsewhere in the scientific world via the Royal Society and similar national and regional scientific bodies. This network was a major contributor to the rapid growth of understanding of reproductive biology that was to allow the full extension of ET to agricultural animals.

The late 1920s and early 1930s saw the beginnings of specific investment in developing ET for use in agriculture on both sides of the Atlantic. For example, the work of a group at the Institut für Allgemeine und Experimentelle Pathologie in Vienna, led by Artur Biedl, achieved a successful pregnancy in rabbits after 70 transfers in 1922 (Biedl et al. 1922). However, two centers in particular are identified with the next key advances in ET, one in Cambridge, Massachusetts, USA, and the other in Cambridge, UK. Cambridge, USA, was the site of one groundbreaking development in embryo transfer technology in 1936 by Gregory Pincus, an outstanding American endocrinologist and scientist. Six years previously he had reported a series of 21 embryo transfers in the rabbit that yielded 3 litters (Pincus 1930), an achievement stemming from his introduction of the use of anesthesia to allow direct exposure of and access to the oviducts and ovaries and a special pipette he had built to facilitate ET. However, the vast majority of these and his subsequent experiments suffered from the lack of knowledge of the need for synchrony between the embryo and the recipient uterus, an appreciation he only made in 1936 when he and his coworker Kirsch recovered blastocysts that had developed following transfer of one- and two-cell embryos to the oviducts of rabbits at estrus, that is, before functional corpora lutea have been established (Pincus and Kirsch 1936). The recognition of the need for synchrony between donor and recipient provided the key to the development of robust and reliable ET for use in livestock breeding programs.

The first steps toward use of ET in livestock breeding had already been made in 1931 by Hartman and his colleagues at the Carnegie Laboratory of Embryology in Baltimore, USA, who harvested bovine two-cell embryos for the first time (Hartman et al. 1931; Miller et al. 1931). This was followed a year later by the first recorded actual transfer of livestock embryos by the group of Berry and Warwick at the Agricultural and Mechanical College in Texas, USA, who used ET to investigate causes of early embryonic loss in sheep and goats (Warwick et al. 1934; Warwick and Berry 1949). To honor this achievement, Dr. Berry became the first recipient of the Pioneer Award of the International Embryo Transfer Society (IETS) in 1982. World War II interrupted progress and development of ET techniques in Europe, but the prevalent food shortage from the war and its aftereffects urged research aimed at improving livestock breeding technologies including ET. In the UK, embryo transfer was identified as critical for the production of high-quality meat from beef cattle produced from dairy herds. This need was an important prompt for the Agricultural Research Council (ARC) Unit of Animal Reproduction at the Huntingdon Road in Cambridge, UK, the remarkable body of work on ET contributed by the Unit from then until its closure in 1986, making it a must go to scientific center in assisted reproductive technologies (ARTs). Among its early achievements were major advances in superovulation and the introduction by Lionel Edward Aston (Tim) Rowson, of nonsurgical collection of embryos in cattle breeds through his development of a catheter for transcervical recovery. As a consequence of the broad-spread interest triggered by these and subsequent developments in ET among breeders of both livestock, specifically cattle, robust and reliable ET protocols are now available for a large number of species (Table 1.4).

Important contributions to embryo transfer technology in other livestock species, such as sheep and pigs, came from the former Soviet Union (USSR) and Poland (Lopyrin et al. 1950, 1951; Kvasnitski 1951). An English translation of the Kvasnitski paper can be found in the proceedings of the conference held in May 2000 in Kiev, now Ukraine, that commemorated the 50th anniversary of the first successful porcine embryo transfer (Kvasnitski 2001).

From the 1970s onward, ET technology developed at a rapid pace through the work at the ARC Unit and other groups operative throughout the world. Important steps in this included the development of robust and reliable superovulation and synchronization protocols based on the better understanding of reproductive endocrinology and physiology, the use of frozen semen, and the implementation of nonsurgical transfer and collection techniques. Important advances were also made in the development of media suitable for the holding and culture of early embryos. Field application of the new technologies was advanced in 1972, through an instruction course on ET technology organized in Cambridge, UK, which brought together

Table 1.4 First successful (with the delivery of live offspring) embryo transfers in different species

Year	Author	Country	Species
1891	Heape	UK	Rabbit
1933	Nicholas	USA	Rat
1934	Warwick et al.	USA	Sheep
1942	Fekete and Little	USA	Mouse
1949	Warwick and Berry	USA	Goat
1951	Willett et al.	USA	Cattle
1951	Kvasnitski	UdSSR (Ukraine)	Pig
*1964	Mutter et al.	USA	Cattle
1968	Chang	USA	Ferret
1974	Oguri and Tsutsui	Japan	Horse
1976	Kraemer et al.	USA	Primate
1978	Steptoe and Edwards	UK	Human
1978	Shriver and Kraemer	USA	Cat
1979	Kinney et al.	USA	Dog

*Transcervical transfer

a group of veterinarians and scientists from around the globe. This group later played a crucial role in forming the International Embryo Transfer Society (IETS) (Carmichael 1980; Schultz 1980), now regarded as the lead scientific forum for the exchange of new ideas on embryo transfer and related technologies. In 2016, the name of the society was changed to "International Embryo Technology Society," to better reflect the importance of the emerging embryo-related techniques such as in vitro fertilization, freezing, or cloning.

Another important step toward practical ET techniques was the report of the first successful freezing of a mammalian embryo, the mouse (Whittingham 1971), an advance based on the demonstration by M.C. Chang, in 1947, of the feasibility of this by his successful transfer of rabbit embryos that had been cooled to 10 °C (Chang 1947). The report of the first successfully frozen/thawed bovine embryos quickly followed (Wilmut and Rowson 1973). This success allowed animal breeders not only to freeze and store valuable gene stock for transfer to appropriate recipients as needed but opened up the way for global exchange of gene stock through frozen embryos as well as sperm. Refinements in freezing protocols have been rapid, due in part to the co-interest in cryopreservation of human tissues. This had led to a number of different freezing protocols now being available for freezing bovine and other livestock embryos. The number of transfers of bovine embryos, both freshly collected and frozen/thawed, increased significantly in the last decade from ~823.200 in 2006 (Thibier 2008) to up to ~965.000 embryos in 2016 (Perry 2017). While ET is widely used in dairy and parts of beef cattle, it is much less applied in pigs (few thousand ETs), small ruminants (few hundred ETS), and horses (few thousand ETs) (Perry 2017). Thus, embryo transfer technology is now an integral part of modern breeding concepts for cattle and is widely applied across the globe. However, while embryo transfer technology allows a better exploitation of the genetic potential of the female germ pool than AI, it is still only used in the top 1–2% of a breeding population.

A major expansion of interest in ET technology followed the landmark achievement in human reproductive medicine with the birth of Louise Brown in 1978 in Oldham, UK, following in vitro fertilization and transfer procedures developed by Robert Edwards, a Cambridge, UK, physiologist, and Patrick Steptoe, a surgeon from Oldham, UK (Edwards and Steptoe 1978). The foundation stones for Edwards' success were laid nearly 20 years earlier in what has been described as a golden age in IVF studies (Bavister 2002). Highlights of this era were the reports of Anne McLaren and John Biggers of successful development and birth of mice cultivated in vitro as early embryos (McLaren and Biggers 1958) and, a year later, MC Chang's findings that in vitro fertilized rabbit eggs could develop normally following transfer to surrogate mothers (Chang 1959).

A prime motivation for Edwards' in vitro fertilization was his interest in addressing the high incidence of infertility in humans, in particular the growing number of women in the post-pill era with infertility due to hydrosalpinx, a blockage in their fallopian tubes that could be traced to a prior reproductive tract infection, most commonly chlamydia. Demonstrating that in vitro fertilization of human oocytes was possible was the first step (Edwards et al. 1969); the next was for Steptoe to use his skills in laparoscopy to develop minimally invasive procedures allowing repeated collection of oocytes that Edwards could fertilize in vitro and reimplant in the uterus thus by-passing the damaged tubes and achieving pregnancy. Their epoch-making achievement was the culminating point of Robert Edwards lifetime of pioneering research in human infertility and earned him the Nobel Prize in 2010 (Johnson 2011).

IVF is now used to address a wide range of fertility issues, and the number of babies born from assisted reproductive technologies (ART) is increasing rapidly: their numbers have more than quadrupled since 1995, and to date, >5 million babies worldwide have been born after ART (ESHRE 2009). ART births constitute 1.5–4.5% of all births in the USA and other countries such as the UK (Sunderam et al. 2018; HFEA 2011). In livestock breeding, the technology initially lagged behind that in human, with the first successful IVF from in vivo matured oocytes in cattle in 1982 (Brackett et al. 1982) and entirely from IVM/IVF/IVC in 1987 (Fukuda et al. 1990) and in the pig in 1985 (Cheng et al. 1986). IVM/IV + IVC have now been refined to a stage that it is possible to repeatedly harvest oocytes by laparoscopic and nonsurgical techniques, mature and fertilizing the harvested oocytes in vitro, followed by culture of the resultant zygotes to the blastocyst stage for transfer to synchronized recipients. These IVM/IVF/IVC procedures are now being widely used for experimental studies and commercially as well, for recovery of valuable gene stock postmortem (usually from abattoirs), and to reduce the generation interval via collection of oocytes from juvenile animals (JIVET). Current global figures revealed a total of ~450.000 entirely in vitro produced bovine embryos that had been transferred to recipients with geographical emphasis in South America (Perry 2017). The application of IVM/IVF/IVC combined with cryopreservation procedures in the introduction of highly productive genotypes is now supporting development of China's dairy herds.

1.3.3   "Dolly" and Beyond

The birth of an ewe named "Dolly" in Scotland in July 1996 opened up a new world of possibilities for animal breeders. Dolly's distinction was the fact that she was the first cloned mammal derived from a fully differentiated adult cell (Wilmut et al. 1997), a fact that challenged the then ruling paradigm that genes not required in the development of specific tissues were lost or permanently inactivated (Weissmann 1893). From the animal breeders' perspective, this was interpreted as limiting any developments in cloning technology to cells from early embryos, that is, before cells become committed to their specific differentiation pathway. This limiting paradigm was a consequence of studies made with amphibian embryos in 1952 by Robert Briggs and Thomas J. King in Philadelphia, USA. Using the amphibian species *Rana pipiens*, and the nuclear transfer procedures they had specifically developed for the purpose, they showed that while normal tadpoles could be obtained after transplanting the nucleus of a blastula cell into the enucleated egg, tadpole development became increasingly restricted as cells underwent differentiation (Briggs and King 1952). This led to the hypothesis that the closer the nuclear donor is developmentally to early embryonic stages, the more successful nuclear transfer is likely to be. Support for this viewpoint came from John Gurdon, an Oxford, UK, based developmental biologist, who used another amphibian, the frog *Xenopus laevis*, as model species. *Xenopus* has some distinct advantages over *Rana pipiens*, because (1) the embryos can be grown to sexual maturity in less than a year, (2) *Rana pipiens* lives more than 4 years, and (3) *Xenopus* frogs can be induced to lay eggs throughout the year after hormonal injections. In contrast, *Rana pipiens* and other frogs are strictly seasonal. Gurdon showed that only with less differentiated donor cells, he could achieve development and developmental rates dropped when more differentiated cells were used as donors (Gurdon 1960, 1962, 2017). This viewpoint prevailed for many years and had a strong influence on the design of experiments in the 1970s and 1980s.

Cloning of mammals became possible when laboratory equipment became available in the late 1960s and early 1970s that allowed micromanipulation of the much smaller mammalian eggs (100–130 μm in diameter, i.e., about one tenth of the diameter of the amphibian egg). The first report on cloning in mammals was by Illmensee and Hoppe (1981) who reported the birth of three cloned mice after transfer of nuclei from the inner cell mass cells of a blastocyst into enucleated zygotes. However, these results could not be repeated, with other researchers finding that development was arrested following the transfer of the nucleus of a zygote or two-cell embryos into an enucleated zygote (McGrath and Solter 1983). The same researchers also found no development when nuclei from donor cells from later development stages were used (McGrath and Solter 1984). This led the authors to conclude that cloning of mammals by nuclear transfer would be biologically impossible, presumably due to the rapid loss of totipotency in developing embryonic cells. The challenge to this viewpoint came only few years later in 1986, from Steen Willadsen, a Danish developmental biologist working in the ARC Unit in Cambridge, UK, through his demonstration that nuclei obtained from blastomeres from cleavage stage ovine embryos could be inserted into enucleated oocytes and viable lambs obtained following transfer to recipient ewes (Willadsen 1986). This major technical advance, together with the later finding that donor cells could even be obtained from the inner cell mass (ICM) of bovine blastocysts (Sims and First 1994), established a base for the following successful embryonic cloning of rabbits, mice, pigs, cows, and monkeys (for review see Niemann et al. 2011).

The possibility of cloning mammals through somatic cells was heralded in 1996/1997 through the publication of two landmark papers by the group at the Roslin Institute, Edinburgh, Scotland, UK. Their initial achievement was the demonstration of the feasibility of deriving donor nuclei from an established cell line derived from a day 13 ovine conceptus and maintained in vitro for several passages (Campbell et al. 1996). This remarkable success they attributed to their synchronizing of the cell cycle of the donor cells through lowering the concentrations of serum in the culture medium, thus causing the cells to exit the cell cycle and hold at the Go stage. Transfer of donor cells from these quiescent cell lines to enucleated matured oocytes and transfer of the reconstructed embryos into synchronized recipient ewes resulted in the birth of two healthy cloned lambs called "Megan" and "Morag." Their achievement encouraged the group to extend their studies to somatic cells derived from mammary epithelial tissue that led to the birth of "Dolly" the following year (Wilmut et al. 1997). The prospect of translation of these findings into animal breeding enterprises was enhanced by "Dolly" living a rather normal life at the Roslin Institute until she had to be euthanized in February 2003 due to a fatal pulmonary disease caused by the adenomatosis virus endemic in Scottish sheep flocks. The significance of this advance is documented through the exhibition of Dolly's preserved remains in the Science and Technology Galleries of the National Museum of Scotland, Edinburgh (Fig. 1.1). Interestingly, Dolly is one of the museum's most popular exhibits and has become a symbol of Scottish national pride (García-Sancho 2015). Important steps into the evolution of somatic cloning are depicted in Table 1.5.

Dolly's birth launched a heated ethical debate worldwide and sparked a series of science fiction stories. Initially, the origin of Dolly from a fully differentiated donor cell was questioned by many scientists. However, in the next 5–10 years, the validity of their claims was proven and the feasibility of somatic cell cloning fully realized and established as an important tool in research. Somatic cloning by somatic cell nuclear transfer (SCNT), resulting in the production of live clones, has now been successfully extended to more than two dozen species, including sheep, cattle, mouse, goat, pig, cat, rabbit, horse, rat, dog, ferret, red deer, buffalo, gray wolf, camel, and very recently nonhuman primates (see Niemann 2016; Liu et al. 2018), and, despite a slow start, has been developed to a stage where it is now being offered commercially in all the important agricultural species, including cattle, pigs, and horses.

The underlying mechanisms that determine success in somatic nuclear transfer are still a subject of active research. One initial hypothesis was that the clones only

Fig. 1.1 Dolly, the first mammal cloned from a somatic cell, can be visited in the Scottish National Museum in Edinburgh

arose from a subpopulation of stem cells (Hochedlinger and Jaenisch 2002). However, this was short lived as evidence built up showing that differentiated somatic cells can successfully be employed in SCNT. The reprogramming of the genome following nuclear transfer causes dramatic changes of the epigenetic landscape of the donor cell consistent with the expression profile of the differentiated cells being abolished and a new, embryo-specific expression profile established to drive embryonic and fetal development (Niemann et al. 2008). It is now known that such epigenetic reprogramming involves the erasure of the gene expression program of the respective donor cell and the reestablishment of the well-orchestrated sequence of expression of the estimated 10,000–12,000 genes critical for early embryonic development. Through Dolly, mammalian development is now established as having high plasticity with significant implications for many areas in the natural sciences and in public debate.

Soon after "Dolly" the sheep was born, the journal "Cloning" was launched in 1999 to cover the emerging new information in this area. The journal was expanded in 2002 and 2010 to include all mechanisms of cellular reprogramming and is now called "Cellular Reprogramming" (Wilmut and Taylor 2018). This reflects the dramatic impact of somatic cell cloning not only on animal breeding but in both the biological and medical sciences. One use is in the derivation of so-called induced

Table 1.5 Important milestones in the development of somatic cloning via somatic cell nuclear transfer (SCNT)

Author	Year	Main findings
Spemann	1938	Embryonic development and early differentiation
Briggs and King	1952	Viable tadpoles from nuclei transplanted from blastula stages in *Rana pipiens*; nuclei are multipotent
Gurdon	1962	Viable tadpoles from intestinal epithelial cells in *Xenopus laevis*; nuclei are multipotent
Gurdon and Uehlinger	1966	Fertile adult frogs from intestinal epithelial cells of feeding tadpoles in *Xenopus laevis*; nucleus is still totipotent
McGrath and Solter	1984	Arrested development of reconstructed mouse embryos; claim: Mammalian cloning is biologically impossible
Willadsen	1986	Successful nuclear transfer-based cloning using embryonic donor cells in sheep (8–16 cells)
Tsunoda et al.	1987	Successful nuclear transfer in mice using 4–8 cell embryos as donors
Prather et al.	1987	Successful cloning of cattle by using 2–32 cell stage embryos as donors
Sims and First	1994	Successful cloning of cattle by using cultured cells from the inner cell mass (ICM) of blastocysts
Campbell et al.	1996	Successful cloning of sheep by using 13-days-old cultured fetal donor cells
Wilmut et al.	1997	Dolly, the sheep, successful cloning from a fully differentiated (mammary epithelial) cell
Cibelli et al.	1998	Successful somatic cloning of cattle using fibroblasts as donors
Wakayama et al.	1998	Successful somatic cloning of mice using adult cells
Many different authors	Since 1998	>24 species have been successfully cloned up to 2018
Liu et al.	2018	Successful cloning of nonhuman primate

Modified from Gurdon (2017)

pluripotent stem cells (iPSCs) in 2006 (Takahashi and Yamanaka 2006), an advance which earned S. Yamanaka the Nobel Prize together with John Gurdon in 2012 and established iPSCs as important tools for derivation of patient-specific therapeutic stem cells and regenerative medicine. In the biological sciences, SCNT has proven to be a research tool of great value in the study of early development and epigenetic mechanisms governing the expression of genes that regulate embryonic and fetal development (Kues et al. 2008; Niemann 2014).

SCNT is now developed to a stage where it has commercial application in major farm animals, including cattle, pigs, and horses. However, its main impact on animal breeding will not be through cloning of existing genomes but through its use as a route allowing the full armory of genome editing tools to be applied to the animal genome, allowing precise modification of existing genes or precise insertion of new genes in the animal genome.

1.4 Genome Editing and Precision Breeding

The demonstration, in 1980, by Jon W. Gordon and Frank Ruddle at Yale University, USA, that it was possible to introduce new and functional genetic material into the germline of laboratory rodents heralded a new era in animal breeding. Called transgenesis, it was achieved by microinjection of foreign DNA into oocytes shortly after fertilization (Hammer et al. 1985). The potential application of this powerful new tool was immediately recognized, and within 5 years the creation of the first genetically modified farm animals, including rabbits, pigs, and sheep, had been achieved (Hammer et al. 1985).

However, the microinjection approach to germline modification proved to be highly inefficient in practice and had other major limitations due the fact that it only allowed additive gene transfer and that the introduced DNA was integrated randomly in the recipient genome and a frequent incidence of mosaicism. These limitations were only overcome with the development of cell-based gene transfer methods realized following the confirmation of the feasibility of using SCNT by the birth of Dolly. SCNT-based procedures were quickly developed that now allow the full application of DNA editing technology to be applied to the somatic cells in culture prior to the introduction of the modified genome into the germline. The introduction of SCNT and its capacity to allow the selection and use of highly defined donor cells dramatically improved the production of genetically modified livestock. As a consequence, a whole new range of useful application models became available not only for rodents and other species used in basic research but for various livestock species with new traits of interest to agricultural and biomedical enterprises (Laible et al. 2015). However, cell-mediated transgenesis was still hampered by the inability to produce animals with targeted genetic modifications. This was at least partly due to the fact that in farm animals, in contrast to laboratory species (mouse and rat), robust and reliable procedures for the establishment of true pluripotent stem cell cultures have not yet been achieved (Nowak-Imialek and Niemann 2012). Primary cells only have a limited lifespan in culture, and being limited to their use in SCNT was not compatible with the high selection needed for targeted mutations, thus severely limiting the extent of the genetic modification that could be achieved.

This situation changed dramatically with the introduction of genome editing technologies based on the use of DNA nucleases (see Petersen and Niemann 2015). These molecular scissors, including zinc finger nucleases (ZFN), transcription activator-like effector nucleases (TALEN), and the CRISPR/Cas (clustered regularly interspaced short palindromic repeats) system, allow precise modifications of the genome. In animals all three nucleases can be applied either via microinjection into early fertilized eggs (zygotes) or after transfection into donor cells that are subsequently used in somatic cloning. Within a few years following their introduction, numerous research groups have described the successful production of genetically modified cattle, pigs, and sheep covering a range of potentially useful genetic modifications, both for agricultural and biomedical application (Petersen and Niemann 2015; Telugu et al. 2017).

For the first time, it became possible to overcome the limitations of the glacially slow classical breeding and selection processes traditionally used in agricultural enterprises. Using the new technologies of genome editing, new phenotypes can be produced and introduced within a single generation (Laible et al. 2015). Furthermore, with the capacity to target and edit individual genes or noncoding sequences in the genome in combination with the use of homologous recombination protocols, to introduce new DNA sequences provides the basis for establishing a whole new world of opportunities for animal breeding enterprises.

To date, only a limited number of products from genetically modified animals have been approved for use through the national supervisory bodies established to monitor and govern the use of these technologies. All were derived by conventional transgenic technologies, including recombinant human antithrombin (ATryn[R]) from goat milk for prophylactic treatment of hereditary antithrombin deficiency within a surgery, recombinant C1 esterase inhibitor from rabbit milk for treatment of hereditary angioedema (HAE) (Ruconest®), and Kanuma (sebelipase alfa®), a recombinant human enzyme that is produced in egg white of hens to treat lysosomal acid lipase deficiency. Pigs and other livestock with enhanced production traits have been developed, but only one has been accepted for commercial use, namely, the AquAdvantage Atlantic salmon from the company Aquabounty. Of concern is that the fish which grows twice the size of the normal Atlantic salmon over the same time period only received official approval from the FDA, the supervisory body in the USA, in 2015, 20 years after its development and after a major regulatory battle. It will be interesting to see how products from animals derived from gene editing will be legalized as similar genetic changes may occur naturally, making it difficult, if not impossible, to identify the origin of the mutation. The recent acceptance in March 2018 of the safety of products derived through gene editing in food plants by the FDA is encouraging. The genomic maps of both plants and farm animals are constantly being refined, and a wealth of new opportunities for genomic editing that majorly expand genetic diversity from a variety of important application perspectives can be confidently anticipated (Petersen and Niemann 2015; Telugu et al. 2017).

1.5   Future Perspectives

Modern animal breeding strategies, mainly based on population genetics, novel molecular tools, and assisted breeding technologies (ARTs) such as AI and ET, have significantly increased the performance of domestic animals. This forms the basis for a regular supply of high-quality animal-derived food and fiber at competitive prices. For example, in both Australia and the USA, Holstein-Friesian dairy bovine milk production increased annually by about 1%, corresponding to 40–80 kg/cow/year, between 1980 and 2010 (Hayes et al. 2013). Gains that played an important part in the reduction are seen in the costs of milk and milk products. Similar gains were achieved in the efficiency of production of other animal-derived food products, such as meat and eggs.

The introduction of precision breeding concepts based on genome editing is an important step in allowing the necessary developments required to address compounding challenges in global food security, environmental sustainability, and animal welfare (Rothschild and Plastow 2014). It is predicted that by the year 2050, the global population will have grown up to 9.5–10 billion people. This growth will take place mainly in developing countries and in major urban areas requiring a dramatic increase in food production, including animal-derived protein. Estimates of future need for meat products indicate that meat production will need to increase by at least 70% to cope with this future demand. As the majority of arable land is already in production, there is a clear challenge to livestock breeders to increase efficiency of food production from both intensive and non-intensive animal production enterprises in a sustainable manner (Telugu et al. 2016). Encouragingly, considerable genetic variation for traits contributing to efficiency improvements in all livestock species still exists (Hayes et al. 2013). Realizing the full potential of these traits and the introduction of new traits will require the precision breeding concepts introduced in this brief history. DNA-based breeding concepts and genome editing are critical for ensuring an efficient and sustainable future for both plant- and animal-based agricultural enterprises. Further development and acceptance of bioengineered products will also be of immense medical importance in the generation of models for human diseases, xenotransplantation, the production of pharmaceutically active proteins, environmental remediation, and regenerative medicine.

The USDA has recently accepted (March 2018) that with precision editing now possible mutations would be indistinguishable from rare but possible natural mutations and stated that it does not and has no plans to regulate gene editing of plants or crops but will still treat plants with introduced foreign genes as GMOs (genetically modified organisms). Experience gained from repeated attempts to gain acceptance of genetically modified meat products suggests that there is still a way to go for even the most subtle gene modifications. The pathway to public acceptance of genome editing technologies in farm animals is probably an indirect one through initial demonstrations of its safety and value in addressing issues of animal welfare, human health, and sustainability. Procedures from genomic editing in animals must be rigorously screened for off-target mutations to avoid any violation of the integrity of the animal. This is now entirely feasible using advanced CRISPR/Cas and similar systems. The value of persisting in seeking to introduce this approach more broadly in the livestock sector has been confirmed by the recent demonstration that genome editing can be used to increase the genetic gain in farm animal breeding in both the short- and medium-term perspective. By applying gene drive concepts using genome editing tools, increasing the allele frequency using gene drive mechanisms would accelerate genetic gain even further and without the risk of increased inbreeding (Gonen et al. 2017). This viewpoint is supported by a recent simulation study that revealed that this approach could be used to refine the increase in genetic gain through accelerating the increase in the frequency of favorable alleles and reducing the time to fix them in germlines; labeling nucleotides, for a more rapid targeting of quantitative traits; and finally increasing the efficiency of converting genetic variation into genetic gain (Gonen et al. 2017), all desirable capacities for inclusion in future breeding concepts.

In summary, researchers and animal breeders now have tools in hand to modify the genome in a previously unprecedented very precise manner. The potential to rapidly increase favorable genes in a given population is an important step toward achieving genetic gain and modulating economically important gene loci. These opportunities need to be exhaustively explored and their potential fully assessed as they are of vital importance to development of the animal enterprises needed to combat the looming challenges to food security from the hyperbolically increasing demands and predicted climatic and environmental uncertainties. These advances need to be carried out in a manner that ensures sufficient transparency and information to the public and decision-makers so that there is a general understanding of the importance and need for full support for initiatives in this area.

References

Bavister BD (2002) Early history of *in vitro* fertilization. Reproduction 124:181–196
Beja-Pereira A, Caramelli D, Lallueza-Fox C et al (2006) The origin of European cattle: evidence from modern and ancient DNA. Proc Natl Acad Sci U S A 103:8113–8118
Betteridge KJ (2003) A history of farm animal embryo transfer and some associated techniques. Anim Reprod Sci 79:203–244
Biedl A, Peters H, Hofstätter R (1922) Experimentelle Studien über die Einnistung und Weiterentwicklung des Eies im Uterus. Z Geburtshilfe Gynäk 84:59–103
Blasco A, Toro MA (2014) A short critical history of the application of genomics to animal breeding. Livest Sci 166:4–9
Brackett BG, Bousquet D, Boice ML, Donawick W, Evans JF, Dressel MA (1982) Normal development following in vitro fertilization in the cow. Biol Reprod 27:147–158
Bradford MW, Bradley DG, Luikart G (2003) DNA markers reveal the complexity of livestock domestication. Nat Rev Genet 4:900–910
Briggs R, King TJ (1952) Transplantation of living nuclei from blastula cells into enucleated frogs' eggs. Proc Natl Acad Sci U S A 38:455–463
Campbell KHS, McWhir J, Ritchie WA et al (1996) Sheep cloned by nuclear transfer from a cultured cell line. Nature 380:64–66
Carmichael RA (1980) History of international embryo transfer society I. Theriogenology 13:3–6
Chang MC (1947) Normal development of fertilized rabbit ova stored at low temperature for several days. Nature 159:602–603
Chang MC (1959) Fertilization of rabbit ova *in vitro*. Nature 179:466–467
Chang MC (1968) Reciprocal insemination and egg transfer between ferrets and mink. J Exp Zool 168:49–60
Cheng WTK, Polge C, Moor RM (1986) In vitro fertilization of pig and sheep oocytes. Theriogenology 25:146 (abstr)
Cibelli J, Stice SL, Golueke PJ et al (1998) Cloned transgenic calves produced from nonquiescent fetal fibroblasts. Science 280:1256–1258
Comizzoli P, Holt WV (2014) Recent advances and prospects in germplasm preservation of rare and endangered species. Adv Exp Med Biol 753:331–356
Comizzoli P, Mermillod P, Mauget R (2000) Reproductive biotechnologies for endangered mammalian species. Reprod Nutr Dev 40:493–504
Connolly J, Colledge S, Dobney K et al (2011) Meta-analysis of zooarchaelogical data from SW Asia and SE Europe provides insight into the origins and spread of animal husbandry. J Archeol Sci 38:538–545

Cran DG, Johnson LA, Miller NGA, Cochrane D, Polge C (1993) Production of bovine calves following separation of X- and Y-chromosome bearing sperm and in vitro fertilization. Vet Rec 132:40–51

Dekkers J (2004) Commercial application of marker- and gene-assisted selection in livestock: strategies and lessons. J Anim Sci 82(E.Suppl):E313–E328

Diamond J (2002) Evolution, consequences and future of plant and animal domestication. Nature 418:700–707

Edwards RG, Steptoe PC (1978) Birth after the reimplantation of a human embryo. Lancet 312:366

Edwards RG, Bavister BD, Steptoe PC (1969) Early stages of fertilization in vitro of human oocytes matured in vitro. Nature 221:632–635

European Society of Human Reproduction (ESHRE) (2009) ART fact sheet. In: Embryology ESHRE. European Society of Human Reproduction (ESHRE), 2009

Fekete E, Little CC (1942) Observations on the mammary tumor incidence of mice born from transferred ova. Cancer Res 2:525–530

Foote RH (1996) Review: dairy cattle reproductive physiology research and management – past progress and future prospects. J Dairy Sci 79:980–990

Foote RH, Bratton RW (1950) The fertility of bovine semen in extenders containing sulfanilamide, penicillin, streptomycin and polymyxin. J Dairy Sci 33(8):544–547

Fukuda Y, Ichikawa M, Naito K, Toyoda Y (1990) Birth of normal calves resulting from bovine oocytes matured, fertilized, and cultured with cumulus cells in vitro up to the blastocyst stage. Biol Reprod 42:114–119

García-Sancho M (2015) Animal breeding in the age of biotechnology: the investigative pathway behind the cloning of Dolly the sheep. HPLS 37:282–304

Garner DL, Seidel GE Jr (2008) History of commercializing sexed semen in cattle. Theriogenology 69:886–895

Gerbault P, Allaby RG, Boivin N et al (2014) Storytelling and story testing in domestication. Proc Natl Acad Sci U S A 111:6159–6164

Gianola D, Rosa GJM (2015) One hundred years of statistical developments in animal breeding. Annu Rev Anim Biosci 3:19–56

Gonen S, Jenko J, Gorjanc G et al (2017) Potential of gene drives with genome editing to increase genetic gain in livestock breeding programs. Genet Sel Evol 49:3. https://doi.org/10.1186/s12711-016-0280-3

Groenen MAM (2016) A decade of pig genome sequencing: a window on pig domestication and evolution. Genet Sel Evol 48:23. https://doi.org/10.1186/s12711-016-0204-2

Groeneveld LF, Lenstra JA, Eding H et al (2010) Genetic diversity in farm animals – a review. Anim Genet 41:6–31

Gurdon JB (1960) The developmental capacity of nuclei taken from differentiating endoderm cells of Xenopus laevis. J Embryol Exp Morphol 8:505–526

Gurdon JB (1962) The developmental capacity of nuclei taken from intestinal epithelium cells of feeding tadpoles. J Embryol Exp Morphol 10:622–640

Gurdon JB (2017) Nuclear transplantation, the conservation of the genome, and prospects for cell replacement. FEBS J 284:211–217

Gurdon JB, Uehlinger V (1966) "Fertile" intestine nuclei. Nature 210:1240–1241

Hammer RE, Palmiter RD, Pursel VG et al (1985) Production of transgenic rabbits, sheep and pigs by microinjection. Nature 315:680–683

Hartman CG, Lewis WH, Miller FW et al (1931) First findings of tubal ova in the cow, together with notes on oestrus. Anat Rec 48:267–275

Hayes BJ, Lewin HA, Goddard ME (2013) The future of livestock breeding: genomic selection for efficiency, reduced emissions intensity, and adaptation. Trends Genet 29:206–214

Heape W (1891) Preliminary note on the transplantation and growth of mammalian ova within a uterine foster mother. Proc R Soc Lond Biol Sci 48:457–459

Heape W (1897) The artificial insemination of mammals and subsequent possible fertilization or impregnation of their ova. Proc R Soc Lond B 61:52–63

Hochedlinger K, Jaenisch R (2002) Monoclonal mice generated by nuclear transfer from mature B and T donor cells. Nature 415:1035–1038

Human Fertilisation and Embryology Authority (HFEA) (2011) Fertility treatment in 2011: trends and figures in UK: Human Fertilisation and Embryology Authority (HFEA)

Illmensee K, Hoppe PC (1981) Nuclear transplantation in Mus musculus: developmental potential of nuclei from preimplantation embryos. Cell 23:9–18

Johnson MH (2011) Robert Edwards: the path to IVF. Reprod Biomed Online 23:245–262

Johnson LA, Flook JP, Hawk HW (1989) Sex selection in rabbits: live births from X and Y sperm separated by DNA and cell sorting. Biol Reprod 41:199–203

Kellis M, Wold B, Snyder M (2014) Defining functional DNA elements in the human genome. Proc Natl Acad Sci U S A 17:6131–6138

Kinney GM, Pennycook JW, Schriver MD et al (1979) Surgical collection and transfer of canine embryos. Biol Reprod 20:96A

Kraemer DC, Moore GT, Kramen MA (1976) Baboon infant produced by embryo transfer. Science 192:1246–1247

Kues WA, Sudheer S, Herrmann D et al (2008) Genome-wide expression profiling reveals distinct clusters of transcriptional regulation during bovine preimplantation development in vivo. Proc Natl Acad Sci U S A 105:19768–19773

Kvasnitski AV (1951) Interbreed ova transplantations (in Russian). Soc Zooteckh 1:36–42 (Anim Breed Abstr 19:224)

Kvasnitski AV (2001) Research on interbreed ova transfer in pigs. Theriogenology 56:1285–1289

Laible G, Wei J, Wagner S (2015) Improving livestock for agriculture – technological progress from random transgenesis to precision genome editing heralds a new area. Biotechnol J 10:109–120

Larson G, Albarella U, Dobney K et al (2007) Ancient DNA, pig domestication, and the spread of the Neolithic into Europe. Proc Natl Acad Sci U S A 104:15276–15281

Liu Z, Cai Y, Wang Y (2018) Cloning of macaque monkeys by somatic cell nuclear transfer. Cell 172:881–887

Lopyrin AI, Loginova NV, Karpov PL (1950) Changes in the exterior of lambs as a result of interbreed embryonic transfer (in Russian). Dokl Adad Nouk SSSR Ser Biol 74:1019–1021 (Anim Breed Abstr 19, Abstr 1262)

Lopyrin AI, Loginova NV, Karpov PL (1951) The effect of changed conditions during embryogenesis on the growth and development of lambs (in Russian). Sov Zootheh 6:83–95 (Anim Breed Abstr 20, Abstr 729)

McGrath J, Solter D (1983) Nuclear transplantation in mouse embryos. J Exp Zool 228:355–362

McGrath J, Solter D (1984) Inability of mouse blastomere nuclei transferred to enucleated zygotes to support development in vitro. Science 226:1317–1319

McHugh DE, Bradley DG (2001) Livestock genetic origins: goats buck the trend. Proc Natl Acad Sci U S A 98:5382–5384

McLaren A, Biggers JD (1958) Successful development and birth of mice cultivated in vitro as early embryos. Nature 182:877–878

Meuwissen THE, Hayes BJ, Goddard ME (2001) Prediction of total genetic value using genome-wide dense marker maps. Genetics 157:1819–1829

Miller FW, Swett WW, Hartman CG et al (1931) A study of ova from the Fallopian tubes of dairy cows, with a genital history of the cows. J Agric Res 43:627–636

Mutter LR, Graden AP, Olds D (1964) Successful non-surgical bovine embryo transfer. AI Digest 12:3

Nicholas JS (1933) Development of transplanted rat eggs. Proc Soc Exp Biol Med 30:1111–1113

Niemann H (2014) Epigenetics of cloned livestock embryos and offspring. In: Cibelli J, Gurdon J, Wilmut I, Jaenisch R, Lanza R, West MD, KHS C (eds) Principles of cloning. Elsevier, Amsterdam, pp 453–463

Niemann H (2016) Epigenetic reprogramming in mammalian species after SCNT-based cloning. Theriogenology 86:80–90

Niemann H, Tian XC, King WA et al (2008) Epigenetic reprogramming in embryonic and foetal development upon somatic cell nuclear transfer. Reproduction 135:151–163

Niemann H, Kues WA, Lucas-Hahn A et al (2011) Somatic cloning and epigenetic reprogramming in mammals. In: Atala A, Lanza R, Thomson JA, Nerem R (eds) Principles of regenerative medicine. Elsevier., ISBN: 978-0123814227, Amsterdam, pp 129–158

Nowak-Imialek M, Niemann H (2012) Pluripotent cells in farm animals: state of the art and future perspectives. Reprod Fertil Dev 25:103–128

Oguri N, Tsutsumi Y (1974) Non-surgical egg transfer in mares. J Reprod Fertil 41:313–320

Ombelet W, van Robays J (2015) Artificial insemination history: hurdles and milestones. Facts Views Vis Obgyn 7:137–143

Orland B (2017) The invention of artificial fertilization in the eighteenth and nineteenth century. Hist Philos Life Sci 39:11

Perry G (2017) 2016 statistics of embryo collection and transfer in domestic farm animals. IETS Newsletter 37:8–18

Petersen B, Niemann H (2015) Molecular scissors and their application in genetically modified farm animals. Transgenic Res 24:381–396

Pincus G (1930) Observations on the living eggs of the rabbit. Proc R Soc Lond B 107:132–167

Pincus G, Kirsch RE (1936) The sterility in rabbits produced by injections of oestrone and related compounds. Am J Phys 115:219–228

Polge C (2007) The work of the animal research station, Cambridge. Stud Hist Philos Biol Biomed Sci 38:511–520

Polge C, Smith AU, Parkes AS (1949) Revival of spermatozoa after vitrification and dehydration at low temperatures. Nature 164(4172):666

Prather RS, Barnes FL, Sims MM et al (1987) Nuclear transplantation in the bovine embryo: assessment of donor nuclei and recipient oocyte. Biol Reprod 37:859–866

Rothschild MF, Plastow G (2014) Applications of genomics to improve livestock in the developing world. Livest Sci 166:76–83

Schafberg R, Swalve HH (2015) The history of breeding for polled cattle. Livest Sci 179:54–70

Schriver MD, Kraemer DC (1978) Embryo transfer in the domestic feline. Am Ass Lab Anim Sci Publ 78–4:12

Schultz H (1980) History of the international embryo transfer society II. Theriogenology 13:7–11

Shendure J, Balasubramanian S, Church GM et al (2017) DNA sequencing at 40: past, present and future. Nature 550:345–353

Sims M, First NL (1994) Production of calves by transfer of nuclei from cultured inner cell mass cells. Proc Natl Acad Sci U S A 91:6143–6147

Spemann H (1938) Embryonic development and induction. Yale University Press, New Haven, CT, pp 373–398

Steptoe PC, Edwards RG (1978) Birth after reimplantation of a human embryo. Lancet ii:366

Sunderam S, Kissin DM, Crawford SB, Folger SG, Boulet SL, Warner L, Barfield WD (2018) Assisted reproductive technology surveillance – United States, 2015. MMWR Surveill Summ 16:1–28

Takahashi K, Yamanaka S (2006) Induction of pluripotent stem cells from mouse embryonic and adult fibroblast cultures by defined factors. Cell 126:663–676

Telugu BP, Donovan DM, Mark B et al (2016) Genome editing to the rescue: sustainably feeding 10 billion global human population. NIB J. https://doi.org/10.2218/natlinstbiosci.1.2016.1743

Telugu BP, Park KE, Park CH (2017) Genome editing and genetic engineering in livestock for advancing agricultural and biomedical applications. Mamm Genome 28:338–347

Thibier M (2008) Data retrieval committee statistics of embryo transfer – year 2007. IETS Newsletter 26:4–9

Tsunoda Y, Yasui T, Shioda Y et al (1987) Full-term development of mouse blastomere nuclei transplanted into enucleated two-cell embryos. J Exp Zool 242:147–151

Venter JC, Adams MD, Myers EW et al (2001) The sequence of the human genome. Science 291:1304–1351

Vishwanath R (2003) Artificial insemination: the state of the art. Theriogenology 59:571–584

Wakayama T, Perry ACF, Zuccotti M et al (1998) Full-term development of mice from enucleated oocytes injected with cumulus cell nuclei. Nature 394:369–374

Wang GD, Xie HB, Penb MS et al (2014) Domestication genomics: evidence from animals. Annu Rev Anim Biosci 2:65–84

Warwick BL, Berry RO (1949) Inter-generic and intra-specific embryo transfers in sheep and goats. J Hered 40:297–303

Warwick BL, Berry RO, Horlacher WR (1934) Results of mating rams to Angora female goats. In: Proc 27th Ann Meet Am Soc Anim Prod, pp 225–227

Weissmann A (1893) The germ-plasm: a theory of heredity. Translated by W. Newton Parker and Harriet Rönnfeldt. Scribner, New York, NY

Whittingham DG (1971) Survival of mouse embryos after freezing and thawing. Nature 233:125–126

Willadsen SM (1986) Nuclear transplantation in sheep embryos. Nature 320:63–65

Willett EL, Black WG, Casida LH et al (1951) Successful transplantation of a fertilized bovine ovum. Science 113:247

Wilmut I, Rowson LE (1973) Experiments on the low-temperature preservation of cow embryos. Vet Rec 92:686–690

Wilmut I, Taylor J (2018) Cloning after Dolly. Cell Reprogram 20:1–3

Wilmut I, Schnieke AE, McWhir J et al (1997) Viable offspring derived from fetal and adult mammalian cells. Nature 385:810–813

Zeder MA, Bradeley DG, Emshwiller E et al (2006) Documenting domestication: new genetic and archaeological paradigms. University of California Press, Berkeley/Los Angeles, CA

Future Agricultural Animals: The Need for Biotechnology

2

G. E. Seidel Jr.

Abstract

Agricultural animals, by definition, must have utility. There are dozens of desirable agricultural phenotypes, even within a species, and they vary according to the hundreds of agricultural environments on our planet. In the course of domestication and husbandry of animals, phenotypes have continually evolved, a process that has accelerated over the past century. Specifying desirable phenotypes of future farm animals has become exceedingly complex and now includes characteristics such as carbon footprint, minimization of greenhouse gases, and modifying methods and products to adapt to wants of consumers and activists, many of whom have no connection with agriculture.

The tools for attaining phenotypic improvements of animals include increasingly powerful biotechnologies, which are sometimes oversold. In some cases the biotechnologies even drive phenotypes, as, for example, sperm of dairy bulls have become more tolerant of cryopreservation since bulls whose semen does not tolerate cryopreservation leave few progeny due to extensive use of artificial inseminations with frozen semen. In any case, biotechnologies are tools, and should be used to benefit mankind as well as animals. There are costs to making any change in animal agriculture (including making no change), and the benefit to cost ratio should be the main consideration in evaluating a change. Benefits, such as many fewer people killed by bulls through use of artificial insemination, and costs, such as discomfort to animals due to confinement, also need to be considered when evaluating biotechnologies. Baggage such as whether the technology was developed by a company vs. nonprofit organization or whether DNA was modified in the laboratory vs. a "natural" mutation should be minor considerations relative to efficacy, minimizing undesirable side effects, and what is best for the animals and the environment.

G. E. Seidel Jr.
Animal Reproduction and Biotechnology Laboratory (ARBL),
Colorado State University, Fort Collins, CO, USA
e-mail: George.Seidel@colostate.edu

© Springer International Publishing AG, part of Springer Nature 2018
H. Niemann, C. Wrenzycki (eds.), *Animal Biotechnology 1*,
https://doi.org/10.1007/978-3-319-92327-7_2

2.1   Introduction

Although I grew up on a farm with a variety of livestock species and currently own a beef cattle ranch, I have attempted to prepare this chapter as a practicing scientist. While this point of view will be primarily scientific, readers may bring many different perspectives to the discussion, including religion, philosophy, conservation, etc.

The task at hand is to anticipate phenotypes of ideal future agricultural animals and the role of biotechnology in achieving those phenotypes. Mankind evolved with animals, and has depended on them in various ways over the ages. Domestication of animals has greatly enriched our evolution, especially over the last 10 millennia, although in many cultures co-domestication is another way of thinking. For example, in dairying cultures, humans evolved to lifelong functioning of lactase for digestion, and in many situations, companion animals consume embarrassing amounts of resources. Even in many agricultural situations, the bond between persons and animals has similarities with interpersonal bonds. An uptick in depression and suicide among owners of dairy herds decimated by catastrophe has been postulated to be due partly to disruption of such bonds. Despite the above considerations, the main emphasis of this chapter will be in the context of agriculture, i.e., the products and services that domestic animals provide, such as food, fiber, power, and by-products.

What phenotypes are we aiming for in agricultural animals? I have listed some of these in Table 2.1. A second discussion is required about the approaches to attain/maintain these phenotypes, and biotechnology is one of the important tools available (Table 2.2). I also summarize trends in Fig. 2.1, which are likely to continue for the foreseeable future. Biotechnology has to be implemented in the context of these trends.

Agricultural animals, with very few exceptions such as game farms, have been domesticated to various degrees from wild animals. There is some pressure based on ethical grounds to reverse domestication of agricultural animals genetically, i.e., make them more like their wild progenitors. While ethical issues are an important consideration in choosing what phenotypes to aim at in selection, these often are more related to our own moral well-being than what is best from the animal's perspective. An example: all animals will die, and nearly all agricultural animals will be killed

Table 2.1 Examples of desirable phenotypes of agricultural animals	Efficient growth, reproduction, milk production, etc.
	Robust health, disease resistance, etc.
	Suitable end-product characteristics, e.g., meat, milk, hides, athletic performance
	Safety to personnel, docility, etc.
	Environmental fit such as tolerance to heat and parasites
	Minimizing environmental impact, e.g., methane and CO_2 production
	Appropriate, even pleasing, phenotypes for their purposes
	Longevity in breeding populations
	Minimizing variability, predictability
	Profitability and sustainability

Table 2.2 Approaches to attaining/maintaining desirable phenotypes

Selective breeding, selective culling, genomic evaluation
Assisted reproductive technologies
Data collection, evaluation, and use in management
Environmental manipulations, such as shelter, cooling/warming practices, etc.
Management practices such as vaccinations, and castration
Pharmacological interventions such as growth implants, bst, prophylactic antibiotics, methane suppressors, etc.

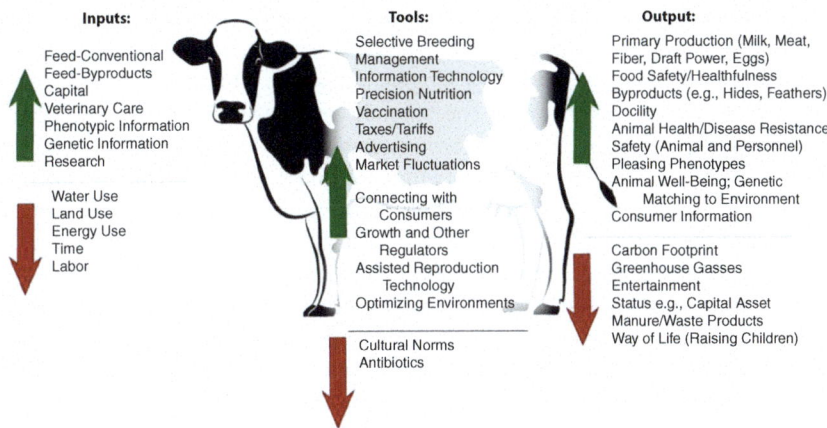

*Sustainability will require optimization of the factors in this figure. The number one requirement for sustainability is profitability. Long term, resources will not be invested in animal agriculture unless profitable, including a "livable wage" for personnel.

Fig. 2.1 Trends per Unit of Animal Products. Sustainability will require optimization of the factors in this figure. The number one requirement for sustainability is profitability. Long term, resources will not be invested in animal agriculture unless profitable, including a "livable wage" for personnel

deliberately, before they live out a "natural" life span. Some methods of killing animals are offensive to people, and this varies greatly from culture to culture. However, if the method is quick, painless, and minimally stressful beforehand, the specific method is of no interest to the animal. The term natural life span is an oxymoron for domesticated animals and especially agricultural animals. If not killed deliberately for production purposes, life spans of domestic animals often greatly exceed those of similar wild animals. Also, I contend that cattle, for example, that are culled for various reasons resulting in beef, die much more humanely than wild counterparts or animals left to die of "old age" without deliberate euthanasia when debilitated. Carnivores such as wolves, buzzards, and mountain lions in North America kill with much more stress and pain than occurs when cattle are killed for beef. One of the costs of managing livestock in more "natural" settings such as mountain pastures or plots for "free-range" chickens is that some will be killed by wild carnivores. Of course, it is the responsibility of those in agriculture to minimize such painful endpoints, both on moral and sustainability grounds.

2.2   Phenotypes

By definition, phenotypes of domestic animals are manipulated by people (although there are also cases of animals farming animals). Even in so-called game farms, outlier animals are culled, and, obviously, subsequent generations result from those animals that in fact reproduced in the environment. Put differently, the phenotypes of the population will change in the direction of those who reproduce, and away from the phenotypes of those not reproducing. Much of Darwin's thinking on evolution was inspired by his observations of which domestic animals reproduce (Darwin 1859).

Later in this paper I will expand on methods of manipulating phenotypes, but for now keep in mind that methods can be genetic, environmental, and managerial plus interactions among these. Several of the phenotypes in Table 2.1 are required even for agricultural animals to exist. The animals must have some minimal level of docility. Animals that routinely kill farmers or routinely escape are not phenotypes that will endure as farm animals. An obvious required trait is reproduction. Also, for the system to survive, it must be profitable. In my opinion, agriculture that is unduly subsidized is more like a zoo than a farm.

After the basic requirements of docility, reproduction, and sustainability are met, farmers have sought to improve animals for traits like health, longevity, product characteristics, and efficiency. All of these must meet minimal standards. For example, unpalatable milk or meat will not suffice, nor wool unsuitable for applications. Farmers with animals that are chronically ill or debilitated will go out of business. There also needs to be some minimal level of efficiency, maximizing the amount of product produced per unit of input such as feed, labor, and especially time.

Over the past few decades attempts to change (improve) phenotypes have concentrated in two broad areas, efficiency and quality of product, while of course retaining minimal levels of nearly all the characteristics in Table 2.1. Results have been spectacular, particularly in efficiency. The amounts of meat and milk produced per unit of feed, labor, environmental impact, time, etc. have increased dramatically (Hume et al. 2011), and this has enabled satisfying the needs/wants of a doubling population with much less than a doubling of inputs such as land use or waste products such as methane production. A particularly important trait is maximizing the number of offspring per breeding female, as this affects profitability, efficiency, sustainability, environmental impact, etc. Also, new traits evolve continually such as suitability of cows to adapt to robotic milking, in which cows enter the milking station at times and frequencies of their own choosing (Jacobs and Siegford 2012).

There, of course, have been some problems/costs associated with the drive for efficiency, for example, lower reproductive success in dairy cows (Lucy 2001), increased dystocia from larger calves in beef cattle, and leg problems in broiler poultry (Paxton et al. 2013). Three points need to be made about this situation:

1. There are costs and benefits to every change, and it is the ratio of benefits to costs that is most important; there will be costs.
2. Costs can be minimized or ameliorated once identified. For example, reproduction in dairy cows is now improving since various reproductive traits are now

included in selection schemes; the nadir in reproduction in North America occurred a decade ago (Garcia-Ruiz et al. 2016). Dystocia has greatly decreased in beef cattle in North America, particularly for primiparous heifers, and there now is almost undue selection pressure for easy calving. Broilers with severe leg problems are unprofitable and morally offensive, and this problem is being countered with genetics, nutrition, etc.

3. Animal agricultural systems are sufficiently complex that it is impossible to anticipate all of the costs (and benefits!) when making phenotypic changes via genetics, environmental manipulations, or other means.

This does not constitute an excuse for not attempting to address these preemptively, but does mean that outcomes should be monitored, problems identified, and changes made to ameliorate or correct or even abandon the planned change. Thus, collecting empirical information is required. Evaluating phenotypic changes via projects at universities, agricultural institutes, government laboratories, etc. is one mechanism to evaluate phenotypic changes effectively.

Many are trapped in thinking that it is best to maintain the status quo, or even aim at phenotypes present decades ago. There are plenty of problems (and benefits) with current and past phenotypes, and doing nothing to improve/correct these cannot be justified ethically.

Recently there has been a shift away from emphasizing direct phenotypic efficiency and concentrating on traits such as health, minimizing pain, and ecological aspects such as carbon footprint, and minimizing production of greenhouse gases (Herrero et al. 2016). Emphasizing these while maintaining efficiency and quality of product is important. This shift in emphasis is due in part to recognizing the importance of nonefficiency traits and in part due to societal demands. Funding agencies also drive research in this direction, and marketing forces have a huge impact. New technologies such as robotic milking, methods to minimize parasites, and approaches to minimize heat stress can be very positive from the perspective of animal welfare.

2.3 Manipulating Phenotypes

Most people, including many scientists, usually first think of using genetics when considering changing phenotypes. As alluded to earlier, huge changes also are made routinely via the environment, particularly via nutrition. Other examples (Table 2.1) are improving health via vaccinations, improving docility via management practices, decreasing birthweight via induced parturition, manipulating health and efficiency via more or less ambulation, and adding methane suppressors to feed (Hristou et al. 2015). There are also innumerable experiments on the interaction of nutrition and/or other management factors with genetics. For example, there appears to be an interaction between genetic background and whether dairy cows get the bulk of their nutrition from grazing or via stored feed (Washburn and Mullen 2014). For efficient beef production, it may be preferable to select replacement heifers managed with limited nutrient intake (Funston and Summers 2013).

There almost always are interactions of genetic propensity for growth or milk production with nutrient availability. For example, beef cattle that produce large amounts of milk survive poorly under semidesert range conditions (which constitute a huge percentage of the worldwide environment for beef cattle); not only do animals end up in poor body condition, they often fail to get pregnant, or pregnancy is greatly delayed under such conditions, whereas other genotypes/phenotypes thrive. Recently, my colleagues and I revived the concept of the single-calf heifer system of producing beef, which at least theoretically is much more efficient than conventional beef production, but has very different optimal phenotypes (Seidel Jr and Whittier 2015). For example, in this system all animals are slaughtered before reaching 30 months of age so longevity is of limited interest, nor is there a need to establish pregnancy during lactation.

Despite these important environmental and managerial aspects of manipulating phenotypes, making genetic changes is quite important and has the huge advantage that, once made, requires minimal inputs; moreover changes are transmitted to future generations. Reproductive biotechnologies are especially valuable in effecting genetic changes (Taylor et al. 2016).

With rare exceptions, genetic changes are almost entirely due to manipulating allelic frequencies. New alleles are constantly being added to the population via mutations, most of which are both recessive and deleterious. However, there are many beneficial recessive alleles.

In any case, the most powerful method of changing allele frequencies to make phenotypic changes has been selective breeding (which is simply choosing the parents of the next generation). Examples of the success of this process abound, such as the more than tenfold differences in mature size and weight among breeds of dogs and horses and similar magnitudes in various traits in cattle, rabbits, poultry, etc. Tools to speed up selective breeding abound, with the spectacularly effective example of artificial insemination in cattle. Of course, one must also consider the less mundane tools such as superovulation, in vitro fertilization, embryo transfer, cloning, sexed semen, etc. which excite the imagination, and while extremely powerful, usually do not measure up to the power of artificial insemination in practice. Of course, these tools are usually superimposed on artificial insemination and its benefits. All of these tools require information such as records, genotypes, etc. to be effective.

A special case is transgenic manipulations, which enable adding specific new alleles at precise locations on chromosomes as opposed to those that occur de novo from spontaneous mutations. So-called genetically modified organisms are of two varieties, those for which intraspecies changes such as making animals polled instead of with horns or moving genes/alleles from heat-tolerant breeds to intolerant breeds. Many contend that these animals should not be designated as genetically modified organisms because identical changes could be made, although slowly, by introgression without any need for molecular manipulations. Modifying animals by using DNA sequences from another species or sequences designed de novo are in a different category and require more stringent testing for safety, etc. However, from a purely logical perspective, most of these deliberate changes are less problematic theoretically than the billions of mutations that occur daily within every species in the natural course of reproduction. In fact, intraspecies alterations also occur naturally as in the *Agrobacterium* genes recently found in sweet potatoes, something that occurred naturally in the course of domestication millennia ago and which mimic exactly in principle what geneticists have done with Bt transgenic crops (Kyndt et al. 2015).

In any case, it would seem that the end product should be the important issue, not how it was produced, and apart from possible negative by-products of the method such as overuse of a resource or creation of an undesirable by-product, it is spectacularly illogical, for example, to discriminate sugar produced from genetically modified vs. nonmodified sugar beets since the sugar is identical in every respect (Oguchi et al. 2009), and the genetically modified beets result in greatly decreased amounts of insecticide and herbicide use, plus a greatly decreased carbon footprint.

Some of the procedures for modifying allele frequencies are more invasive to animals than others, for example, those requiring surgery such as embryo transfer to the oviduct, although even this can be accomplished laparoscopically. Nearly all procedures are invasive in some respect; for example, even simple selective breeding requires identifying specific animals, and as a practical matter, this requires branding, tattooing, or ear tagging as examples. Most people do not object to momentary pain to animals; even vaccinating animals requires restraint and usually an injection with a hypodermic needle.

2.4 Marketing

One of the forces that interferes with use of scientific or even logical consideration of phenotype is marketing forces. For example, retailers use various means to differentiate their products from those of others with terms such as non-GMO, local, natural, organic, bst-free, etc. These labels rarely have a scientific basis in terms of a safer or more efficacious product, although sometimes there is a component of animal welfare such as cage-free eggs. Often the labels are misleading or nonsensical, such as labeling animal products as gluten-free. These marketing forces can be extremely powerful, and often override any scientific considerations. While most of the world is stuck with the consequences of marketing forces, these should not override decisions of policy makers. For example, supplying poor people with nutritious food is often driven by policy makers who are not at all poor, and their biases creep into decision-making.

2.5 Biotechnologies

Individual biotechnologies are discussed in other chapters of this volume, and I will have embarrassingly little to say about them; I have written about them in some detail previously (Seidel Jr 1991, 2015). One issue is simply defining biotechnology. For example, I consider selective breeding, ovulation synchronization, and artificial insemination to be biotechnologies; they are extremely powerful and efficacious. In the context of this chapter, I do not consider vaccination, growth implants, and tools such as ultrasonography as biotechnologies, although they could be argued either way, depending on context. Use of SNP chips for selective breeding (Meuwissen et al. 2013) is another very powerful tool that could be classified either way.

Evaluating individual biotechnologies is complicated by interactions and multiple benefits/costs. For example, sexed semen is useless without artificial insemination or in vitro fertilization. Artificial insemination and superovulation plus embryo transfer require selective breeding information to be efficacious in most contexts, although there are nongenetic benefits such as reduced venereal diseases; interestingly, these technologies even synergize with selective breeding by providing robust data on individuals (Soller 2015). There are the biotechnology-specific phenotypes to deal with such as fertility of cryopreserved semen with artificial insemination, responses to superovulation, accuracy of sexed semen, etc. Determining normalcy of offspring from cloning, transgenics, sexed semen, etc. represents another set of biotechnology-specific endpoints.

Despite these complexities, agriculturists and the rest of the world are stuck with continuing to use biotechnologies if we are to feed the growing world population without destroying the environment. One can argue that there should be less animal and more plant agriculture, but severe reduction in the animal component would waste resources that end up producing food such as by-products fed to animals, inedible plants eaten by grazing ruminants, etc. (Wilkinson 2011). Also, animals are important culturally, and we would be less human (and likely less humane) without animals. An ethical issue would be having to allocate resources to either animals for food vs. animals for companionship. Hopefully we can continue to accommodate both; increased use of biotechnology will be needed to do so.

References

Darwin C (1859) On the origin of species by means of natural selection, or the preservation of favored races in the struggle for life. John Murray, London

Funston RN, Summers AF (2013) Epigenetics: setting up lifetime production of beef cows by managing nutrition. Annu Rev Anim Biosci 1:339–363

Garcia-Ruiz A, Cole JB, Van Raden PM et al (2016) Changes in genetic selection differentials and generation intervals in US Holstein dairy cattle as a result of genomic selection. Proc Natl Acad Sci U S A 113:E3995–E4004

Herrero M, Henderson B, Havlik P et al (2016) Greenhouse gas mitigation potentials in the livestock sector. Nat Clim Change 6:452–461. https://doi.org/10.1038/NCLIMATE2925

Hristou AN, Oh J, Giallongo F et al (2015) An inhibitor persistently decreased enteric methane emission from dairy cows with no negative effect on milk production. Proc Natl Acad Sci U S A 112:10663–10668

Hume DA, Whitelaw CBA, Archibald AL (2011) The future of animal production: improving productivity and sustainability. J Agric Sci 149:9–16

Jacobs JA, Siegford JM (2012) The impact of automatic milking systems on dairy cow management, behavior, health, and welfare. J Dairy Sci 95:2227–2247

Kyndt T, Quispe D, Zhai H et al (2015) The genome of cultivated sweet potato contains Agrobacterium T-DNAs with expressed genes: an example of a naturally transgenic food crop. Proc Natl Acad Sci U S A 112:5844–5849

Lucy MC (2001) Reproductive loss in high producing dairy cows: where will it end? J Dairy Sci 84:1277–1293

Meuwissen T, Hayes B, Goddard M (2013) Accelerating improvement of livestock with genomic selectors. Annu Rev Anim Biosci 1:221–237

Oguchi T, Onishi M, Chikagawa Y et al (2009) Investigation of residual DNAs in sugar from sugar beet (Beta vulgaris L.). J Food Hyg Soc Japan 50:41–46

Paxton H, Daley MA, Corr SA, Hutchinson JR (2013) The gait dynamics of the modern broiler chicken: a cautionary tale of selective breeding. J Exp Biol 216:3237–3248

Seidel GE Jr (1991) Embryo transfer: the next 100 years. Theriogenology 35:171–180

Seidel GE Jr (2015) Lessons from reproductive technology research. Annu Rev Anim Biosci 3:467–487

Seidel GE Jr, Whittier JC (2015) Beef production without mature cows. J Anim Sci 93:4244–4251

Soller M (2015) If a bull were a cow, how much milk would he give? Annu Rev Anim Biosci 3:1–17

Taylor JF, Taylor AH, Decker JE (2016) Holsteins are the genomic selection poster cows. Proc Natl Acad Sci U S A 113:7690–7692

Washburn SP, Mullen KAE (2014) Genetic considerations for various pasture-based dairy systems. J Dairy Sci 97:5923–5938

Wilkinson JM (2011) Re-defining efficiency of feed use by livestock. Animal 5:1014–1022

Artificial Insemination in Domestic and Wild Animal Species

3

Dagmar Waberski

Abstract

Artificial insemination (AI) is the key technology in livestock production for achieving genetic progress and maintenance of genetic diversity. It is also a basic tool for advanced assisted reproductive technologies in animal species. This article reviews the state-of-the-art and current development in AI, including its principle steps, i.e., collection, evaluation, and preservation of semen, as well as various insemination strategies. Opportunities for this first-generation biotechnology are illustrated in domestic and wild animal species against the background of emerging molecular techniques.

3.1 Introduction: Artificial Insemination as the Key Technology in Animal Reproduction

Artificial insemination (AI) is the key technology in livestock production and is critically important for the maintenance of genetic diversity. Moreover, it is fundamental for the use of many other assisted reproductive technologies (ARTs) in domestic and wild animals. In addition, AI has been recognized as the least-invasive, low-cost, and most promising biotechnology for companion animals, non-domestic animals, and endangered species (Durrant 2009). AI comprises the collection and preservation of semen and its manual or instrumental transfer into the female reproductive tract. The basic steps of AI and semen use for other ARTs are illustrated in Fig. 3.1. In the public perception, artificial insemination is often misinterpreted as

D. Waberski
Unit for Reproductive Medicine of Clinics/Clinic for Pigs and Small Ruminants,
University of Veterinary Medicine, Hannover, Germany
e-mail: dagmar.waberski@tiho-hannover.de

© Springer International Publishing AG, part of Springer Nature 2018
H. Niemann, C. Wrenzycki (eds.), *Animal Biotechnology 1*,
https://doi.org/10.1007/978-3-319-92327-7_3

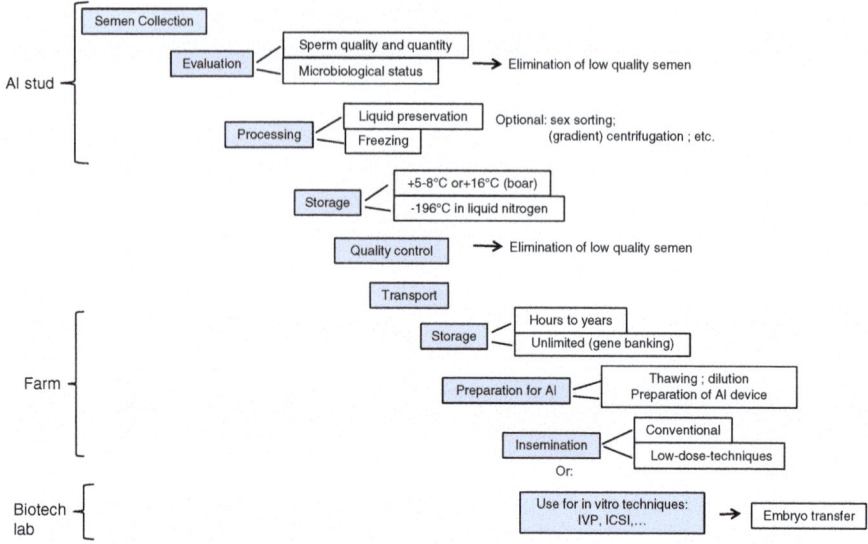

Fig. 3.1 Work flow from semen collection to use for insemination and other assisted reproductive techniques

"artificial fertilization." These terms should be strictly differentiated from each other because of significant different meanings.

As a first generation and to date the most widely used biotechnology, AI evolved during the last century to routine technique in countries with significant livestock production. In today's agricultural industry in the developed countries, AI is used in about 80% of dairy cattle and more than 90% of breeding sows. Efficiency of AI together with progress in second- and third-generation biotechnologies will be important to overcome the increasing energy demand from animal sources created by the anticipated increase in world population from 7.6 billion people in 2017 to 9.8 billion by 2050 (United Nations 2017). Initially, in 1940, the development of commercial AI was triggered by the worldwide threat of venereal diseases in cattle, such as trichomoniasis and brucellosis which led to a dramatic loss of food production and decrease of the global economy. To date, the prevention of direct contact between female and males with artificial breeding remains crucial for prevention of epizootic diseases.

The major technological breakthrough for the application of AI in livestock production was the development of efficient semen preservation methods, particularly the use of frozen-thawed semen, thereby allowing the geographically and timely independent use of male genetics distributed in multiple semen doses. More than 1000 frozen semen doses can be produced from a single bull ejaculate. This has enabled the enormous acceleration of genetic progress, especially when AI was combined with modern reproductive technologies, and more recently genomic selection. Albeit semen cryopreservation seems to be well established in farm animal species, with exception of the pig, the sperm-intrinsic cold shock sensitivity still limits

efficiency of AI. For this reason, and for the great importance of satisfying requirements of semen quality to be used for gene banking, current and emerging technologies relating to freezing technology will be discussed below in a more detailed
manner. Another major contribution to progress in AI was the establishment of AI
management strategies on the farm. With the possibility of sonographic detection of
ovulation, it quickly became apparent that the time interval between insemination
and ovulation is a key factor for successful AI. Strategies for proper timing of insemination, therefore, will be another main topic described in this chapter. A more recent
milestone was the introduction of sex-sorted sperm for use in AI. This technology
stimulated the development of low-dose AI techniques to compensate for the limited
availability and reduced quality of flow cytometric-sex-sorted sperm. Current and
future opportunities of AI technology will be illustrated in domestic and wild mammalian species with major advances or application in AI practice.

3.2   Semen Collection

3.2.1   Artificial Vagina

In domestic farm animals, semen usually is collected from genetically superior,
healthy sires, which have successfully passed breeding soundness and health inspec-
tions. For other species where legislative regulations do not limit trade, e.g., wild or
exotic animals, a clinical andrological examination, including a spermiogram,
should precede semen collections for AI purpose in order to exclude males with
hereditary defects or fertility problems. Various semen collection methods are available. Species-specific mating behavior, prospects for training effects, animal welfare, and effects on quality and quantity of the collected ejaculate may contribute to
the decision about which collection technique is preferable. The goal of any semen
collection is to obtain a complete, high-quality ejaculate, without compromising
safety of the personnel and the animals, including the female teasers.

Male reproductive behavior is initiated by sexual stimuli and can be divided into
three distinct phases with a characteristic sequence of reflexes:

- Precopulatory phase: *excitatio sexualis—erectio penis—emissio penis*
- Copulatory phase: *ascensus—circumplectio—adjustatio penis—immissio penis*
  frictio (species dependent, e.g., stallion, boar) or *propulsus* (e.g., bull, ram,
  buck)—*ejaculatio*
- Postcopulatory phase: *descensus—relaxatio penis—calmatio sexualis*

Ideally, a semen collection method has to be incorporated into the physiological
mating behavior without disturbing the reflex chain. This is accomplished in domestic ruminants, horses, and camelids by use of an artificial vagina (AV), consisting
typically of a rubber tube with a soft non-spermicidal inner liner and an attached
semen collection vessel. In these and other species, thermo receptivity of the *glans
penis* and its sensitivity to pressure are critical for the initiation of ejaculation.

Fig. 3.2 Semen collection
in a stallion at a dummy
using an artificial vagina

Consequently, the artificial vagina must mimic the temperature and pressure conditions of the natural vagina. These conditions are achieved by filling warm water between the outer rubber tube and the inner smooth rubber line. To enhance the stimulatory effect of the AV, the temperature is set slightly above body temperature (41 °C). Environmental conditions are also important for successful semen collection. Sires in AI centers usually are trained to mount artificial dummies (Fig. 3.2), thereby allowing risk-free, easy, and hygienic semen collections. Additional exposure to female teasers, preferentially in estrus, will enhance sexual stimulation, especially in horses. This shortens the reaction time, which is defined as the time interval from presentation of the mounting partner until first mounting, and presents a measure for the *libido sexualis*. Further stimulation may be achieved by false mounting, i.e., mounting without allowing *immissio penis* and subsequent steps of the mating cascade. In contrast to other production animals, semen collection in boars does not require the use of an AV. Instead, the simulation of cervical contractions by rhythmic digital pressure using the "gloved hand technique" or by automatic semen collection systems is essential to initiate ejaculation. This mimics natural mating conditions, where the corkscrew-like tip of the penis is anchored into the helical cervix uteri of the sow during the 5–15 min of the ejaculation phase.

3.2.2   Electroejaculation

For species or individual animals where the use of AV is not possible, for example, due to handling conditions, e.g., zebu (*Bos indicus*) or Bali bulls (*Bos javanicus*) under field conditions in the tropics, or physical conditions, e.g., leg disease, electroejaculation (EE) is the method of choice (Sarsaifi et al. 2013). A sine-wave electrostimulator connected to a rectal probe with three embedded electrodes is used to administer pulse stimuli to autonomic nerves in the pelvic plexus, the lumbar and sacral region involved in reflexes ultimately leading to ejaculation. A lack of response to electrical stimuli may however occur. Indication for use of this technique should be strict because of several concerns. First, EE is associated with pain and therefore consideration of animal welfare is mandatory (Palmer 2005). Animals can be therefore sedated or anesthetized, and the technique should be applied by experienced operators as gently as possible with respect to the pulse frequency and length of rest intervals between wave pulses. Retrograde ejaculation or urine contamination resulting from the use of EE renders the semen unusable. In addition, EE affects the seminal plasma composition and yields ejaculates of larger volumes and lower sperm concentration owing to the direct stimulation of the accessory glands. Nevertheless, reports on Zebu and Brown Swiss bulls (León et al. 1991) as well as in Guirra rams (Marco-Jiménez et al. 2005) suggest that sperm quality is not affected by collection via electrical stimulation, although an increased retention of cytoplasmic droplets due to disrupted removal of droplets as would occur during natural ejaculation conditions cannot be ruled out. Motility of frozen-thawed sperm previously collected by EE was significantly reduced by an average of 10% compared to that collected with an AV in the bull study, whereas post-sperm viability parameters were not impaired in EE semen in Guirra rams. Noteworthy, the birth of offspring of an endangered wild equid subspecies, the Persian onager, from AI using a frozen-thawed electroejaculated semen was reported (Schook et al. 2013), although use of EE in equids is discouraged due to their temperament and the likelihood of trauma to the animal, the operator, or the handlers (Cary et al. 2004).

3.2.3 Transrectal Massage

Alternatives to EE should be considered due to animal welfare concerns or when animals do not respond to EE. Manual transrectal massage (TM) in the area of the ampullae seminal vesicles, prostate, and pelvic urethra was effective in 80% of range beef bulls and from 95% of yearling beef bulls accustomed to handling (Palmer 2005). However, semen quality was reduced compared to EE, presumably due to a higher contamination caused by the inability of penis emission. Semen has been collected successfully from elephant bulls (Schmitt and Hildebrandt 1998) and large ungulates by TM, but massage requires a male sufficiently accustomed to handling and restraint (Durrant 2009).

3.2.4 Collection of Epididymal Spermatozoa

In case of severe illness or accident of a valuable animal, terminal or postmortem recovery of Spermatozoa from the *cauda epididymis* is an applicable tool for germplasm rescue. The retrograde flushing method after cannulating the vas deferens is superior to the flotation technique in excised epididymal tissue pieces due to lower risk of contamination and better spermatozoa quality after freezing (Martinez-Pastor et al. 2006). Compared to ejaculated spermatozoa, cauda epididymal spermatozoa have not been in contact with seminal plasma and are more resistant to stressors associated with cryopreservation, e.g., osmotic imbalances and cold shock. Successful AI using frozen epididymal spermatozoa has been reported for several domestic species (incl. sheep, Ehling et al. 2006; cattle, Bertol et al. 2016; horse, Morris et al. 2002; pig, Holtz and Smidt 1976) and a few endangered species (see Ref. in Durrant 2009).

In pigs and likely in other species as well, fertility of epididymal spermatozoa can be improved adding seminal plasma to the thawing solution, presumably by activation of spermatozoa motility and prevention of premature capacitation (Okazaki et al. 2012). The health of the male, storage duration and conditions of the isolated genital organs postmortem, and the use of freezing protocols specifically adapted for epididymal spermatozoa are critical factors of cryopreservation success. Pregnancies were reported in four out of ten mares after deep uterine AI with frozen-thawed epididymal spermatozoa when the epididymis had been stored at 5 °C up to 48 h, although a significant decline in spermatozoa motility compared to 24-h storage period was observed (Stawicki et al. 2016). Storage of epididymides for approximately 18 h at ambient temperature (18–20 °C) maintains spermatozoa quality and in vitro fertilization capacity of Zebu bull spermatozoa, although low pregnancy levels were achieved even after 30 h of storage (Bertol et al. 2016). Prospects for successful AI using frozen-thawed epididymal spermatozoa are high, even with inferior spermatozoa quality, when AI timing is optimized and low-dose AI techniques are used.

3.3   Semen Evaluation

3.3.1   Basic Spermatology

The goal of domestic farm animal semen evaluation for eventual AI is primarily to eliminate ejaculates with poor fertility potential and to determine sperm concentration in order to calculate the appropriate dilution with the extenders for the intended number of spermatozoa per AI dose. Semen analysis in AI centers is performed within a few minutes after collection on each ejaculate in a production routine. Typically, standard semen parameters are established and consist of visual assessment for contaminants, determination of semen volume, assessment of sperm concentration, and microscopic analysis of sperm motility, followed by an estimation of sperm morphological characteristics of individual spermatozoa. Additionally, in most cases, sperm motility and membrane integrity are reevaluated after a brief storage period or cryopreservation. This is necessary since high-quality native semen does not guarantee a high preservation capacity of the spermatozoa. Evaluation of digitalized microscope images with computer-assisted sperm analysis (CASA) has become increasingly popular in AI laboratories allowing kinematic analyses and detection of distinct morphological abnormalities (reviewed by Amann and Waberski 2014). Recognition of structural integrity, e.g., membrane integrity, can be incorporated in CASA systems equipped with fluorescence modules. Overall, the benefit of CASA in AI centers is its potential to provide objective measurements of sperm motion. With increasing computational power, the introduction of digital holographic imaging of unlabeled (Di Caprio et al. 2015) or fluorescently labeled (Su et al. 2016) spermatozoa into routine semen analyses is anticipated. Such high-throughput computational imaging of sperm in deep (ca. 100 µm) chambers and statistical quantification of large pools of 3D trajectory data offer new possibilities for the characterization of swimming patterns in sperm subpopulations, possibly adding a new tool for the assessment of fertility potential.

Under field conditions in the tropics or the wilderness, standardized semen evaluations may be much more challenging due to lack of available laboratories and/or sophisticated equipment, in addition to sometimes harsh environmental conditions. Lens-free digital (Tseng et al. 2010) or simplified lens (Kobori et al. 2016) microscopes equipped with sample temperature control attached to smartphones or tablet computers may result in easy-to-use, low-cost field-compatible devices. Appropriate software apps would then give instantaneous information regarding semen quality and spermatozoal kinematics or enable telediagnosis by experts in remote laboratories.

3.3.2 Advanced Spermatology

The tendency to increase AI efficiency by lowering sperm numbers per AI dose or the use of "stressed" spermatozoa, e.g., after sex-sorting and/or cryopreservation, requires additional measures to estimate the fertility potential. The high variability in semen quality observed in animals held under extreme field conditions, e.g., Zebu bulls, or the use of genome-selected young bulls during their first months of collection requires subtle semen evaluation procedures in order to determine their suitability for liquid storage or cryopreservation. In addition, advanced sperm assays could help to explain causes of subfertility of individual males with normal spermatological reports, denoted as "idiopathic sub- or infertility." A plethora of sperm assays are available and more are being developed. A comprehensive review of current and potential future methods, including their potential for fertility prognosis, can be found elsewhere (Rodriguez-Martinez 2014). Current methods primarily involve flow cytometric assessment of structural and functional sperm integrity. The simultaneous assessment of plasma membrane and acrosome integrity by a combination of two different fluorophores, e.g., propidium iodide, a membrane impermeable DNA stain, and FITC-labeled peanut agglutinin with its ability to bind acrosomal galactose residues, has become the most widely used application of flow cytometry in spermatology. Plasma membrane integrity often is equated with sperm viability, albeit this trait is far from being predictive of the fertility potential of individual spermatozoa. To take into account the pronounced vulnerability of sperm membranes to cold shock and senescence in vitro, flow cytometric analysis of 10,000 spermatozoa per sample within a few seconds has evolved as an important tool for the assessment of storage effects. The potential of flow cytometry has increased enormously in recent years due to the large numbers of commercially available fluorophore probes targeting virtually all compartments of the spermatozoon. Chromatin structure, formation of reactive oxygen species, lipid peroxidation, and apoptotic-like changes, all, for example, associated with cryopreservation and in vitro aging (Hossain et al. 2011; Martínez-Pastor et al. 2010; Petrunkina and Harrison 2011; Peña et al. 2016), can be assessed. Flow cytometers equipped with at least three lasers in combination with four and more fluorescent markers allow the simultaneous multicolor analyses of traits of individual cells. Therefore, flow cytometry can visualize the heterogeneity of a semen sample by differentiating subpopulations within the cohort of "viable," i.e., plasma membrane intact sperm. Ideally, such analyses would be performed at different time points after incubating sperm in in vitro capacitating media to consider dynamic changes in subpopulations as expected to occur in vivo (Petrunkina et al. 2007). Multicolor flow cytometry in conjunction with advanced computational data analysis may refine the identification of sperm subpopulations responding to capacitation conditions or cryopreservation (Ortega-Ferrusola et al. 2017a). High-end cytometry will remain restricted to specialized andrology laboratories and mainly for research purposes. Meanwhile low-end flow cytometry has become commonplace in AI laboratories, for example, for analyses of membrane integrity of frozen-thawed spermatozoa. Innovation in fluorescence technology is appealing but requires increased human input to setup, performance, and maintenance of the instruments and preparation of semen samples and selection of fluorescence dyes and ultimately in data interpretation (Petrunkina and Harrison 2013). Thus, AI centers may benefit from external specialized laboratories for advanced semen tests, especially for selection of young males, in situations of unexplained subfertility or for quality control of semen processing.

3.3.3   The Future: "-Omics" in Spermatology

In addition to spermatology, cytogenetic or molecular screening is used to detect chromosomal abnormalities affecting fertility of males. Reciprocal translocations are associated with increased embryonic mortality causing significant economic loss, especially in the prolific porcine species (Popescu et al. 1984). It is estimated that approximately 50% of boars with low fertility are carriers of this abnormality, even though they have a normal phenotype and semen profile (Rodríguez et al. 2010).

Current advances in molecular biology related to "-omic sciences" open completely new possibilities for developing male fertility biomarkers based on the analysis of the transcriptome and proteome, both in spermatozoa and seminal plasma. With increasing knowledge of the functionality of spermatozoa, proteomic tools can be complemented with flow cytometry, allowing rapid assays to investigate sperm function at a single-cell level (Ortega-Ferrusola et al. 2017b).

Utilization of molecular diagnostic tools for semen evaluation may go beyond the estimation of fertilizing capacity. Large numbers of datasets have recently been generated and analyzed with the aim to gain a better understanding of the epigenetic mechanisms in sperm and their function postfertilization (Casas and Vavouri 2014). There is now evidence that noncoding microRNAs (miRNA) in the fertilizing spermatozoa, albeit present in small quantity, influence preimplantation development and even the phenotype of the offspring (Jenkins and Carrell 2011). Moreover, it has been shown that traumatic stress in early life alters miRNA expression in mouse sperm leading to behavioral and metabolic responses in the progeny (Gapp et al. 2014). Future "-omic" analysis of semen components should consider seminal plasma proteomes, not only due to their potential to serve as fertility marker but also in light of the recent finding that seminal fluid influences the metabolic phenotype of offspring in mice (Bromfield et al. 2014). At present, genome-wide association studies are being employed to identifying genomic regions and genes associated with semen traits in bulls (Hering et al. 2014a, 2014b), boars (Diniz et al. 2014), and stallions (Gottschalk et al. 2016). Despite the genetic complexity of most sperm traits, SNP markers for poor semen quality have great potential for marker-assisted selection in early life and thus will be of economic value.

3.3.4 Prognosis of Fertility

Despite all the opportunities ahead provided by advanced cellular and genetic semen analysis, any expectation to improve prediction of fertility should remain realistic. Most analyses do not take into account the highly variable interaction of spermatozoa with cells and fluids in the female reproductive tract that are critical for sperm selection, survival, capacitation, and the acrosome reaction (reviewed by Amann et al. 2018), thus rendering assessment of male fertility as imprecise. Recently developed standardizable bioassays, such as sperm migration in microfluidic devices (Suarez and Wu 2016) and complex 3D cultures of female genital tissue (Ferraz et al. 2017; Xiao et al. 2017), have implications for novel sperm diagnosis, at least at research level. Moreover, herd fertility management, particularly the timing of insemination relative to ovulation, has a predominant influence on the AI results, whereas the influence of semen quality often is overestimated. As an example, a long-term study analyzing records from 165,000 inseminated sows in 350 farms with semen doses from 7429 boars revealed that less than 7% of the total variation in farrowing rate and litter size was boar and semen related (Broekhuijse et al. 2012). However, since the impact of a subfertile male on herd fertility may be higher, the primary goal of every semen evaluation is to identify males with a reduced fertility potential. In production animals, such selection must consider the heterogeneity of sperm in a given semen sample and the number of sperm in the AI dose. Some, but not all, sperm defects may be compensated by higher sperm numbers (Saacke 2000), but AI efficiency may not be compromised, at least in livestock breeding.

3.4 Semen Preservation

3.4.1 Aims and Principles

Freshly ejaculated spermatozoa are activated by seminal plasma ingredients and gradually lose their fertilization capacity within the first hours after collection. Therefore, in a situation where breeding partners are not at the same location,

Table 3.1 Sperm numbers per dose used for conventional artificial insemination in domestic animal species

	Liquid-preserved semen	Frozen-thawed semen
Bull	5×10^6	15×10^6
Boar	2×10^9	5×10^9
Stallion	200×10^6 progressive motile	800×10^6 progressive motile
Ram	$50–100 \times 10^6$	150×10^6
Dog	Total sperm-rich fraction	$100–200 \times 10^6$ progressive motile
Camel	$80–150 \times 10^6$	$150–300 \times 10^6$

effective semen preservation is mandatory. The primary goal is to maintain the fertilizing potential of spermatozoa over a long period of time in a pathogen-free milieu. In addition, efficiency of male gamete use should be increased by dilution with semen extenders and producing multiple insemination doses.

The principle features of semen preservation are:
– Nutrition and protection of spermatozoa: energy substrates, buffers, osmolytes, membrane protectants (e.g., antioxidants), cryoprotectants in semen extenders.
– Concentration of spermatozoa and removal of seminal plasma (freezing only).
– Microbial control: antibiotics in semen extenders, according to national legislative standards, serve as a second line of defense against bacteria causing decreased fertility or pregnancy loss. The first defense is always periodic, routine health evaluations of the stud males.
– Portioning spermatozoa into AI doses and mechanical protection: filling in plastic semen tubes or bags (5–100 ml) for liquid storage or in plastic straws (0.25 or 0.5 ml) for frozen storage.
– Prevention of cold shock: controlled cooling regime to storage temperature.
– Immobilization and reduction of sperm metabolism at low storage temperatures, i.e., liquid semen at 16 °C (boar) or 5 °C (most other species), frozen semen at −196 °C in liquid nitrogen.

The number of sperm per semen dose differs between species and depends on the type of preservation method (liquid/frozen); see Table 3.1.

3.4.2 Semen Cryopreservation

Cryopreservation is the preferred preservation method allowing indefinite storage of male gametes, international trade of superior genetics, and screening for pathogens prior to use. For biobanking, storage in the frozen state is indispensable. However, there are limitations specifically related to spermatozoa. The distinct composition and thermotropic phase behavior of membrane lipids render spermatozoa susceptible to cold shock, the extent of which varies between species. The vast majority of commercially marketed bull semen is cryopreserved, whereas less than 1% frozen semen is used in pig AI. Boar spermatozoa are especially sensitive to chilling damage even at supra-zero temperature which has been attributed to their relatively high content of polyunsaturated fatty acids and a low sterol-to-phospholipid ratio (Parks and Lynch 1992). Differences in membrane lipid composition are associated with freezing-relevant biophysical properties such as the permeability for water and the cryoprotectant glycerol or osmotic tolerance limits (Holt 2000). Breed and male-to-male differences within species discriminate sires into "good," "average," and "poor freezers" pointing to a genetically determined variation as shown by amplified restriction fragment length polymorphism (AFLP) technology in Large White boars (Thurston et al. 2002). This could render sperm freezability as a promising candidate for marker-assisted selection. Even in those domestic animal species regarded as more cryotolerant, e.g., horse and dog, cryopreservation is lethal to a significant proportion (circa 30–50%) of spermatozoa directly after thawing. Moreover, there is evidence from various sperm function tests that the surviving sperm population experience sublethal damage resulting in abnormal sperm transport, altered sperm-oviduct interaction, and shortened survival in the female reproductive tract (Watson 1995). Consequently, sperm injury must be compensated by typically doubling the number of sperm per AI dose compared to liquid-preserved semen and by intense AI management to optimize the time and technique of insemination. Damage to cells during freezing and thawing strongly depends on the cooling rate: slow cooling (about 5 °C/min) results in dehydration due to hypertonic conditions induced by extracellular ice formation, whereas rapid cooling (>100 °C/min) leads to the formation of damaging intracellular ice crystals (Mazur 1963). Typically, sperm diluted in freezing extenders are slowly cooled from room temperature to 5 °C at a rate of approximately −0.1 °C to −0.3 °C/min, followed by freezing at a rate of −10 to −60 °C/min down to a temperature of −80° or −120 °C, after which samples are plunged into liquid nitrogen (−196 °C). Adaptation of cooling velocity to a medium rate is not sufficient to overcome freezing/thawing injury because of membrane phase transitions and resulting disturbance of cell homeostasis which begins already at supra-zero temperatures. The high degree of structural and functional variability in spermatozoa renders calculations of optimal cooling rates inaccurate. Even though knowledge of biophysical membrane properties of sperm from domestic animals has increased, to date cryopreservation protocols remained mostly empirical with only minor modifications over decades.

3.4.2.1 Cryoprotectants

Many efforts have been invested in development and testing of cryoprotectants. Sperm cryopreservation requires the use of cryoprotectants, preferentially those with minimal cell toxicity, high efficiency, and low risk of introducing contaminants. The action of cryoprotectants even increases the biological complexity of the cellular response to cooling and freezing. Still, the most widely used cryoprotectants are glycerol and dimethyl sulfoxide (DMSO). These membrane-permeable agents exert their effect mainly by inhibition of intracellular ice formation. Cryoprotective actions vary for different agents and are influenced by concentration, presence of other solutes, and the exposure temperature of the spermatozoa. Non-membrane-permeable cryoprotectants include osmotically active molecules such as disaccharides, e.g., sucrose and trehalose, and osmotically inactive macromolecules, e.g., polyvinylpyrrolidone (PVP), hydroxyethyl starch, and dextran. These promote cell dehydration, lower the freezing point, and increase the viscosity of media, thus together inhibiting ice crystal formation. The mode of action of cryoprotectants for sperm preservation has been comprehensively reviewed (Holt and Penfold 2014; Sieme et al. 2016). Typically, freezing extenders for domestic livestock species semen contain egg yolk for the protection of membranes as the most vulnerable sperm compartment. Because of its animal origin with the risk for transmission of diseases and its undefined mode of action, efforts continue to replace egg yolk by synthetic or plant lipoproteins, particularly soybean extract (Layek et al. 2016). Addition of antioxidants to extenders is a further approach for the reduction of cryoinjury caused by the formation of reactive oxygen species (ROS) during cooling and thawing (reviewed by Amidi et al. 2016). This strategy also applies for liquid-preserved semen stored under hypothermic conditions.

3.4.2.2 Semen Freezing in Rare and Endangered Species
Reviews of the actual state of the art in sperm cryopreservation, including alternative gonadal tissue preservation, from rare and endangered species have been recently published (Comizzoli and Holt 2014; Spindler et al. 2014; Comizzoli 2015). Biobanks are expanding, for example, currently 800,000 semen samples from 20,000 individuals of 14 different mammalian and nonmammalian species are stored as part of the National Animal Germplasm Program (2016) in the USA. Taxon-inherent seminal traits and sensitivity of spermatozoa to cryopreservation limit the adaptation of freezing protocols from related domestic species. Despite the absence of specific membrane biophysical data, slightly modified standard freezing procedures with glycerol as the cryoprotectant yield acceptable post-thaw results in many different species. In exotic species, however, successful AI with frozen semen has only rarely been reported, especially due to the lack of knowledge of the corresponding female reproductive physiology.

3.4.2.3 Alternative Freezing Strategies

Directional Freezing
Directional freezing techniques shall reduce cryoinjury by preventing uncontrolled ice nucleation. This is accomplished by a multi-thermal gradient device consisting of one warm (+5 °C) and one cold (−50 °C) block with a gap in between to create a temperature gradient. Straws containing semen are moved with precise velocity from the warm to the cold block and then transferred to an even cooler collection chamber (−100 °C), thus avoiding the effect of "supercooling," a process which would damage the spermatozoa due to sudden and fast formation of ice crystals. This technique can now be applied to larger volumes (up to 12 ml) which has been successfully used in domestic animals and wildlife species as well (Arav and

Saragusty 2016). Use of directional frozen semen led to first reported pregnancies after AI with frozen-thawed semen in rhinoceros (Hermes et al. 2009) and elephants (Hildebrandt et al. 2012). More in vitro and in vivo trials and comparisons with conventional freezing could explore the full potential for the use in AI practice.

Vitrification

Ultra-rapid freezing (2000 °C/min) by direct exposure of extended semen samples (with or without cryoprotectants) to liquid nitrogen was successfully applied with human sperm and that from other species. Using this "vitrification" technique, samples instantly reach a glasslike state without formation of deleterious ice crystals. Progress has been made toward an increase of sample volume (currently up to 0.5 ml) and protection against contamination by liquid nitrogen (Isachenko et al. 2005, 2011). Ice crystal formation must be avoided by careful temperature control during rewarming of the sample in the devitrification process. Because of its relative simplicity, low cost, and "field-friendliness", vitrification is of future interest for preservation of semen from endangered or wild species. Details of this and other alternative freezing protocols can be found in the handbook *Cryopreservation and Freeze-Drying Protocols* (Wolkers and Oldenhof 2015).

Freeze-Drying

Sperm freeze-drying would allow easy and low-cost semen storage at supra-zero temperature (4 °C–21 °C) without the need for liquid nitrogen. This method involves a multistep process including primary and secondary drying and two phase transitions to arrive at completely dried samples. It basically follows the principles of anhydrobiosis occurring in nature. To date, freeze-drying is deleterious to most sperm components including DNA, thus precluding its use for AI or IVF (Keskintepe and Eroglu 2015; Gil et al. 2014). However, despite DNA fragmentations, freeze-dried sperm can be successfully used for intracytoplasmic sperm injection (ICSI; Wakayama and Yanagimachi 1998) and therefore is an option for germplasm banking.

3.4.3 Liquid Semen Preservation

Preservation of semen in the liquid state, i.e., at temperatures above 0 °C, is preferred in species with cold-shock-sensitive sperm (e.g., porcine), in males with poor semen quality but high genetic value, or in sires with the best genetics to extend the number of insemination doses. It also may be beneficial for sperm stressed by sex-sorting (Xu 2014). The main advantages of liquid preservation are (1) low cost, (2) avoidance of chilling injury thus preserving higher sperm quality, and (3) low carbon footprint. Sperm numbers reaching the site of fertilization and longevity of the oviductal sperm population are higher, making AI management more flexible compared to cryopreserved sperm. In practice, AI with lower numbers of liquid-preserved spermatozoa often results in higher fertility results. The main drawbacks of liquid semen preservation are the limited life span in vitro (typically a few days) and the higher risk of bacterial growth, at least if the semen is stored at temperatures above 4 °C. In most domestic animal species (bovine, equine, canine), spermatozoa are chilled to temperatures between 4 and 10 °C for the purpose of restricting the metabolic rate during storage, therefore reducing the depletion of ATP and the production of detrimental by-products, e.g., ROS. Similar to cryopreserved semen, animal-derived compounds, e.g., milk and egg yolk, are incorporated into most species-specific extenders to protect sperm membranes from chilling injury. Storage at room temperature would circumvent the need for specific membrane stabilizers and increase the economic use of an ejaculate due to a higher sperm quality; however this has been demonstrated to limit the fertile life span to approximately 12 h in stallion sperm. In this case, the major concern is that the ongoing oxidative phosphorylation production of significant quantities of ROS results in compromised sperm function (Gibb and Aitken 2016).

Several antioxidants and membrane stabilizers such as synthetic surfactants or exogenous lipids presented to the plasma membranes in microvesicles or pre-loaded cyclodextrins have been studied to date. These may not "over-stabilize" surface membranes, thereby maintaining an active, well-balanced oxidative system, which is essential for capacitation (Leahy and Gadella 2011). In boars, liquid semen is traditionally stored between 16 and 18 °C, thus taking into account the critical lower-limit temperature of 15 °C for irreversible loss of membrane integrity and function. Effective long-term extenders (boar semen) allow storage up to 7 days.

In addition, the risk of generating and spreading multiresistant bacteria by the use of critically important antibiotics in boar semen extenders came recently into focus. This is particularly crucial in pigs because of the comparatively large volume of the insemination doses (60–100 ml) used worldwide in sow herds. The development of alternative antimicrobial approaches in pig AI includes rigorous sanitary measures during semen processing (see Fig. 3.3), the search for substitutes for conventional antibiotics, and the development of sperm quality-compatible concepts for hypothermic storage below 10 °C (Schulze et al. 2015, 2016).

In domestic camelids, semen is mostly stored at 4 °C for a maximum of 48 h in ruminant semen extenders. To date, cryopreservation success is poor in this species, presumably due to the high viscosity of the ejaculate, the low ejaculate volume, and the low sperm concentration (Skidmore et al. 2013; Tibary et al. 2014). Even though liquid semen preservation generally is less harmful to sperm function compared to cryopreservation, sperm handling may alter the sperm surface by dilution effects and shearing forces that may cause the removal of protective extracellular matrix components originating from the seminal plasma, e.g., decapacitating factors (Leahy and Gadella 2011). Excessive dilution may therefore decrease fertilization rates, albeit absolute sperm number in the insemination dose is sufficiently high. Moreover, it must be considered that lipid phase transitions in sperm of many mammalian species occur in a temperature range between 30 °C and 10 °C causing leakage of solutes across membranes (Drobnis et al. 1993). To prevent cold shock

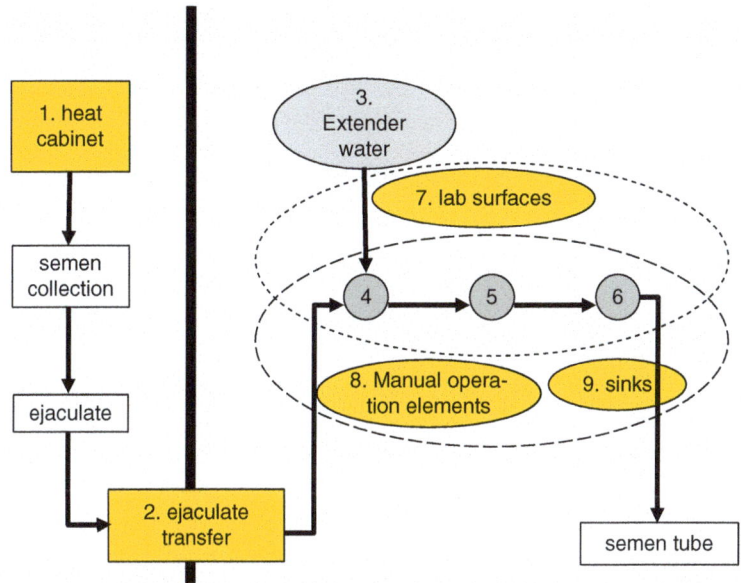

Fig. 3.3 The workflow and nine hygienic critical control points in semen processing. Yellow: Highest risk of bacterial contamination as assessed in two consecutive audits in 24 boar AI centers. 4, extender; 5, inner face of dilution tank lids; 6, semen dyes. Modified from Schulze et al. (2015)

damage, isothermic dilution and slow cooling to the desired storage temperature are crucial.

3.5 Insemination Management

3.5.1 Timing of Insemination

The optimal timing of insemination relative to ovulation is crucial for the success of AI, especially if semen of lower quality is used. Noteworthy, if insemination takes place in a narrow time window of 4 h prior to ovulation, fertility results using cryopreserved and liquid-stored boar semen do not differ (Waberski et al. 1994), even though boar spermatozoa are regarded as particularly sensitive to cooling stress. Species-specific optimal AI timing is illustrated in Fig. 3.4. In most spontaneous-ovulating domestic species, e.g., horse, pig, sheep, and goat, insemination shortly before ovulation will yield the highest pregnancy results, since the life span of spermatozoa is limited to 12 h with frozen semen and between 12 and 36 h (species dependent) with liquid-stored semen. Exceptions are some canids, where ovulation of primary oocytes of domestic dogs and farmed foxes requires a postovulatory maturation period between 2 and 3 days before they are capable of being fertilized (Thomassen and Farstad 2009). Additionally, oocytes and nonfrozen spermatozoa

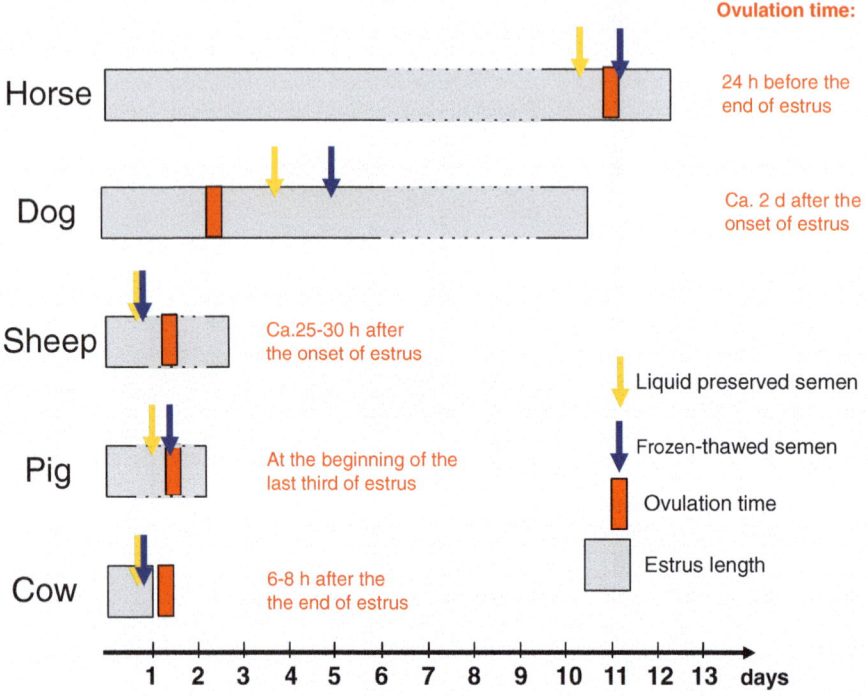

Fig. 3.4 Illustration of typical estrus duration, time of ovulation (typed in red), and optimal insemination time in domestic animal species with spontaneous ovulation. Dotted lines in bars indicate variation in estrus duration. Timing of AI must be adapted accordingly

of these species maintain fertilizing capacity for several days allowing a more flexible, postovulatory AI timing. In most other mammalian species, postovulatory insemination usually has low success rates due to rapid aging of oocytes within the first 4–8 h after ovulation associated with a loss of fertilization capacity or early embryonic death (Hunter 2003). In addition to the senescence of gametes, species-specific differences in the duration of sperm transport to the oviductal sperm reservoir may be important for optimum AI timing. In cows, sustained functional sperm transport to the site of fertilization in the oviduct requires a minimum of 6 h following insemination (Hunter and Wilmut 1983). In addition, high pregnancy rates in cows have been recognized as a compromise between early insemination, resulting in low fertilization rates (due to sperm aging) but good embryo quality, and late insemination characterized by high fertilization rates but low embryo quality due to reduced selection pressure of sperm present only a short time in the oviductal reservoir (Saacke et al. 2000).

Timing of insemination may be even more challenging in induced (reflex) ovulating species, such as felids and camelids. In these species, neural signals from copulatory stimuli trigger hypothalamic secretion of gonadotropin-releasing hormone (GnRH) as the inducer for an endocrine and paracrine cascade leading to ovulation. Interestingly, in camelids and in spontaneous ovulators, including cattle, horse, and pig, a highly conserved ovulation-inducing factor (OIF) was found in seminal plasma. In Bactrian camels, llamas, and alpacas, OIF triggers ovulation by release of pituitary LH (for review see Adams and Ratto 2013). Recently, it was demonstrated that llamas can be induced to ovulate by "insemination" of seminal plasma in the absence of copulation and that copulation alone cannot induce ovulation in the absence of seminal plasma (Berland et al. 2016), whereas the role of OIF in spontaneous ovulators remains to be elucidated. This brings the concept of "facultative-induced ovulators" proposed by Jöchle (1975) into new light. The porcine species provides an example, where seminal plasma was shown to advance spontaneous ovulation by a locally active mechanism (Waberski et al. 1995). Given the complexity of physiological processes necessary for ensuring the fusion of gametes in a state of full fertilizing competence and the limited knowledge on regulation of female reproduction by seminal fluid, future research is necessary to incorporate such knowledge into improved AI strategies.

3.5.1.1 Estrus Detection and Ovulation

In spontaneously cycling animals, detection of estrus is crucial for proper timing of AI because in most species ovulation occurs at a relatively fixed time point during or after estrus. Estrus, also known as "positive standing reflex," is defined as the period where the female tolerates mounting of the male. In modern farming and in wild life, mounting activity is often difficult to observe. Indirect measures are therefore used as signs for the approaching estrus, e.g., edema and hyperemia of the vulva, vaginal mucus, restlessness, and other changes of behavior. Manual provocation of the standing reflex by massage of the lumbar-sacral region in cows or by the "back pressure test" in sows is helpful to identifying females in estrus. Exposure of the female to a male teaser is important because this stimulates the onset of estrus and is essential for early and timely estrus detection, especially in nulliparous females. Insufficient stimuli during estrus detection leads to apparent shorter estrus length and hence inaccurate prediction of ovulation time (Langendijk et al. 2000).

Since the degree of estrus expression has low heritability and varies individually between females and even within female from one estrus period to the other (Roelofs et al. 2010), a careful estrus detection in each cycle is crucial. Health status and environmental conditions are further factors influencing estrus behavior. The interval and intensity of estrus detection are of utmost importance for reliable and early estrus detection. Generally, monitoring for estrus activity is recommended twice daily at an interval of 8–12 h.

Continuous monitoring for estrus behavioral symptoms clearly would be advantageous. A battery of electronic monitoring systems is commercially available, especially for use in large dairy cow farms, including radiotelemetric mount sensors attached to the lumbar-sacral region of the cows which record mounting activity and frequency by a herd mate, telemetric pedometers strapped to the cow's leg, or collars with digital signal processing chips recording physical activity, different modes to measure estrus-associated increased body temperature, electronic "noses" to identify pheromonal odor in vaginal mucus, and online progesterone monitoring systems based on applying a milk sample to an electrochemical biosensor and reading the electrical response (reviewed by Fricke et al. 2014; Mottram 2016). Precise prediction of ovulation time is as yet not achievable, but implementation of these systems in AI programs may increase estrus detection rates in dairy farms. Without doubt, the most reliable tool to estimate ovulation time is manual palpation and ultrasound examination of ovaries to determine size, shape, and the maturity of the Graafian follicles. These methods are also routinely applied in mares; however, they are not practical in most other species.

In wild and exotic species, information of the female cycle has primarily been derived from noninvasive monitoring of estrogen and progestagen metabolites from fecal samples. Comparative studies in related wild and domestic species have revealed metabolic and endocrine differences affecting ovarian activity and estrus behavior. Moreover, captivity may alter reproductive physiology of wild populations, including seasonality of estrus behavior as demonstrated in cheetahs and the maned wolf (Comizzoli et al. 2009). Due to a variety of erroneous assumptions, the simple technology transfer of AI programs from domestic to wild species often has led to failure of breeding programs (Durrant 2009). Most wild animals, e.g., wild felids, in ex situ populations do not reliably exhibit overt signs of estrus, and behavioral indices of sexual receptivity are inconsistent or too difficult (or dangerous) to identify. As a result, the most effective means for timing of AI in these and rare species is to stimulate ovarian activity using exogenous gonadotropins (Howard and Wildt 2009).

3.5.1.2 Induction of Ovulation for Fixed-time Insemination

As an alternative to the time-consuming and often inaccurate estrus detection, insemination can be timed after hormonal treatment to synchronize follicular growth, corpus luteum regression, and ovulation. With increasing herd sizes and advanced knowledge of ovarian physiology, fixed-time artificial insemination (FTAI) programs have been introduced in dairy cows since the late 1990s. In beef cattle, FTAI can successfully increase the AI rate, which is currently less than 10% in the USA. Numerous reports describe treatment protocols, influencing factors, and outcome from FTAI in dairy and beef cattle (for recent reviews see Bisinotto et al. 2014; Bó et al. 2016; Colazo and Mapletoft 2014; Wiltbank and Pursley 2014). In countries where estradiol is banned for treatment of livestock (in North America, Europe, New Zealand), modified GnRH-based protocols published by Pursley et al. (1995) are used. GnRH injected at a random stage of the cycle promotes ovulation of the dominant follicle and induces a new follicular wave. Injections of PGF2α 7 days later cause regression of all corpora lutea. Forty-eight hours later, cows are given a second injection of GnRH to induce ovulation of the new dominant follicle. Insemination is then performed 24 h later irrespective of the presence of estrous signs. Modifications of this protocol, known as "Ovsynch," including a second injection of PGF2α and intravaginal progesterone application through controlled internal drug release (CIDR) inserts, have been adopted. In the USA, research projects for optimizing FTAI protocols in beef cattle are being reinforced by the Beef Reproduction Task Force, and annual updates of recommendations are published

(Johnson et al. 2011). Limitation to use FTAI programs remains their yet relatively poor accuracy to induce ovulation of competent oocytes and consumer's acceptance of hormonal treatments in production animals. Ovsynch-based FTAI protocols are also applied to other domestic species, such as goats, sheep, water buffalo, and yaks (reviewed in Wiltbank and Pursley 2014).

Similarly, estrus synchronization protocols are commonly used in small ruminant AI in order to decrease variation in onset of estrus or for fixed-time AI (reviewed by Romano 2013). In pigs, efficient fixed-timed AI protocols were developed and widely used in large sow units in former East Germany (Brüssow et al. 1996; Hühn et al. 1996). The goal was to synchronize all reproductive events so that periodic and group-wise insemination would allow the hygienically advantageous "all-in-all-out" system and would improve the work flow in the barn without the need for estrus detection. Key elements of these protocols were the synchronization of estrus by the use of the steroid progestin altrenogest in gilts, group weaning in sows, stimulation of follicular development using equine chorionic gonadotropin (eCG), and induction of ovulation using human chorionic gonadotropin (hCG) or GnRH analogues. At present, variations of FTAI protocols, including the vaginal administration of the GnRH agonist triptorelin (Stewart et al. 2010), are being investigated with the aim to reduce the number of inseminated sperm per cycle, thus allowing a wider use of boars of high genetic merit and a reduction of labor in sow farms (Knox 2014; de Rensis and Kirkwood 2016). This would be possible by reducing the number of inseminations per cycle and by the use of lower sperm numbers per dose with a single AI performed close to ovulation. In domestic horses, timed-single AI is often desired in spontaneous cycles, especially if expensive frozen semen is used. However, in mares, as in other species, repeated injections of hCG may induce an immune response resulting in failure to induce ovulation (Roser et al. 1979; Swanson et al. 1995). Alternatively, ovulation can be successfully induced with repeated injections of the GnRH analogue buserelin or using a short-term subcutaneous implant releasing the GnRH analogue deslorelin (Jöchle 1994; Squires et al. 1994; Hemberg et al. 2006).

In many wild and endangered species, the most consistent AI protocols still require induction and/or synchronization of ovulation using exogenous gonadotropins due to the difficulty to detect estrus and unknown ovulation times. However, hormonal protocols used in related domestic species can be ineffective or cause ovarian hyperstimulation and abnormal oocyte/embryo development (Pukazhenthi and Wildt 2004). In Przewalski's horses and felids, estrous cycles can be synchronized and ovulatory follicles developed by administering altrenogest in combination with PGF2α, thus offering strategies for the use of AI in critically endangered species (Howard and Wildt 2009; Collins et al. 2014). Encouraging results have been reported using eCG/hCG injections or GnRH agonists injected or implanted for the induction of ovulation in felids, canids, camelids, and other species (canids, Asa et al. 2006; Johnson et al. 2014; felids, Graham et al. 2006; Howard and Wildt 2009). Exogenous hormonal induction of ovulation seems to be more successful in induced ovulating species, e.g., camel, most felids, and the maned wolf, because ovulation normally would not occur in the absence of gonadotropins or gonadotropin-releasing hormones (Howard and Wildt 2009).

3.5.2   Insemination Techniques

3.5.2.1 Conventional Insemination

Successful insemination in most species relies on sperm deposition into the uterus, especially when using frozen semen. In comparison to fresh semen, frozen semen may show impaired sperm transport and a shorter sperm survival time in the female reproductive tract. Uterine semen deposition is relatively easy to perform in large females, where either the cervix can be positioned by manual transrectal guidance (cow, camel) or a transvaginal manual or instrumental insemination with visual ultrasound monitoring (mare) is possible. A flexible plastic pipette with an inserted straw of semen and a thin steel plunger is positioned in the cranial part of the cervix in order that the semen can then be released into the *corpus uteri*. In pigs, single-use plastic catheters with spiral tips are "screwed" into the cervix, and the semen is then slowly released from storage tubes or bags into the uterus, thus mimicking the natural mounting procedure. AI techniques are more challenging in species with complex anatomical vaginae and/or cervices, e.g., caprine, elephants, and rhinoceros. The goat cervix can be penetrated and the inseminating dose deposited into the uterus in approximately 25%–60% of multiparous females. In ewes, intrauterine insemination by the transcervical approach requires special restraint systems and insemination devices with a bent tip (reviewed in Cseh et al. 2012; Romano 2013). AI in elephants with frozen semen has been successful with the guidance of custom-made insemination catheters by endoscopy and ultrasonography to the distal vagina at the cervical os (Hildebrandt et al. 2012). Similarly, transcervical AI in domestic dogs and cats can be performed using vaginal endoscopy (Romagnoli and Lopate 2014; Zambelli et al. 2015) and is especially recommended for frozen semen.

3.5.2.2 Laparoscopic Insemination

Laparoscopic artificial insemination through the abdominal wall allows semen to be placed directly into the lumen of the uterine horns close to the uterotubal junction, thus overcoming the hindered sperm transport through tortuous folds and crypts, semen backflow, and sperm phagocytosis. This technique became routine in commercial sheep operations using frozen semen, whereas effectivity (safety and success) in other species is still under debate (Vazquez et al. 2008). When using frozen sex-sorted sperm or when only small numbers of spermatozoa are available, laparoscopic AI either into the tip of the uterine horns or into the ampulla of the oviduct is an option, for example, in pigs (del Olmo et al. 2014). After transcervical or even vaginal insemination, offspring have been reported in programs aimed at the conservation of wildlife species, but in most cases laparoscopic AI is more promising. As an example, laparoscopic intrauterine artificial insemination has been successfully used to enhance the dissemination of founder descendants in wild carnivores where male offspring from wild-caught individuals were underrepresented (Comizzoli et al. 2009). However, in addition to the general risks of this invasive approach, it must be considered that anesthesia associated with laparoscopy might affect ovulation, sperm transport, and subsequent establishment of pregnancy, as reported in felids (Howard and Wildt 2009).

3.5.2.3 Low-Dose Insemination

In the last decade, different insemination techniques, which permit the use of lower sperm numbers, have evolved in several species, primarily livestock (c.f. cow, López-Gatius 2000; horse, Samper and Plough 2010; pig, Vazquez et al. 2008; Bortolozzo et al. 2015; small ruminants, de Graaf et al. 2007; camel, Skidmore et al. 2013). Low-dose insemination is preferred for several reasons: (1) the emerging use of sex-sorted sperm in various species due to the to-date limited sorting efficiency and the harvest of sperm with poorer quality, (2) the increased efficiency of the use of males with the consequent reduction of fixed costs and reduction of the female population (pig, Gonzalez-Peña et al. 2014), (3) the acceleration of genetic progress by frequent use of higher indexing sires (pig, Knox 2016), (4) the increase in the availability of semen from males in high demand (horse, Samper and Plough 2010), (5) the use of low-quality semen from genetically valuable males, or (6) the use of

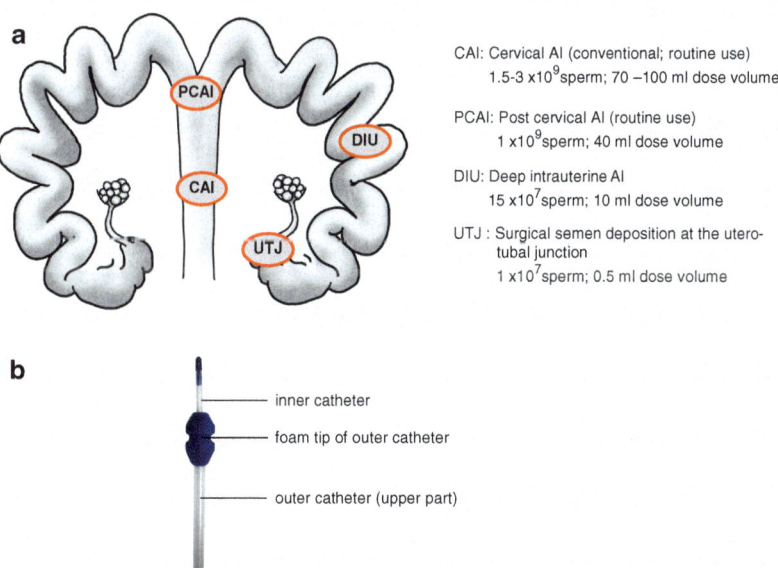

CAI: Cervical AI (conventional; routine use)
     $1.5\text{-}3 \times 10^9$ sperm; 70–100 ml dose volume

PCAI: Post cervical AI (routine use)
      1×10^9 sperm; 40 ml dose volume

DIU: Deep intrauterine AI
     15×10^7 sperm; 10 ml dose volume

UTJ : Surgical semen deposition at the utero-
      tubal junction
      1×10^7 sperm; 0.5 ml dose volume

inner catheter

foam tip of outer catheter

outer catheter (upper part)

Fig. 3.5 (**a**) Schematic drawing of the porcine female genital tract with different sites of semen deposition. Insemination closer to site of fertilization (oviduct) results in a gradual decrease of minimum sperm number and volume of semen dose required for insemination success. Semen deposition deeply intrauterine or at the uterotubal junction allows the use of sex-sorted sperm but is not yet common practice in pigs. (**b**) Insemination catheter used for postcervical insemination (PCI) in sows. It consists of an outer catheter whose foam tip gets introduced into the cervix. This location corresponds to traditional cervical insemination (CAI). For PCI a second inner catheter is inserted and moved approximately 10 cm toward the uterine body

epididymal spermatozoa, usually only available in limited numbers after a terminal semen collection. Insemination of sperm in low numbers requires semen deposition close to the site of fertilization at an optimal time relative to ovulation. In cattle, horses, and camels, however, nonsurgical techniques are performed by rectally guided deep uterine horn insemination ipsilateral to the ovary with a preovulatory follicle. In cows and camels, limitations of these methods are the requirement for gentle palpation of the ovaries by well-trained inseminators to avoid diagnostic errors and manual induction of premature ovulation. Deep insemination in mares has been extended to hysteroscopic AI using a long [approximately 1.5 m] flexible endoscope which is inserted vaginally and rectally guided through the uterine horn to the oviductal papilla. A small volume of semen is then slowly deposited onto the papilla through delivery systems introduced in the working channel of the endoscope (Morris et al. 2000). In sows after their first parity, postcervical AI has become routine in some countries allowing a threefold reduction of sperm numbers compared to conventional AI (Fig. 3.5a). A double catheter system is used where the inner catheter is gently pushed forward toward the uterine body (Fig. 3.5b). Deep intrauterine insemination is challenging in this species because of the long (about 2 m) convoluted uterine horns and the impossibility of rectal guidance but can be achieved using a specially designed flexible catheter (Martinez et al. 2002). This allows a 20–60-fold reduction in the number of spermatozoa inseminated but is inconvenient for field use. Insemination of very low doses of semen to date would be necessary for the use of sex-sorted spermatozoa in pigs but requires laparoscopic insemination or alternative assisted reproductive techniques as discussed elsewhere in this book.

Conclusion

Artificial insemination being the most traditional and widest used biotechnology in animal reproduction currently evolves with new trends in multidisciplinary fields, including basic research on reproductive physiology, computational power, and cytometric engineering. Powered by emerging knowledge on proteomics and genomics and yet available technology to modify genetic codes, new goals appear to be achievable. Overall, artificial insemination targets to increase the efficient use of germplasm of high genetic males and to provide strategies for long-term survival of non-domestic and endangered species. New developments in technology must be sustainable and compatible with animal welfare, and, at least in production animals, cost-effective for AI industry and farmers.

References

Adams GP, Ratto MH (2013) Ovulation-inducing factor in seminal plasma: a review. Anim Reprod Sci 136:148–156

Amann RP, Saacke RG, Barbato GF, Waberski D (2017) Measuring male-to-male differences in fertility or effects of semen treatments. Annu Rev Anim Biosci 6:255–286. https://doi.org/10.1146/annurev-animal-030117-014829 [Epub ahead of print]

Amann RP, Waberski D (2014) Computer-assisted sperm analysis (CASA): capabilities and potential developments. Theriogenology 81:5–17

Amidi F, Pazhohan A, Shabani Nashtaei M, Khodarahmian M, Nekoonam S (2016) The role of antioxidants in sperm freezing: a review. Cell Tissue Bank 17(4):745–756

Arav A, Saragusty J (2016) Directional freezing of sperm and associated derived technologies. Anim Reprod Sci 169:6–13

Asa CS, Bauman K, Callahan P, Bauman J, Volkmann DH, Jöchle W (2006) GnRH-agonist induction of fertile estrus with either natural mating or artificial insemination, followed by birth of pups in gray wolves (Canis lupus). Theriogenology 66:1778–1782

Berland MA, Ulloa-Leal C, Barría M, Wright H, Dissen GA, Silva ME, Ojeda SR, Ratto MH (2016) Seminal plasma induces ovulation in llamas in the absence of a copulatory stimulus: role of nerve growth factor as an ovulation-inducing factor. Endocrinology 157:3224–2332

Bertol MA, Weiss RR, Kozicki LE, Abreu AC, Pereira JF, da Silva JJ (2016) In vitro and in vivo fertilization potential of cryopreserved spermatozoa from bull epididymides stored for up to 30 hours at ambient temperature (18 °C–20 °C). Theriogenology 86:1014–1021

Bisinotto RS, Ribeiro ES, Santos JE (2014) Synchronisation of ovulation for management of reproduction in dairy cows. Animal 8(Suppl 1):151–159

Bó GA, de la Mata JJ, Baruselli PS, Menchaca A (2016) Alternative programs for synchronizing and resynchronizing ovulation in beef cattle. Theriogenology 86:388–396

Bortolozzo FP, Menegat MB, Mellagi AP, Bernardi ML, Wentz I (2015) New artificial insemination technologies for swine. Reprod Domest Anim 50(Suppl 2):80–84

Broekhuijse ML, Soštarić E, Feitsma H, Gadella BM (2012) The value of microscopic semen motility assessment at collection for a commercial artificial insemination center, a retrospective study on factors explaining variation in pig fertility. Theriogenology 77:1466–1479

Bromfield JJ, Schjenken JE, Chin PY, Care AS, Jasper MJ, Robertson SA (2014) Maternal tract factors contribute to paternal seminal fluid impact on metabolic phenotype in offspring. Proc Natl Acad Sci U S A 111:2200–2205

Brüssow KP, Jöchle W, Hühn U (1996) Control of ovulation with a GnRH analog in gilts and sows. Theriogenology 46:925–934

Cary JA, Madill S, Farnsworth K, Hayna JT, Duoos L, Fahning ML (2004) A comparison of electroejaculation and epididymal sperm collection techniques in stallions. Can Vet J 45:35–41

Casas E, Vavouri T (2014) Sperm epigenomics: challenges and opportunities. Front Genet 5(330):1–5

Colazo MG, Mapletoft RJ (2014) A review of current timed-AI (TAI) programs for beef and dairy cattle. Can Vet J 55:772–780

Collins CW, Monfort SL, Vick MM, Wolfe BA, Weiss RB, Keefer CL, Songsasen N (2014) Oral and injectable synthetic progestagens effectively manipulate the estrous cycle in the Przewalski's horse (Equus ferus przewalskii). Anim Reprod Sci 148:42–52

Comizzoli P, Crosier AE, Songsasen N, Gunther MS, Howard JG, Wildt DE (2009) Advances in reproductive science for wild carnivore conservation. Reprod Domest Anim 44(Suppl 2):47–52

Comizzoli P (2015) Biobanking efforts and new advances in male fertility preservation for rare and endangered species. Asian J Androl 17:640–645

Comizzoli P, Holt WV (2014) Recent advances and prospects in germplasm preservation of rare and endangered species. Adv Exp Med Biol 753:331–356

Cseh S, Faigl V, Amiridis GS (2012) Semen processing and artificial insemination in health management of small ruminants. Anim Reprod Sci 130:187–192

de Graaf SP, Evans G, Maxwell WM, Downing JA, O'Brien JK (2007) Successful low dose insemination of flow cytometrically sorted ram spermatozoa in sheep. Reprod Domest Anim 42:648–653

del Olmo D, Parrilla I, Sanchez-Osorio J, Gomis J, Angel MA, Tarantini T, Gil MA, Cuello C, Vazquez JL, Roca J, Vaquez JM, Martinez EA (2014) Successful laparoscopic insemination with a very low number of flow cytometrically sorted boar sperm in field conditions. Theriogenology 81(2):315–320

De Rensis F, Kirkwood RN (2016) Control of estrus and ovulation: fertility to timed insemination of gilts and sows. Theriogenology 86:1460–1466

Di Caprio G, Ferrara MA, Miccio L, Merola F, Memmolo P, Ferraro P, Coppola G (2015) Holographic imaging of unlabelled sperm cells for semen analysis: a review. J Biophotonics 8:779–789

Diniz DB, Lopes MS, Broekhuijse ML, Lopes PS, Harlizius B, Guimarães SE, Duijvesteijn N, Knol EF, Silva F (2014) A genome-wide association study reveals a novel candidate gene for sperm motility in pigs. Anim Reprod Sci 151:201–207

Durrant BS (2009) The importance and potential of artificial insemination in CANDES (companion animals, non-domestic, endangered species). Theriogenology 71:113–122

Drobnis EZ, Crowe LM, Berger T, Anchordoguy TJ, Overstreet JW, Crowe JH (1993) Cold shock damage is due to lipid phase transitions in cell membranes: a demonstration using sperm as a model. J Exp Zool 265:432–437

Ehling C, Rath D, Struckmann C, Frenzel A, Schindler L, Niemann H (2006) Utilization of frozen-thawed epididymal ram semen to preserve genetic diversity in Scrapie susceptible sheep breeds. Theriogenology 66:2160–2164

Ferraz MAMM, Henning HHW, Stout TAE, Vos PLAM, Gadella BM (2017) Designing 3-dimensional in vitro oviduct culture systems to study mammalian fertilization and embryo production. Ann Biomed Eng 45:1731–1744 Erratum in: Ann Biomed Eng. 2016 Nov 28

Fricke PM, Carvalho PD, Giordano JO, Valenza A, Lopes G Jr, Amundson MC (2014) Expression and detection of estrus in dairy cows: the role of new technologies. Animal 8(Suppl 1):134–343

Gapp K, Jawaid A, Sarkies P, Bohacek J, Pelczar P, Prados J, Farinelli L, Miska E, Mansuy IM (2014) Implication of sperm RNAs in transgenerational inheritance of the effects of early trauma in mice. Nat Neurosci 17:667–669

Gibb Z, Aitken RJ (2016) The impact of sperm metabolism during in vitro storage: the stallion as a model. Biomed Res Int 2016:9380609

Gil L, Olaciregui M, Luño V, Malo C, González N, Martínez F (2014) Current status of freeze-drying technology to preserve domestic animals sperm. Reprod Domest Anim 49(Suppl 4):72–81

Gonzalez-Peña D, Knox RV, Pettigrew J, Rodriguez-Zas SL (2014) Impact of pig insemination technique and semen preparation on profitability. J Anim Sci 92:72–84

Gottschalk M, Metzger J, Martinsson G, Sieme H, Distl O (2016) Genome-wide association study for semen quality traits in German Warmblood stallions. Anim Reprod Sci 171:81–86

Graham LH, Byers AP, Armstrong DL, Loskutoff NM, Swanson WF, Wildt DE, Brown JL (2006) Natural and gonadotropin-induced ovarian activity in tigers (Panthera tigris) assessed by fecal steroid analyses. Gen Comp Endocrinol 147:362–370

Hemberg E, Lundeheim N, Einarsson S (2006) Successful timing of ovulation using deslorelin (Ovuplant) is labour-saving in mares aimed for single ai with frozen semen. Reprod Domest Anim 41:535–537

Hering DM, Olenski K, Kaminski S (2014a) Genome-wide association study for poor sperm motility in Holstein-Friesian bulls. Anim Reprod Sci 146:89–97

Hering DM, Oleński K, Ruść A, Kaminski S (2014b) Genome-wide association study for semen volume and total number of sperm in Holstein-Friesian bulls. Anim Reprod Sci 151:126–130

Hermes R, Göritz F, Saragusty J, Sós E, Molnar V, Reid CE, Schwarzenberger F, Hildebrandt TB (2009) First successful artificial insemination with frozen-thawed semen in rhinoceros. Theriogenology 71:393–399

Hildebrandt TB, Hermes R, Saragusty J, Potier R, Schwammer HM, Balfanz F, Vielgrader H, Baker B, Bartels P, Göritz F (2012) Enriching the captive elephant population genetic pool through artificial insemination with frozen-thawed semen collected in the wild. Theriogenology 78:1398–1404

Holt WV (2000) Fundamental apsects of sperm cryobiology: the importance of species and individual differences. Theriogenology 53:47–58

Holt WV, Penfold LM (2014) Fundamental and practical aspects of semen cryopreservation. In: Chenoweth P, Lorton SP (eds) Animal andrology. CABI Oxfordshire, UK, pp 76–99

Holtz W, Smidt D (1976) The fertilizing capacity of epididymal spermatozoa in the pig. J Reprod Fertil 46:227–229

Hossain MS, Johannisson A, Wallgren M, Nagy S, Siqueira AP, Rodriguez-Martinez H (2011) Flow cytometry for the assessment of animal sperm integrity and functionality: state of the art. Asian J Androl 13:406–419

Howard JG, Wildt DE (2009) Approaches and efficacy of artificial insemination in felids and mustelids. Theriogenology 71:130–148

Hühn U, Jöchle W, Brüssow KP (1996) Techniques developed for the control of estrus, ovulation and parturition in the east German pig industry: a review. Theriogenology 46:911–924

Hunter RHF (2003) Physiology of the Graafian follicle and ovulation. Cambridge University Press, Cambridge

Hunter RHF, Wilmut L (1983) The rate of functional sperm transport into the oviducts of mated cows. Anim Reprod Sci 5:167–173

Isachenko V, Isachenko E, Montag M, Zaeva V, Krivokharchenko I, Nawroth F, Dessole S, Katkov II, van der Ven H (2005) Clean technique for cryoprotectant-free vitrification of human spermatozoa. Reprod Biomed Online 10:350–354

Isachenko V, Maettner R, Petrunkina AM, Mallmann P, Rahimi G, Sterzik K, Sanchez R, Risopatron J, Damjanoski I, Isachenko E (2011) Cryoprotectant-free vitrification of human spermatozoa in large (up to 0.5 mL) volume: a novel technology. Clin Lab 57:643–650

Jenkins TG, Carrell DT (2011) The paternal epigenome and embryogenesis: poising mechanisms for development. Asian J Androl 13:76–80

Jöchle W (1975) Current research in coitus-induced ovulation: a review. J Reprod Fertil Suppl 22:165–207

Jöchle W (1994) Control of ovulation in the mare with Ovuplant TM. A short-term release implant (STI) containing the GnRH analogue deslorelin acetate: studies from 1990 to 1994 (a review). J Equine Vet Sci 14:632–644

Johnson SK, Funston RN, Hall JB, Kesler DJ, Lamb GC, Lauderdale JW, Patterson DJ, Perry GA, Strohbehn DR (2011) Multi-state beef reproduction task force provides science-based recommendations for the application of reproductive technologies. J Anim Sci 89:2950–2954

Johnson AE, Freeman EW, Colgin M, McDonough C, Songsasen N (2014) Induction of ovarian activity and ovulation in an induced ovulator, the maned wolf (Chrysocyon brachyurus), using GnRH agonist and recombinant LH. Theriogenology 82:71–79

Keskintepe L, Eroglu A (2015) Freeze-drying of mammalian sperm. Methods Mol Biol 1257:489–497

Kobori Y, Pfanner P, Prins GS, Niederberger C (2016) Novel device for male infertility screening with single-ball lens microscope and smartphone. Fertil Steril 106:574–578

Knox RV (2014) Impact of swine reproductive technologies on pig and global food production. Adv Exp Med Biol 752:131–160

Knox RV (2016) Artificial insemination in pigs today. Theriogenology 85:83–93

Langendijk P, Soede NM, Bouwman EG, Kemp B (2000) Responsiveness to boar stimuli and change in vulvar reddening in relation to ovulation in weaned sows. J Anim Sci 78:3019–3026

Layek SS, Mohanty TK, Kumaresan A, Parks JE (2016) Cryopreservation of bull semen: evolution from egg yolk based to soybean based extenders. Anim Reprod Sci 172:1–9

Leahy T, Gadella BM (2011) Sperm surface changes and physiological consequences induced by sperm handling and storage. Reproduction 142:759–778

León H, Porras AA, Galina CS, Navarro-Fierro R (1991) Effect of the collection method on semen characteristics of zebu and European type cattle in the tropics. Theriogenology 36:349–355

López-Gatius F (2000) Site of semen deposition in cattle: a review. Theriogenology 53:1407–1414

Marco-Jiménez F, Puchades S, Gadea J, Vicente JS, Viudes-de-Castro MP (2005) Effect of semen collection method on pre- and post-thaw Guirra ram spermatozoa. Theriogenology 64:1756–1765

Martinez EA, Vazquez JM, Roca J, Lucas X, Gil MA, Parrilla I, Vazquez JL, Day BN (2002) Minimum number of spermatozoa required for normal fertility after deep intrauterine insemination in non-sedated sows. Reproduction 123:163–170

Martinez-Pastor F, Garcia-Macias V, Alvarez M, Chamorro C, Herraez P, de Paz P, Anel L (2006) Comparison of two methods for obtaining spermatozoa from the cauda epididymis of Iberian red deer. Theriogenology 65:471–485

Martínez-Pastor F, Mata-Campuzano M, Alvarez-Rodríguez M, Alvarez M, Anel L, de Paz P (2010) Probes and techniques for sperm evaluation by flow cytometry. Reprod Domest Anim 45(Suppl 2):67–78

Mazur P (1963) Kinetics of water loss from cells at subzero temperatures and the likelihood of intracellular freezing. J Gen Physiol 47:347–369

Morris LH, Hunter RHF, Allen WR (2000) Hysteroscopic insemination of small numbers of spermatozoa at the uterotubal junction of preovulatory mares. J Reprod Fertil 118:95–100

Morris L, Tiplady C, Allen WR (2002) The in vivo fertility of cauda epididymal spermatozoa in the horse. Theriogenology 58:643–646

Mottram T (2016) Animal board invited review: precision livestock farming for dairy cows with a focus on oestrus detection. Animal 10:1575–1584

National Animal Germplasm Program 2016 (http://nrrc.ars.usda.gov/A-GRIN/main_webpage/ars?record_source=US) Accessed 28 Oct 2016

Okazaki T, Akiyoshi T, Kan M, Mori M, Teshima H, Shimada M (2012) Artificial insemination with seminal plasma improves the reproductive performance of frozen-thawed boar epididymal spermatozoa. J Androl 33:990–998

Ortega-Ferrusola C, Anel-López L, Martín-Muñoz P, Ortíz-Rodríguez JM, Gil MC, Alvarez M, de Paz P, Ezquerra LJ, Masot AJ, Redondo E, Anel L, Peña FJ (2017a) Computational flow cytometry reveals that cryopreservation induces spermptosis but subpopulations of spermatozoa may experience capacitation-like changes. Reproduction 153:293–304

Ortega-Ferrusola C, Gil MC, Rodríguez-Martínez H, Anel L, Peña FJ, Martín-Muñoz P (2017b) Flow cytometry in Spermatology: a bright future ahead. Reprod Domest Anim 52:921–931

Palmer CW (2005) Welfare aspects of theriogenology: investigating alternatives to electroejaculation of bulls. Theriogenology 64:469–479

Parks JE, Lynch DV (1992) Lipid composition and thermotropic phase behavior of boar, bull, stallion, and rooster sperm membranes. Cryobiology 29:255–266

Peña FJ, Ortega Ferrusola C, Martín Muñoz P (2016) New flow cytometry approaches in equine andrology. Theriogenology 86:366–372

Petrunkina AM, Waberski D, Günzel-Apel AR, Töpfer-Petersen E (2007) Determinants of sperm quality and fertility in domestic species. Reproduction 134:3–17

Petrunkina AM, Harrison RA (2011) Cytometric solutions in veterinary andrology: developments, advantages, and limitations. Cytometry A 79:338–348

Petrunkina AM, Harrison RA (2013) Fluorescence technologies for evaluating male gamete (dys)function. Reprod Domest Anim 48(Suppl 1):11–24

Popescu CP, Bonneau M, Tixier M, Bahri I, Boscher J (1984) Reciprocal translocations in pigs. Their detection and consequences on animal performance and economic losses. J Hered 75:448–452

Pukazhenthi BS, Wildt DE (2004) Which reproductive technologies are most relevant to studying, managing and conserving wildlife? Reprod Fertil Dev 16:33–46

Pursley JR, Mee MO, Wiltbank MC (1995) Synchronization of ovulation in dairy cows using PGF2alpha and GnRH. Theriogenology 44:915–923

Rodríguez A, Sanz E, De Mercado E, Gómez E, Martín M, Carrascosa C, Gómez-Fidalgo E, Villagómez DA, Sánchez-Sánchez R (2010) Reproductive consequences of a reciprocal chromosomal translocation in two Duroc boars used to provide semen for artificial insemination. Theriogenology 74:67–74

Rodriguez-Martinez H (2014) Semen evaluation and handling: emerging techniques and future development. In: Chenoweth P, Lorton SP (eds) Animal andrology. CABI Oxfordshire, UK, pp 509–549

Roelofs J, López-Gatius F, Hunter RH, van Eerdenburg FJ, Hanzen C (2010) When is a cow in estrus? Clinical and practical aspects. Theriogenology 74:327–344

Romano JE (2013) Assisted reproductive techniques in small ruminants. Clin Theriogenol 5:293–310

Romagnoli S, Lopate C (2014) Transcervical artificial insemination in dogs and cats: review of the technique and practical aspects. Reprod Domest Anim 49(Suppl 4):56–63

Roser JF, Kiefer BL, Evans JW, Neely DP, Pacheco DA (1979) The development of antibodies to human chorionic gonadotrophin following its repeated injection in the cyclic mare. J Reprod Fertil Suppl 27:173–179

Saacke RG, Dalton JC, Nadir S, Nebel RL, Bame JH (2000) Relationship of seminal traits and insemination time to fertilization rate and embryo quality. Anim Reprod Sci 60-61:663–677

Samper JC, Plough T (2010) Techniques for the insemination of low doses of stallion sperm. Reprod Domest 45(Suppl 2):35–39

Sarsaifi K, Rosnina Y, Ariff MO, Wahid H, Hani H, Yimer N, Vejayan J, Win Naing S, Abas MO (2013) Effect of semen collection methods on the quality of pre- and post-thawed Bali cattle (Bos javanicus) spermatozoa. Reprod Domest Anim 48:1006–1012

Schook MW, Wildt DE, Weiss RB, Wolfe BA, Archibald KE, Pukazhenthi BS (2013) Fundamental studies of the reproductive biology of the endangered persian onager (Equus hemionus onager) result in first wild equid offspring from artificial insemination. Biol Reprod 89(41):1–13

Schmitt DL, Hildebrandt TB (1998) Manual collection and characterization of semen from Asian elephants (Elephas maximus). Anim Reprod Sci 53:309–314

Schulze M, Ammon C, Rüdiger K, Jung M, Grobbel M (2015) Analysis of hygienic critical control points in boar semen production. Theriogenology 83:430–437

Schulze M, Dathe M, Waberski D, Müller K (2016) Liquid storage of boar semen: current and future perspectives on the use of cationic antimicrobial peptides to replace antibiotics in semen extenders. Theriogenology 85:39–46

Sieme H, Oldenhof H, Wolkers WF (2016) Mode of action of cryoprotectants for sperm preservation. Anim Reprod Sci 169:2–5

Skidmore JA, Morton KM, Billah M (2013) Artificial insemination in dromedary camels. Anim Reprod Sci 136:178–186

Spindler R, Keeley T, Satake N (2014) Applied andrology in endangered, exotic and wildlife species. In: Chenoweth P, Lorton SP (eds) Animal andrology. CABI Oxfordshire, UK, pp 450–473

Squires EL, Moran DM, Farlin ME, Jasko DJ, Keefe TJ, Meyers SA, Figueiredo E, McCue PM, Jochle W (1994) Effect of dose of GnRH analog on ovulation in mares. Theriogenology 41:757–769

Stawicki RJ, McDonnell SM, Giguère S, Turner RM (2016) Pregnancy outcomes using stallion epididymal sperm stored at 5 °C for 24 or 48 hours before harvest. Theriogenology 85:698–702

Stewart KR, Flowers WL, Rampacek GB, Greger DL, Swanson ME, Hafs HD (2010) Endocrine, ovulatory and reproductive characteristics of sows treated with an intravaginal GnRH agonist. Anim Reprod Sci 120:112–119

Su TW, Choi I, Feng J, Huang K, Ozcan A (2016) High-throughput analysis of horse sperms' 3D swimming patterns using computational on-chip imaging. Anim Reprod Sci 169:45–55

Suarez SS, Wu M (2016) Microfluidic devices for the study of sperm migration. Mol Hum Reprod 23(4):227–234

Swanson WF, Horohov DW, Godke RA (1995) Production of exogenous gonadotrophin-neutralizing immunoglobulins in cats after repeated eCG-hCG treatment and relevance for assisted reproduction in felids. J Reprod Fertil 105:35–41

Thomassen R, Farstad W (2009) Artificial insemination in canids: a useful tool in breeding and conservation. Theriogenology 71:190–199

Thurston LM, Siggins K, Mileham AJ, Watson PF, Holt WV (2002) Identification of amplified restriction fragment length polymorphism markers linked to genes controlling boar sperm viability following cryopreservation. Biol Reprod 66:545–554

Tibary A, Pearson LK, Anouassi A (2014) Applied andrology in camelids. In: Chenoweth P, Lorton SP (eds) Animal andrology. CABI Oxfordshire, UK, pp 418–449

Tseng D, Mudanyali O, Oztoprak C, Isikman SO, Sencan I, Yaglidere O, Ozcan A (2010) Lens free microscopy on a cellphone. Lab Chip 10:1787–1792

United Nations, Department of Economic and Social Affairs, Population Division (2017). World population prospects: the 2017 revision, key findings and advance tables. Working Paper No. ESA/P/WP.241. https://esa.un.org/unpd/wpp/publications/Files/WPP2017_KeyFindings.pdf. Accessed 6 June 2018

Vazquez JM, Roca J, Gil MA, Cuello C, Parrilla I, Caballero I, Vazquez JL, Martínez EA (2008) Low-dose insemination in pigs: problems and possibilities. Reprod Domest Anim 43(Suppl 2):347–354

Waberski D, Weitze KF, Gleumes T, Schwarz M, Willmen T, Petzoldt R (1994) Effect of time of insemination relative to ovulation on fertility with liquid and frozen boar semen. Theriogenology 42:831–840

Waberski D, Südhoff H, Hahn T, Jungblut PW, Kallweit E, Calvete JJ, Ensslin M, Hoppen HO, Wintergalen N, Weitze KF, Töpfer-Petersen E (1995) Advanced ovulation in gilts by the intra-uterine application of a low molecular mass pronase-sensitive fraction of boar seminal plasma. J Reprod Fertil 105:247–252

Wakayama T, Yanagimachi R (1998) Development of normal mice from oocytes injected with freeze-dried spermatozoa. Nat Biotechnol 16:639–641

Watson PF (1995) Recent developments and concepts in the cryopreservation of spermatozoa and the assessment of their post-thawing function. Reprod Fertil Dev 7:871–891

Wiltbank MC, Pursley JR (2014) The cow as an induced ovulator: timed AI after synchronization of ovulation. Theriogenology 81:170–185

Wolkers FW, Oldenhof (eds) (2015) Cryopreservation and freeze-drying protocols, 3rd edn. Springer, New York

Xiao S, Coppeta JR, Rogers HB, Isenberg BC, Zhu J, Olalekan SA, McKinnon KE, Dokic D, Rashedi AS, Haisenleder DJ, Malpani SS, Arnold-Murray CA, Chen K, Jiang M, Bai L, Nguyen CT, Zhang J, Laronda MM, Hope TJ, Maniar KP, Pavone ME, Avram MJ, Sefton EC, Getsios S, Burdette JE, Kim JJ, Borenstein JT, Woodruff TK (2017) A microfluidic culture model of the human reproductive tract and 28-day menstrual cycle. Nat Commun 28:14584

Xu ZZ (2014) Application of liquid semen technology improves conception rate of sex-sorted semen in lactating dairy cows. J Dairy Sci 97:7298–7304

Zambelli D, Bini C, Cunto M (2015) Endoscopic transcervical catheterization in the domestic cat. Reprod Domest Anim 50:13–16

Technique and Application of Sex-Sorted Sperm in Domestic Farm Animals

4

Detlef Rath and Chis Maxwell

Abstract

The Food and Agriculture Organization of the United Nations has recognised that the production of pre-sexed livestock by sperm or embryo sexing as a useful breeding tool to increase production efficiency, especially for traits that are sex-related. In this chapter, we briefly explain sex determination in mammals, review approaches to identifying X and Y chromosome-bearing sperm and their practical implications for semen handling and artificial insemination (AI) and compare their importance and success in the main farm animal species. The problems associated with current technology for sperm sexing, as reflected in the damage caused to mammalian sperm are then considered, followed by an assessment of the potential for replacing this technology by other methods.

In mammals, the most efficient method to bias sex ratios in offspring is to separate X and Y chromosome-bearing sperm by flow cytometry before insemination. Numerous other techniques purporting to alter the sex ratio have been proposed or discussed. None of these were able to produce significant separation of fertile X and/or Y sperm populations or were not repeatable. Only quantitative methods, which differentiate between X and Y sperm on the basis of total DNA and then apply flow cytometric sorting, have been able to separate the two sperm populations with high accuracy. Sperm are labelled with a DNA fluorescent dye. After recognition and electric charging, droplets containing single sperm are deflected and pushed into a collection medium from which they are further processed. This set-up allows the identification and selection of individual sperm into populations with sort purities above 90% of the desired characteristics.

D. Rath (✉)
Institute of Farm Animal Genetics, Friedrich-Loeffler-Institut, Neustadt-Mariensee, Germany
e-mail: rath@tzv.fal.de

C. Maxwell
Faculty of Veterinary Science, University of Sydney, Sydney, NSW, Australia
e-mail: chis.maxwell@sydney.edu.au

© Springer International Publishing AG, part of Springer Nature 2018
H. Niemann, C. Wrenzycki (eds.), *Animal Biotechnology 1*,
https://doi.org/10.1007/978-3-319-92327-7_4

A critical point is the orientation of sperm in front of a UV laser, requiring modifications of a standard flow cytometer. A specially designed nozzle assembly hydrodynamically focusses the sperm-containing laminar core stream by means of a sheath fluid and the specific geometrics of the internal assembly parts.

Sperm sorting requires special liquid media. For example, a system based on TRIS extender has been developed for bull and ram semen. Besides TRIS and other ingredients, the medium contains antioxidant scavengers to combat reactive oxygen species (ROS) and the Hoechst dye 33342. Porcine semen is handled in a similar way, except that the sample fluid is based on TRIS-HEPES. The sample fluid for stallion semen is generally based on skim milk, INRA 96 or Kenney's modified Tyrode (KMT). Sorted samples are collected in tubes pre-filled with collection medium. The composition of this medium is, in most cases, a TEST-yolk extender, supplemented with seminal plasma in order to decapacitate the collected sperm.

In the animal industries, changing the sex ratio of offspring can increase genetic progress and productivity. Animal welfare can be improved, for example, by decreasing obstetric difficulties in cattle and minimising environmental impacts by eliminating the unwanted sex. Sexed sperm has been most widely applied in the dairy industry, and it is likely that this will continue, dependent on the market situation. For US dairy farmers, milk production and the sale of surplus calves and cull cows are as important as the production of replacement heifers on-farm. Outside the USA, at least in Europe and Australia, the demand for sexed sperm is potentially high for milk producers to optimise herd management. In these countries, the genetically superior cows will be bred with X chromosome-bearing sperm to produce genetically superior females with high milk yield and for (female) pregnant heifer export to other countries. Besides AI, embryo transfer (ET) can be performed after insemination with sex-sorted sperm. The combination of sex-sorted sperm with in vitro embryo production (IVEP) is advantageous, but much more difficult than ET, and depends on species, individual semen donor and composition of media used for in vitro maturation, in vitro fertilisation (IVF) and in vitro culture.

Commercialisation of sex-sorted ram sperm has, to date, been restricted by the dearth of commercial sorting facilities in Australia and New Zealand, although sheep are the only species in which sex-sorted frozen-thawed sperm have been shown to have comparable, if not superior, fertility to that of non-sorted frozen-thawed controls. Moreover, there has been little incentive to take up the technology due to low rates of adoption of genetic improvement programmes and/or artificial breeding technology.

In pigs, apart from economic benefits from faster growth rates, sex-sorted sperm would provide major welfare advantages through the elimination of surgical castration. However, the current method of individual sperm sorting is not efficient enough to satisfy the potential demands of the porcine AI industry, due to the high number of sperm required for each insemination. For special applications, such as building up nucleus herds or for research, sexed boar sperm can be utilised in combination with specially adapted insemination strategies. A significant reduction in the total sperm dose, maintaining fertility, can be achieved if porcine semen is deposited deep in the uterus in front of the utero-tubal junction or directly into the oviduct. Only very few sperm are required for IVF using in vivo or in vitro matured oocytes. Transferring both gametes into the oviduct at the same time (gamete intrafallopian transfer – GIFT) can be used as an alternative to IVF. Even fewer sperm are required for intracytoplasmic sperm injection (ICSI) than for all other IVF methods. However, to date, these methods require laparoscopy or laparotomy for insemination, embryo or gamete transfer, which are not practicable as alternatives to castration.

In horses the preferred gender depends on the breed and range of use. Stallion sperm have a low sorting index and their sortability varies, not only among stallions but also among ejaculates. Additionally, the freezability of stallion sperm varies widely. Insemination with sex-sorted sperm has to be performed by hysteroscopy deep into the uterine horn, limiting the technology to high-value animals.

The sex-sorting process can cause sperm damage. The main sources of damage are incubation with the fluorescent stain and exposure to the UV laser, mechanical forces and electrical charge.

Future sorting methods may avoid the need to identify quantitative differences between X and Y chromosome-bearing sperm. This would require a specific marker related to only one sex. A promising system is based on gold nanoparticles, which can be functionalised with DNA probes. After internalisation of the probe into the sperm head, the Y chromosome-bearing sperm can be identified due to their strong plasmon resonance, which is more stable than fluorescent dyes. Non-invasive coupling of a specific DNA probe with the intact DNA double strand by triplex binding and accumulation of nanoparticles has been achieved, but to date internalisation of the gold nanoparticles requires further research. Another promising new method promotes the naturally occurring genomic variations by gene editing. It is not a question of if, only when these methods will be ready for the market and replace the existing sexing techniques.

4.1 Introduction

Along with various reproductive strategies, different ways to balance sex ratios have evolved in the animal kingdom. In mammals, sex is determined by an almost equal distribution of two different sex chromosomes, named X and Y, located by meiotic segregation in the sperm head. Random chances of fertilising the X chromosome-bearing oocytes guarantee a balance of sexes in the mammalian population. However, there is some new evidence that the female may have an impact on the final sex of their offspring, by selectively modifying the oviductal environment in response to the presence of X or Y sperm (Alminana et al. 2014) or by epigenetic mechanisms adapting sex ratios to the needs of a population and to environmental challenges (Boklage 2005).

Since ancient times, both scientists and mystics have tried to uncover the mechanisms of sex determination and the reasons for balanced sex distributions. However, reliable scientific investigations awaited the invention of the light microscope, and it took several centuries to discover the biological and genetic mechanisms by which sperm contribute to sex determination. Many but not all details have been elucidated, and scientists have begun to transfer this knowledge into technical methods that allow the sex ratio to be changed, often referred to as 'sperm sexing'. Apart from avoiding specific sex-related diseases in humans, these techniques provide a powerful tool for managing farm animal breeding.

Future agricultural strategies will have to provide sufficient food for an increasing world population. Food production must be at a price that is affordable by consumers and profitable enough to provide farmers with a balanced income, to allow sufficient investment to fulfil farming as well as animal welfare regulations and to establish a sustainable food production chain. In this context, the Food and Agriculture Organization of the United Nations has recognised that the production of pre-sexed livestock by sperm or embryo sexing, when combined with other biotechnologies, is a useful breeding tool to increase efficiency, especially for traits that are sex-related (De Cecco et al. 2010; Niemann et al. 2011). For example, as only cows and not bulls produce milk, sperm sexing has been promoted in dairy cattle, and X chromosome-bearing sperm are preferred for insemination (Seidel 2003b). Such techniques allow farmers to produce an optimal ratio of males and females in their production systems, which is of particular advantage when combined with a genomic selection programme.

In this chapter, we will briefly explain sex determination in mammals, review approaches to identifying X and Y chromosome-bearing sperm and their practical implications for semen handling and artificial insemination (AI) and compare their importance and success in the main farm animal species. The problems associated with current technology for sperm sexing, as reflected in the damage caused to mammalian sperm, will then be considered, followed by an assessment of the potential for replacing this technology by other methods.

4.2   Natural Sex Determination

In mammals, sex is determined at fertilisation by the sex chromosomes, X and Y. These are equally distributed among sperm, whereas the oocyte always carries an X chromosome. In the resulting zygote, the 'XX chromosome' combination determines a female and the 'XY chromosome' combination a male. The biological differences between males and females are set genetically during embryo development.

After fertilisation with sperm carrying either sex chromosome, primordial germ cells (PGCs) develop and start migration within the first weeks of foetal development across the hindgut to the genital ridge, an undefined gonad, which may differentiate into either a testis or an ovary. PGCs originating from fertilisation with a Y chromosome-bearing sperm carry the gene region SRY (Jacobs and Ross 1966) with a length of 35 KB (Sinclair et al. 1990), coding for the protein 'testis-determining factor' (TDF). This protein is the primary signal to engrave the male phenotype on the genital ridge and initialise the development of Sertoli cells. Other Y chromosomal as well as autosomal genes participate in testicular development (Eggers and Sinclair 2012), and several transcription factors control the coordinated process to the mature gonad (Eggers et al. 2014), such as SOX 9 which has to be present downstream and is up-regulated by SRY for testicular development (Hanley et al. 2000; Mittwoch 2013). Thus, the factors influencing sex determination tend to be transcriptional regulators. The Y chromosome has several other pivotal functions in spermatogenesis, and the removal of these genes in the AZF regions causes distinct pathological testis phenotypes (Krauz and Casamonti 2017).

Sex differentiation, on the other hand, occurs once the gonad has developed and is induced by gonadal products. Secreted hormones and their receptors, therefore, largely establish phenotypic sex (Byskov 1986; Eggers and Sinclair 2012). The testis starts to produce testosterone and anti-Mullerian hormone (AMH) early in foetal development. Testosterone induces the masculine differentiation of the brain, the sex (Wolffian) duct and secondary sex characteristics, whereas AMH suppresses the development of the female sexual (Mullerian) duct system. Fertilisation with X chromosome-bearing sperm maintains the female characteristics of the sexual organs with ovaries, the formation of the Mullerian duct and the female secondary sex characteristics (Byskov 1986).

4.3 Techniques to Identify Sex-Related Characteristics of Sperm

In mammals, the most efficient way to bias sex ratios in offspring is to separate X and Y chromosome-bearing sperm before insemination. Over the past 90 years numerous techniques purporting to alter the sex ratio have been proposed or discussed (for reviews see Windsor et al. 1993; Klinc and Rath 2005). None of these methods were able to produce statistically significant separation of fertile X and/or Y sperm populations or were not repeatable (Pinkel et al. 1985; Johnson 1988; Johnson and Clarke 1988; examples in Table 4.1). One of the more promising alternatives was the use of interferometry to detect volume differences between the heads of X and Y sperm, but the technique has not yet reached the level of efficiency that would allow practical application (van Munster 2002).

4.3.1 Sperm Sorting by Quantitative Flow Cytometry

Only quantitative methods, which differentiate between X and Y sperm on the basis of total DNA and then apply flow cytometric sorting, have been able to separate the two sperm populations with high accuracy. The X chromosome carries more DNA than the Y chromosome (Moruzzi 1979), whereas the autosomes of both kinds of sperm have identical DNA content. The DNA difference is widely species-specific

Table 4.1 Examples of physical methods proposed for identification and separation of sperm

Criterion	Status	References
Velocity	Unproven	Ericsson et al. (1973), Dmowski et al. (1979), Beernink and Ericsson (1982); Beal et al. (1984)
Density	Unproven	Bhattacharya (1962); Bhattacharya et al. (1966), Rohde et al. (1975), Ross et al. (1975), Schilling and Thormaehlen (1977), Shastry et al. (1977), Kaneko et al. (1983), Vidal et al. (1993), Pyrzak (1994), Wang et al. (1994a, b), Flaherty et al. (1997), Kobayashi et al. (2004), Koundouros and Verma (2012)
Electrical surface charge	Not reproducible	Sevinc (1968), Shirai et al. (1974), Shirai and Matsuda (1974), Shishito et al. (1975), Uwland and Willmes (1975), Engelmann et al. (1988), Blottner et al. (1994), Manger et al. (1997)
Immunologically relevant structures Surface proteins	Not reproducible	Bennett and Boyse (1973), Erickson et al. (1981), Hancock et al. (1983), Pinkel et al. (1985), Ali et al. (1990), Hendriksen et al. (1993), Sills et al. (1998), Blecher et al. (1999)
Volume/ interferometry	Small differences not at practical stage	van Munster (2002)
Semen deposition site in the uterus	Unproven	Zobel et al. (2011)
Interval insemination: ovulation	No effect	Rorie (1999), Rorie et al. (1999), Roelofs et al. (2006)

with some breed variation (Garner 2006). Gledhill et al. (1976) commenced the first experiments on flow cytometrical sperm analysis, but it was Fulwyler (1977) who developed a technical solution for asymmetric cells to orient them in front of a laser by hydrodynamic focussing.

In the procedure subsequently developed by Johnson and Pinkel (1986), sperm are labelled with a DNA fluorescent dye. After co-incubation with the dye, the cells are hydrodynamically focussed in a flow cytometer into a discontinuous droplet stream. The stream passes an interrogation point, where a UV laser beam is projected on it, illuminating the flat surface of the sperm head and exciting the fluorescent dye. The orthogonal set-up of two fluorescence detectors requires a precise orientation of the sperm head in front of the laser to resolve the small quantitative DNA difference of 2.3–7.5% (Garner 2006) between the X and Y chromosome-bearing sperm (Figs. 4.1 and 4.2). Before the droplets disrupt from the discontinuous stream, the last hanging drop is charged according to the DNA content of the sperm it encloses. The droplets then pass an electrostatic field (3000 V) and are deflected according to their charge. The deflected or sorted cells are pushed into a collection medium from where they are distributed to further preservation steps (Johnson and Welch 1999). This set-up allows the identification and selection of individual sperm into populations with sort purities above 90% of the desired characteristics.

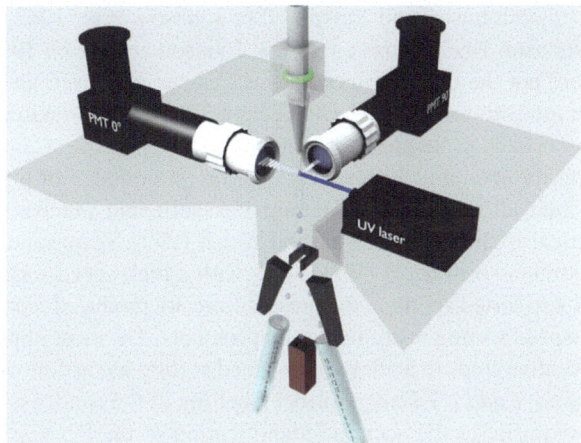

Fig. 4.1 The principle of a flow cytometer modified for sperm sorting. The UV laser-based system requires several modifications to optimise high-speed sperm sorting. The essential elements are the replacement of the usual forward scatter diode by a photo multiplier tube (PMT 0°) and its associated optical lens, a laser with beam shape optic optimally focussed on the flat surface side of the sperm head and a specially designed orientation nozzle assembly. (By courtesy of Roberto Mancini)

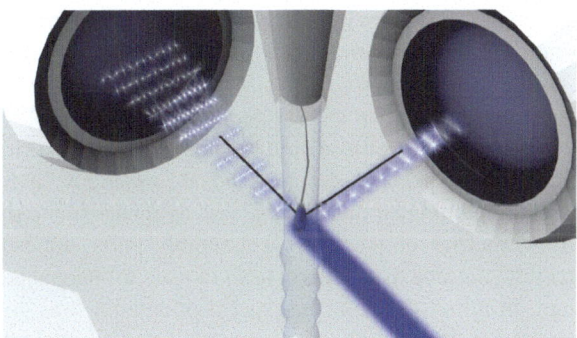

Fig. 4.2 The principle of sperm orientation within the core stream of a modified sperm sorter. The small quantitative difference in DNA content between X and Y chromosome-bearing sperm requires an orthogonal orientation relative to the laser beam. The orientation is accomplished by hydrodynamic focussing caused by the nozzle assembly design and the differential pressure of the core stream and the sheath fluid. The fluorescence signal of the small rim of the sperm head creates an optical breaking effect that is independent of the DNA content. Its recognition by the 90° PMT identifies the position relative to the laser beam. The DNA difference is measured as the emission signal of the excited fluorochromes Hoechst 33342 by the 0° PMT. The signals of the PMTs are digitised and presented as a dot plot on a computer screen. (By courtesy of Roberto Mancini)

The technique described above requires numerous modifications to a standard flow cytometer. Detailed instrument modifications, protocols and technical improvements to the sexing method have been well documented in various articles and reviews (Fulwyler 1977; Stovel et al. 1978; Dean et al. 1978; Pinkel et al. 1982;

Garner et al. 1983; Johnson and Pinkel 1986; Johnson et al. 1987; Johnson and Clarke 1988; Johnson 1997; Rens et al. 1998; Johnson and Welch 1999; Rath et al. 2009). These are not the main subjects of this chapter. However, an outline of the most important innovations that have allowed practical application in domestic farm animals, and associated problems, is pertinent.

A major step for economic production was the introduction of high-speed flow cytometers, which allowed the production of a sufficient number of sex-sorted sperm for practical application (Johnson and Welch 1999). Sharpe and Evans (2009) reported a maximum sort rate of 8000 cells/s with a high-speed sorter under ideal conditions. All sex-sorted sperm at the present time are produced regardless of species with high-speed sorting instruments and protocols. The most important modifications and inventive steps to achieve high-speed sorting are as follows:

Fluorescence dye and UV laser: Correct labelling of the condensed sperm chromatin is a prerequisite to the accurate identification of the DNA size differences between X- and Y-bearing cells. Only very few fluorescent dyes are able to pass through the intact sperm membrane into the nucleus. Hoechst 33342 (bis-benzimide) has been shown to represent the DNA content of sperm precisely, without affecting their integrity (Johnson et al. 1987; Garner 2006), as it binds to the AT-rich regions in the minor groove of the DNA helix (for review, see Rath and Johnson 2008). The dye is apparently not genotoxic, although it is known to be mutagenic and may affect embryo development. Moreover, the fate of the Hoechst dye, once transported by the sperm into the oocyte and thereby into the embryo and offspring, is little understood (Garner 2009). For excitation the dye requires continuous, or at least quasi-continuous, wave UV laser light above 100 mW. The laser beam has to illuminate each sperm with a specifically designed beam shape optic, which projects it into a vertical ellipse onto the flat side of the sperm head.

Sperm orientation: A specially designed nozzle assembly hydrodynamically focusses the sperm-containing laminar core stream by means of the sheath fluid and the asymmetric geometrics of the internal assembly parts (Johnson and Pinkel 1986). This forces the flat side of the sperm head into an orthogonal position relative to the laser beam. In a first development, sperm orientation was generated by an assembly carrying a bevelled sample injection needle and an orientation nozzle tip that promoted the alignment of the cells in front of the laser (Rens et al. 1998). With this assembly inserted into a high-speed flow cytometer, sort rates of 12–15 million sperm per hour became a reality and were the prerequisite for commercial application of the technology (Johnson and Welch 1999). While this so-called 'HISON' orientating nozzle had a double torsional elliptic shape (Rens et al. 1998), most commercial sorters today work with a single torsion nozzle (Cytonozzle; XY, Inc., Fort Collins, CO, USA) a further refinement that provides better sperm orientation. Recently, an updated version of the original double torsional nozzle assembly was developed, with an improved internal geometry designed to optimise the efficiency of sperm orientation. In this assembly, the spatula-like shape and the double-phased edges of the injection tube now amplify the hydrodynamic focussing, rather than the ceramic nozzle tip (Rath et al. 2013).

Replacement of the forward diode: In order to detect the light emitted from the Hoechst 33342 dye, the forward diode of the standard flow cytometer can be replaced by a sensitive fluorescence-detecting photomultiplier tube (PMT) (Dean et al. 1978; Stovel et al. 1978; Pinkel et al. 1982; Garner et al. 1983). Through a (50×) microscopic lens, the 0° PMT recognises the emitted light from the flat side of the labelled sperm head. The emission signal of dye is related to the total DNA content of the sperm head. If correctly oriented, the 0° PMT signal and the corresponding 90° PMT orientation signal display both sperm populations as distinct separated areas in a dot-plot presentation or as separated histograms (Fig. 4.3).

Sort purity and sorting parameters: Sort purity depends on many technical set-up parameters and adjustments of the sorter. Sort regions drawn on the dot-plot presentation, identifying specific cell populations, provide the command to send the related droplet charge to the stream (Johnson et al. 1989; Johnson 1991; Welch and Johnson 1999). The charge is transmitted to the discontinuous fluid stream at the time point when the droplet, with the corresponding sperm cell, detaches from the stream. Thereby, a free droplet is produced that carries the individually recognised sperm with a DNA-content-correlated electrical charge. This precise time point has to be set up as the so-called drop delay before sorting. The free droplet then passes an electrostatic field of around 3000 V and is deflected to either side depending on its charge. Sorted cells are pushed into a collection medium (Johnson and Welch 1999) and then separation from the sheath fluid by centrifugation or continuous filtering.

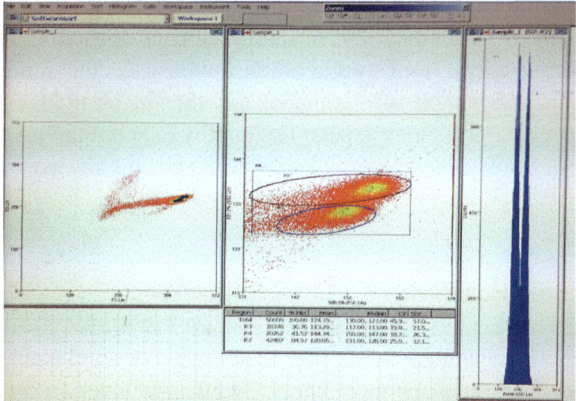

Fig. 4.3 Dot plot presentation of the PMT signals. The dot plots (right dot plot zoomed) show the sperm emission signals from the sperm head rim (orientation, *X* axes) and the relative DNA content (*Y* axes). This information received from the two PMTs is used to draw overlaid sort regions (R2 and R3), which determine the signal to send an electric charge to the last hanging droplet of the discontinuous fluid stream. After disintegration, the free-flowing charged droplets pass an electro-static field, which deflects them to either side, or, if they have not been identified, they continue undeflected into the waste stream

4.4   Semen Processing

4.4.1   Sample Preparation

After collection of the ejaculate and assessment of semen quality, aliquots of raw semen are adjusted to 50–100 million sperm/ml. For most farm animal species, specific extenders have been developed serving as 'sample fluid' for the incubation of sperm with the DNA dye. The sample fluid further provides the material for the core stream during sorting. Whereas some groups prefer a TALP-based medium for sorting and a TRIS-based cooling/freezing system for ruminant semen, it can be advantageous not to change the buffer system during the sorting/freezing process. Therefore, a system based on TRIS extender only has been developed for bull and ram semen (Sexcess® Klinc and Rath 2005). Besides TRIS and other ingredients, the medium contains antioxidant scavengers to combat reactive oxygen species (ROS) and 10–25 µl/ml of the Hoechst 33342 made of a stock solution with 5 mg/ml dye. Semen and fluorescent dye are co-incubated for 30–90 min at 34 °C. Porcine semen is handled in a similar way, except that the sample fluid comprises TRIS-HEPES-buffered extender (modified Androhep™: Johnson 1991; Waberski et al. 1994) and the incubation temperature is set to 30 °C. The sample fluid for stallion semen is generally based on skim milk, INRA 96 or Kenney's modified Tyrode (KMT) (Heer 2007; Clulow et al. 2008, 2012).

In all cases, the sample fluid containing the labelled sperm is filtered after the incubation period through a 50 µm nylon filter into a 5 ml pressure tube (maximum diameter 14.9 mm). Food dye (FD#40) is added in order to identify those sperm with damaged membranes resulting from sorting. FD#40 consists of relatively large molecules, which only enter the heads of sperm with damaged membranes. The dye eliminates defective sperm from sorting because it bleaches the fluorescence signal of the DNA stain. The sample tube is placed into the sample holder, and the liquid is pushed into it under pressure (50 psi). The quality of sperm in the collected sample would benefit from lower pressure (Suh et al. 2005), but this would be in conflict with existing patents.

4.4.2   Post Sort Handling

Sorted samples are collected in tubes pre-filled with collection medium. The composition of this medium is, in most cases, TEST-yolk extender as described by Johnson et al. (1987), which benefits from the inclusion of 2% seminal plasma. The latter component stabilises the membranes of those sperm that have undergone capacitation-like changes during flow sorting.

Most of the volume of the collected sample originates from the sheath fluid, necessary to realise the hydrodynamic focussing of the sperm heads in the nozzle assembly. For ruminant semen, a TRIS-buffered salt solution is preferred, containing at least one additional energy source. Boar sperm are able to tolerate simple PBS, supplemented with at least one antioxidant, as a sheath fluid.

After sorting about 8 ml of sample into each collection tube, they are centrifuged and the supernatant discarded. The remaining sperm pellet is extended with an appropriate, species-specific, medium for liquid preservation or freezing.

4.5   The Importance of Sexing Techniques in Different Farm Animals

In the animal industries, changing the sex ratio of offspring can promote faster genetic progress and higher productivity and support animal welfare, for example, by decreasing obstetric difficulties in cattle and minimising environmental impacts by eliminating the unwanted sex.

Several reviews have been published on the commercial use of sexed bovine sperm, especially in the USA (Amann 1999; Seidel 2003a, b; Garner 2006; Garner and Seidel 2008), but they are helpful guidelines for other countries too. Maxwell et al. (2004) reviewed the situation in other species; and specifically for pigs, there are reviews by Johnson et al. (2005) and Rath et al. (2015). Sex-sorted semen is in high demand, but the range of applications varies widely. This depends on species, products, production lines, economic interests, market requirements, breeding programmes, local specialities and other factors. While the current technique based on modified flow cytometry separates X and Y chromosome-bearing sperm with high accuracy, it is limited because each sperm cell has to be characterised and sorted individually. Therefore, its commercial utilisation differs significantly among species because of different characteristics such as site of semen deposition, length of the oestrous cycle, demand for sperm numbers and the sortability and freezability of sperm.

4.5.1   Cattle

Johnson et al. (1989) performed the first trials that successfully produced offspring from sex-sorted rabbit sperm. The original generation of flow sorters was not very efficient, resulting in low output and purity of sorted sperm compared with their present-day counterparts. Consequently, it was necessary to utilise the sorted sperm for IVF and embryo transfer to produce the first calves (Cran et al. 1993, 1995). Later Seidel et al. (1997), using a newly developed high-speed flow cytometer (MoFlo), obtained the first calves after AI of heifers with sorted liquid-stored sperm, and later with frozen sperm (Seidel et al. 1999). High-speed flow cytometry has been applied most easily in the ruminant species (bovine, 2 million/AI, Seidel et al. 1997; and ovine, 1–5 million sperm/AI, de Graaf et al. 2007c), especially as the sorting index (131) is more suitable for high-throughput sorting than in other species (Garner 2006). By the year 2000, high-speed sorting had been commercially introduced in the UK. However, field data indicated that fertility was still highly variable and depended on bull effects. Such effects were not necessarily due to sorting but may have been related to high dilution and reduced compensatory mechanisms among sperm. This was especially the case in farms with moderate fertility, where limited quality of the sorted sperm became more apparent. Modified insemination protocols may help to improve pregnancy rates. Unilateral intrauterine horn inseminations in heifers with preovulatory follicles seemed to be advantageous under hot conditions (Chang et al. 2017), whereas Kurykin et al. (2016) found no differences in pregnancy rates after intra-cornual (44.9%) or conventional insemination (48.4%). Insemination closer to the expected ovulation yielded higher chances of pregnancy in Jersey cows (Bombardelli et al. 2016).

At the present time, due to decreasing production costs as a consequence of better instrumentation and optimised maintenance of sperm quality, sexed sperm has become more widely applied in the dairy industry. It is likely that this will continue, dependent on the market situation. As long as the demand and price for heifers remain high, a profitable sale of sexed semen can be expected. However, if the prices for milk, heifers and cull cows decrease, feed costs increase and prices remain low for conventional semen, the demand for sexed semen may disappear.

For US dairy farmers, milk production and the sale of surplus calves and cull cows are important, whereas the sex of the calf is relatively unimportant, except for reduced possibilities of dystocia from male compared with female calves (Hohenboken 1999) and a slightly higher milk yield when cows have heavier (male) calves (Quesnel et al. 1995). More important is the production of replacement heifers on-farm, which avoids the need for foreign animals to enter the herd and improves the genetic value of the herd by purchasing semen from highly selected bulls from the seed-stock industry. Genetic gain will be passed on to milk producers, as sexed semen helps to maximise the genetic merit of breeding stock by increasing the rate of selection and reducing the costs of genetic improvement (Hohenboken 1999).

In the past, genetic improvement programmes required heifers for test inseminations. In modern genomic selection programmes, bulls already have a proven genetic status as calves or even at their embryonic stage. In consequence, the turnover of young bulls has increased significantly, requiring more specifically selected elite bull mothers which can be produced with X chromosome-bearing sperm, while Y sperm can be reserved for sire production, encouraged by the monthly published figures on breeding values. Sexed sperm are most efficiently used in the in vitro production of matured 'OPU oocytes', which have been characterised for high breeding value before embryo transfer or storage. Female embryos can be transferred to recipients for provision directly to milk producers, whereas the male embryos serve as a source of superior AI bulls.

Other than in the USA, where heifer replacement within the herd is one of the main reasons for using sexed sperm, at least in Europe and Australia, the demand for sexed sperm is potentially high in milk-producing farms to optimise herd management. As female replacement is here not of such importance, the genetically superior cows will be bred with X chromosome-bearing sperm to produce genetically superior females with high milk yield and for (female) pregnant heifer export to other countries. The remaining females could be bred with Y sperm of a beef breed to optimise sales revenue, which is suboptimal if dairy breed bulls are fattened.

However, the demand is mainly influenced by the milk price rather than by the consumption of milk. Ettema (2007) presented calculations for the dairy industry in Denmark as an example for the European market. He used a model, which included the price for heifers, replacement costs, price per beef calf, the price of sexed semen, conception rates with sexed semen, replacement rates, the sex ratio of the sperm and the incidence of dystocia and stillbirths. The main negative factors were lower fertility, high cost of equipment, personnel cost and investment in intellectual property. In consequence of Ettema's analysis, a net return to assets from the use of sexed semen in a breeding programme would not be expected earlier than 3–4 years after implementation.

For intensive beef production, Seidel and Whittier (2015) proposed a programme for heifer fattening using sexed sperm, based on the principles described by Bourdon and Brinks (1987a, b, c). Without a sire, all AI is performed in this system with X sperm on heifers only. Female offspring are raised and inseminated again with X sperm, and after delivery the heifer offspring are finally fattened and culled. Because only young females exist on the farm, more beef is produced per feeding unit, less water is necessary and less CO_2 and methane are produced. As no older animals exist, losses related to illness are minimised, and treatment costs are low. However, compared with a bull-fattening system, daily weight increases are lower, which has to be compensated for by running more heifers. According to Seidel and Whittier (2015), the heifer system is superior as no mating cows and bulls exist on the farm. Dystocia is more likely in heifers than adult cows, but as only female calves are born, the difficulties are negligible. The biggest disadvantage is that the system of heifer replacement is not completely self-maintaining. Presumably such systems for beef production are more likely to be adopted if milk prices fall below profitable margins and would be useful for farmers who wish to move from dairy production to fattening with limited investment.

Besides AI, embryo transfer (ET) can be performed after insemination with sex-sorted sperm, and the embryo donors may be hormonally stimulated to increase the number of offspring and hence selection differentials. Schenk et al. (2006) reported no difference in embryo production or quality after insemination with sex-sorted compared with unsorted sperm. The combination of sex-sorted sperm with in vitro embryo production (IVEP) is advantageous, but much more difficult than ET, and depends on aspects like species, individual semen donor and composition of media used for in vitro maturation, in vitro fertilisation (IVF) and in vitro culture. Moreover, factors such as origin of gametes, status of the sorting protocol and instrumentation as well as treatment and storage of gametes and embryos and the liquid or frozen status of the derived embryos, to name a few, have an impact on the resulting number of offspring. For in vitro blastocyst production, Inaba et al. (2016) observed reduced competence of oocytes fertilised by X sperm, rather than any effect on sperm fertilising ability. However, the occurrence of this phenomenon varied among bulls. Accordingly, published data from IVEP with sexed sperm vary, and they are partly contradictory.

The number of sperm required for IVF of bovine oocytes differs depending on whether they have been subjected to sex sorting or not. Reasons for this include a change in the propensity of the sperm membrane to undergo the acrosome reaction, resembling a partial 'capacitation' (Moce et al. 2006), and the elimination of DNA-damaged sperm during sex sorting. Thus, Lu and Seidel (2004) found that the heparin concentration in the medium had to be optimised for each bull and increasing sperm dose from 0.5 (1500) to 1.5 (4500) and 4.5×10^6 sperm per ml (13,500 sperm per oocyte) increased cleavage but not blastocyst rates. Barcelo-Fimbres and Seidel (2004) obtained the best IVF and embryo development rates using 1×10^6 sperm per ml (2667 sperm/oocyte). The resulting cleavage and blastocyst rates did not differ between X and Y chromosome-bearing sperm (Cran et al. 1993; Barcelo-Fimbres et al. 2011), whereas others have reported that male IVF embryos grow faster than their female equivalents (Xu et al. 1992).

Not all bovine ejaculates are suitable for sorting and subsequent IVEP. In some experiments, an unexplained decrease in blastocyst production has been reported with sorted sperm compared to controls (Lu et al. 1999), and Xu et al. (2006) could only use one third of the available bulls. From the latter samples, however, more than 33% of sexed IVF embryos developed into blastocysts and, after vitrification, 40% of recipients became pregnant after ET. Similarly, the source of oocytes is important. Palma et al. (2008) identified structural changes of organelles like mitochondria, rough endoplasmic reticulum (ER) and the nuclear envelope after IVF with sex-sorted compared with unsorted sperm. This is in agreement with studies on the mRNA expression pattern of the important developmental genes, glucose-3 transporter (Glut-3), glucose-6-phosphate dehydrogenase (G6PD), X-inactive specific transcript (X-ist) and heat shock protein 70.1 (Hsp), in day 7 and 8 bovine IVP embryos produced with sexed sperm (Morton et al. 2007). Lopez et al. (2013) fertilised oocytes, derived from ovum pick-up, in vitro with sorted frozen-thawed sperm or with non-sorted frozen-thawed sperm from the same ejaculate. In this study, gamete co-incubation, either short (4–12 h) or long (18–24 h), had no effect on monospermy, pronuclear formation or syngamy. This contradicts earlier reports (Maxwell et al. 2004; Rath et al. 2009; Carvalho et al. 2010) and suggests that many of the improvements in sorting and post-sort semen preservation, as well as oocyte handling, might have compensated for the former differences between sorted and unsorted sperm. These data are almost in agreement with those published by Trigal et al. (2012), except that they confirmed a bull-related effect as seen in earlier studies (Lu and Seidel 2004; Xu et al. 2006). However, there were no differences in embryo survival after vitrification, nor in pregnancy rates, between sorted and unsorted semen. The bull effect on IVEP may be related to the capacitation status of sperm after sorting, requiring the heparin concentration to be adjusted for the sperm of each bull in order to obtain the maximum number of competent embryos (Blondin et al. 2009).

4.5.2   Sheep

Commercialisation of sex-sorted ram sperm has, to date, been restricted by the dearth of commercial sorting facilities in those countries where the sheep population is highest, namely, Australia and New Zealand. Moreover, there has been little incentive to take up the technology due to low rates of adoption of genetic improvement programmes and/or artificial breeding technology in some sectors of the industry. This is both surprising and disappointing, as sheep are the only species in which sex-sorted frozen-thawed sperm have been shown to have comparable, if not superior, fertility to that of non-sorted frozen-thawed controls. The use of very low numbers of sperm for laparoscopic insemination of sheep has resulted in the most efficient utilisation of sex-sorted sperm, with the highest levels of fertility, of any species, whether inseminated in superovulated ewes, at very low doses in non-superovulated ewes, or even when the semen has been frozen twice, both before and after sex sorting (reviewed by de Graaf et al. 2009).

The first lamb from sex-sorted spermatozoa was produced by ICSI with a fresh sperm (Catt 1996). Lambs were then produced after sex sorting and laparoscopic insemination using either 10 million non-frozen (Cran 1997) or 2–4 million frozen-thawed spermatozoa (Hollinshead et al. 2002). Two years later, offspring were produced by IVF of oocytes aspirated from hormone-stimulated prepubertal lambs (Morton et al. 2004). After modification of the sexed sperm treatment protocols, de Graaf et al. (2007b) were able to report superior fertility rates to non-sorted sperm when inseminated by laparoscopy (1 or 5 million sorted motile sperm). In another trial, it was shown for the first time for any species that 'reverse sorting' (sorting of previously frozen-thawed sperm) is capable of producing offspring of predicted sex following AI (de Graaf et al. 2006). The fertility of sex-sorted frozen-thawed ram sperm was shown, in a number of subsequent studies, to be equal to unsorted sperm when used for laparoscopic insemination or intrauterine insemination in superovulated ewes and subsequent embryo transfer of morula and blastocysts (de Graaf et al. 2007a, 2009). Furthermore, IVF data showed that sex-sorted sperm elicit equal or greater cleavage and blastocyst rates than their non-sorted counterparts (de Graaf et al. 2009; Beilby et al. 2011).

It seems that sheep are an exception to the general rule that sex sorting reduces the fertility of mammalian sperm. They withstand the stress caused by different treatments such as incubation with the Hoechst dye, flow sorting and post-sort treatments for long-term storage in liquid nitrogen, and they have been shown to possess characteristics predictive of longer fertilising lifespan in the female reproductive tract, compared with unsorted sperm (de Graaf et al. 2009). Presumably, besides physiological characteristics, it is likely that protection by liquid media has been optimised for this species.

4.5.3 Pigs

The first piglets were born from oviductal insemination with sexed sperm only 2 years after the first offspring from sex-sorted rabbit sperm (Johnson 1991). Pig producers would benefit from the use of sexed sperm, either as a fresh or frozen semen product, to obtain more female piglets. As a commercial product, aside from the economic benefits from faster growth rates, there would be major welfare advantages through the elimination of surgical castration. However, until now, sexed sperm have not been commercially available. This is because the current method of individual sperm sorting is not efficient enough to satisfy the potential demands of the porcine AI industry, due to the high number of sperm required for each insemination. Additionally, unlike in the bovine industry, pig AI is based on liquid semen and has no existing infrastructure for the application of frozen semen on farm.

For special applications, such as building up nucleus herds or for research, sexed boar sperm can be utilised in combination with specially adapted insemination strategies. During natural mating, the boar deposits the semen through the cranial part of the cervix and into the uterine body. Selection and binding of sperm occur to a large extent in the uterine horns, and much of the sperm and semen volume is ejected by retrograde flow through the vagina (Viring and Einarsson 1981; Steverink et al. 1997, 1998; Matthijs et al. 2000, 2003). In conventional AI, the semen is deposited at the same location as at natural mating by the boar, and it is subjected to the same selection processes during transport to the oviduct. The normal insemination dose varies, therefore, between 1.5 and 3 billion sperm for liquid-stored semen and 5 billion for frozen-thawed semen. If this were considered in the context of sex sorting, it would theoretically take at least half a day, and require eight sorting machines, to produce one dose of liquid sex-sorted sperm. An important logistical consideration is the storage time of semen before and after sorting. Alkmin et al. (2016) proposed a method for storage of semen for up to 24 h before sorting, which would allow its transport to a central sorting unit. There, sorted sperm could be encapsulated in barium alginate capsules, allowing controlled release of sperm into the female genital tract after post-sorting storage (Spinaci et al. 2016).

Because of the selection and binding of sperm in the porcine uterus, a significant reduction in the total sperm dose, to as low as 1×10^8, can maintain fertility compared with controls if the semen is deposited deep in the uterus in front of the uterotubal junction (UTJ) (Martinez et al. 2001, 2006; Grossfeld et al. 2005; Vazquez et al. 2003, 2005, 2008a, b). Under research conditions, even a very low number of sperm (1 million) was sufficient to produce pregnancies (Krueger et al. 1999). A further significant reduction in the sperm dose can be made if inseminated directly into the oviduct either by laparotomy (Polge et al. 1970; Salamon and Visser 1973), using as little as 200,000 sex-sorted or unsorted sperm per oviduct (Rath et al. 1993), or by laparoscopy (Vazquez et al. 2008b; Roca et al. 2011; del Olmo et al. 2014).

A possible approach to improve fertility after low dose insemination in pigs would be to reduce the losses of sperm during their uterine migration, by interrupting the processes of sperm binding to the uterine wall. While little is known about the physiological importance of such sperm binding, its reduction could lead to fertility similar to that obtained with deep intrauterine AI, while allowing farmers to still use standard insemination tools. Intact sperm bind transitionally to the uterine wall, whereas most of the retrograde flow contains the less viable sperm. Moreover, when sperm bind to the uterine wall, the expression pattern of inflammatory and anti-inflammatory genes changes in uterine epithelial cells, indicating a very specific interaction. This change in gene expression might either reflect the epithelial cells acting as a transient sperm reservoir, which could be important for late ovulating sows, or it might be a priming signal that prepares the uterus for the implantation of embryos (Taylor et al. 2008, 2009a, b, c; Junge et al. 2010, 2011, 2012). However, as pregnancy rates are not reduced when semen is mechanically deposited in front of the UTJ, a biochemical prevention of sperm binding would presumably not affect embryo implantation or pregnancy rates. Bergmann et al. (2012a, b) showed that the in vitro interaction of porcine sperm with monolayers of uterine epithelial cells (UEC) is mediated by lectin-like proteins located on the sperm surface and carbohydrate residues on the UEC. The glycan ligand involved in this binding was identified as sialic acid. With saturation of the ligands before insemination, sperm would no longer bind to the UEC, potentially increasing the number of sperm reaching the UTJ. However, it is not completely clear which further physiological functions, besides a selection process, may be related to these sperm-UEC interactions.

As current technology requires the identification and sorting of individual sperm, the efficiency, even of the latest generation of sorters, can hardly fulfil the commercial demand for sexed sperm in pigs. As opposed to AI, only very few sperm are required for IVF using in vivo or in vitro matured oocytes. The first embryos from IVF with sexed boar sperm were produced some 23 years ago at the USDA in Beltsville, USA (Rath et al. 1993). In these early experiments, mature cumulus-oocyte complexes were collected from superovulated prepubertal gilts shortly before ovulation, with an average cleavage rate after IVF of 56.2%. Offspring were subsequently born after IVF with sex-sorted sperm employing either in vivo matured oocytes (Rath et al. 1997) or, later, in vitro matured oocytes (Abeydeera et al. 1998; Rath et al. 1999). In parallel, a method was developed for gamete intrafallopian transfer (GIFT) as an alternative to IVF, especially for those cases where laboratory equipment was limited and did not allow fertilisation in vitro. For GIFT, matured oocytes and sorted sperm were placed, in two segments, into a 0.5 ml plastic straw, which at the open end was equipped with a smooth silicon tube, and both gametes were simultaneously transferred into the oviducts of peri-ovulatory gilts. Recipient follicles were aspirated to avoid fertilisation of their oocytes by the transferred sperm. Comparing GIFT with unsorted and sorted sperm, 50% and 48% of reflushed blastocysts had 25–80 cells, respectively (Rath et al. 1994a, b).

Fewer sperm are required for intracytoplasmic sperm injection (ICSI) than for all other IVF methods, as only a single cell is required for microinjection into the ooplasm of a matured oocyte. Oocyte activation is induced in many cases by the injection itself but can be supported by $CaCl_2$ as a medium supplement. ICSI has been used successfully to produce male offspring employing only a single sexed sperm per oocyte (Probst and Rath 2003). However, with the exception of GIFT, all these in vitro techniques can only be commercialised if the adjunctive embryo transfer can be performed non-surgically. The prerequisite is an efficient system for the in vitro production of morula or blastocyst stage embryos. Krisher and Wheeler (2010) have developed a mostly automated system for IVEP in microfluidic chips, and Roca et al. (2003, 2006, 2011) built and tested uterine ET equipment, which is already manufactured for commercial use. Therefore, the medium-term future application of sex-sorted sperm in pigs will be in combination with the named biotechniques, at least for specific breeding purposes.

4.5.4 Horses

As horses are multipurpose animals, the preferred gender depends on the breed and range of use. Little is known about the real market demand for sexed equine sperm. Using data from an unofficial survey by the Royal Association of the Friesian Horse Studbook, KFPS, Samper et al. (2012) reported about 52% of owners would make use of sexed sperm (63% female; 29% male), whereas 35% would not. Females are preferred for polo ponies and cutting horses and of course males as reining horses. However, more important especially in standard breeds is the decision of the owner, who demands a specific sex of a foal, either a colt out of a particular mare to produce a stallion or a filly to provide a brood mare replacement.

The anatomy and physiology of the female genital tract provide more challenges to the production of offspring from horses than from any other species. Similar to pigs, the insemination dose has to be much higher than in ruminants, and the UTJ is hardly traversable with insemination devices, even under hysteroscopic control. As opposed to ruminants, where sperm are mainly selected in the cervix, which encloses the AI device when placed into the uterine body or horn, stallions deposit their ejaculate into the uterus, where the AI device must be placed also. Therefore, it is necessary to either bypass the uterus or, in order to minimise the necessary sperm dose, place the inseminate very deep into the uterine horn.

Stallion sperm have a low sorting index of 59 (Garner 2006), and their sortability varies, not only among stallions but also among ejaculates (Rath and Sieme 2003; Clulow et al. 2008). Additionally, the freezability of stallion sperm, that is, their ability to survive freezing and thawing, varies widely (Vidament 2005). Early research on sex sorting of stallion sperm showed that it was even less efficient than in other species. Therefore, the first inseminations that resulted in the birth of a foal had to be made surgically. Buchanan et al. (2000) performed the first successful non-surgical inseminations in mares with 25 million sperm per ml. Lindsey et al. (2005) made it possible to store semen at 18 °C before sorting and insemination, with the aid of a video hysteroscope, deep into the uterine horn, obtaining first cycle pregnancies in 72% of such inseminations (Morris et al. 2000). Better pregnancy rates have been reported from hysteroscopic insemination of low numbers of sex-sorted spermatozoa, compared to rectally guided deep-uterine insemination (Lindsey et al. 2005).

Several groups have investigated ways to improve the quality of sex-sorted stallion sperm. Minor improvements were made, for example, by using cushioned centrifugation to both ameliorate the stress to the sperm resulting from post-sort reconcentration and to select more viable cells (Knop et al. 2005; Mari et al. 2015). Such cushioning agents as Puresperm® have been shown also to enrich the proportion of morphologically normal sperm with high progressive motility and to improve their mitochondrial membrane potential, compared with untreated controls (Heer 2007). The high dilution of the stallion sperm, which occurs during sex sorting, has a major impact on their post-sort quality (Gibb et al. 2013; da Silva et al. 2016a, b) as it does in other species (Klinc et al. 2007). This is exacerbated by the loss of seminal plasma that protects sperm against ROS, leading to mitochondrial damage

(Michl 2014) as well as DNA fragmentation (Gibb et al. 2013; da Silva et al. 2016a, b), and capacitation-like changes (Maxwell and Johnson 1997; da Silva et al. 2013). The loss of seminal plasma can be partially prevented by co-incubating the sperm and Hoechst dye prior to sorting (da Silva et al. 2014).

Long-term storage of sex-sorted stallion sperm in liquid nitrogen has been investigated by Clulow et al. (2008, 2012), in which semen was treated with two different extenders for labelling and sorting (KMT and Sperm TALP) and frozen after sorting in two different media (INRA 82® and a modified EDTA-lactose extender). The most successful sex-sorting protocols used KMT as the staining and incubation medium, while either INRA 82® or lactose-EDTA could be employed as cryo-diluents. After shipment of the sexed-frozen semen from Germany to Australia, one filly was born after hysteroscopic insemination.

Samper et al. (2012) summarised the results of inseminations with sex-sorted sperm and subsequent embryo flushing and transfer. From 173 deep intrauterine inseminations with fresh sex-sorted sperm, 109 embryos were recovered and produced 60.4% pregnancies after ET. Insemination with a high dose of sperm resulted in only a 50% pregnancy rate, although pregnancies were obtained from all stallions used. This was less than might be expected from other studies (Gibb et al. 2012). Conversely, the fertility of sex-sorted frozen-thawed sperm was low (0–16%), with an increased incidence of embryonic death compared with the fresh sex-sorted sperm.

The only large-scale study, conducted in Argentina, presented the best fertility yet obtained using sexed sperm in horses, where a prerequisite was excellent management of both the sorting procedure and of the inseminations and embryo transfer (Panarace et al. 2014). In this experiment, conducted over 3 breeding seasons, mares were inseminated at 838 oestrous cycles, of which 435 (52%) yielded viable embryos, and 81.5% of these embryos resulted in a pregnancy when transferred singly to recipients. These results bode well for the large-scale application of sperm-sexing technology in horses, but to date commercial application remains based on short-term liquid storage of the sex-sorted sperm.

4.6 Sperm Damage Caused by Sex Sorting

The great challenge in sorting mammalian sperm by flow cytometry is to maintain their fertilising ability until they reach the mature oocyte in the oviduct of the inseminated female. During natural mating semen does not come into contact with the external environment and finds, in the female reproductive tract, optimal conditions of temperature, pH and osmotic pressure, to name a few important factors. Conventional semen preservation already causes stress to the sperm, when they are processed and stored short term in a liquid state or even longer in a deep-frozen state in liquid nitrogen. In addition to this stress, sex sorting has the potential to cause further harm to each sperm cell. After insemination of rabbits with sex-sorted sperm, McNutt and Johnson (1996) found increased foetal mortality during early pregnancy. Cran et al. (1993) reported a reduction in both blastocyst development and pregnancy rates after bovine IVF with flow-sorted sperm. It is amazing that so many offspring have been born without major genetic or phenotypic malformations, although insemination doses only contain as few as 1–2 million sorted sperm in bovine AI. Nevertheless, in most species, sorted sperm doses are less fertile than unsorted doses, and in the case of bull sperm, based on day 56 non-return rates, nearly two-thirds of this decline in fertility (8.6%) may be due to the low dose and a third (5.0%) to the process of sorting itself (Frijters et al. 2009).

4.6.1 Incubation with the Fluorescent Stain and Exposure to the UV Laser

Observing sperm through the various stages of the sorting process, a first stress factor is labelling and incubation with the fluorescent dye. The negative effects of the Hoechst stain are dependent on the co-incubation medium and the species (Downing et al. 1991; Guthrie et al. 2002). The most sensitive to the dye (60 μM) are boar sperm, whereas bull (90 μM) and human sperm (900 μM) easily withstand much higher concentrations of stain. In terms of mitochondrial function, Watkins et al. (1996) described a dose-dependent impact of Hoechst 33342 on the tail beat frequency of human sperm, and Spinaci et al. (2005) measured a loss of mitochondrial membrane potential after staining and sorting of boar sperm. Moreover, incubation with Hoechst 33342 increased the rate of spontaneous lipid peroxidation with negative effects on the motility of bull sperm (Klinc and Rath 2007), which were not independent from extender composition but were partly compensable (Mancini et al. 2013). Nevertheless, despite these findings of effects of dose of the stain at the level of sperm function and ultrastructure, at the concentrations necessary for a differentiation of the X and Y boar sperm populations, there appear to be no adverse effects on their motility or fertilising capacity after insemination (Vazquez et al. 2002).

In the subsequent stages of processing, however, Hoechst dye in combination with the energy released from the UV laser (150–200 mW) could affect DNA integrity. In early studies, when higher dye concentrations were used than at present, the results did not completely exclude a combined effect of incubation with stain and exposure to the laser on chromosome integrity (Libbus et al. 1987). When applied to somatic cells, Hoechst 33342 and UV light are toxic and mutagenic (Durand and Olive 1982; Sinha and Hader 2002). However, in more recent studies, no effect was found on DNA methylation, when tested for IGF2 and IGF2 receptor genes in bulls (Carvalho et al. 2012). In human sperm, no increase was found in the incidence of endogenous nicks in any sperm after UV and fluorochrome exposure, compared with controls without exposure, nor after the sorting procedure in the flow cytometer (Catt et al. 1997). This may be due mainly to the ultrashort UV exposure time of cells in the sperm sorter. Pamila et al. (2004) inseminated sows with sex-sorted sperm and investigated the lymphocytes of newborn piglets. No increase was observed in genotoxic effects based on the frequency of the mutagenic index, nor was there evidence of any phenotypic abnormalities. Moreover in pigs, Guthrie et al. (2002) compared the effect of UV laser power on embryos produced with sexed boar sperm and did not find detrimental effects on embryo development between 25 mW and 125 mW of laser power. However, the higher power was advantageous in maintaining high resolution and separation of sperm. Schenk and Seidel (2007) also tested the effect of Hoechst 33342 dye and UV laser power on the integrity of bovine sperm. However, an examination of the various steps in the sorting process indicated that mechanical damage, rather than Hoechst 33342 staining or laser exposure, was responsible for most of the decreased viability of the sex-sorted sperm (Garner and Suh 2002).

Alterations to the fine structure of the sperm tail and mitochondria have been noted by a number of researchers after sex sorting of mammalian sperm. In a comprehensive ultrastructural study on bull sperm, Michl (2014) found that co-incubation with Hoechst 33342, exposure to the laser, increased amplitude of the piezo crystal and exposure to the electrostatic field, all caused direct quantitative changes in the sperm mitochondrial conformation from orthodox to condensed. These changes led to a reduction in matrix volume and an increase in electron density. In parallel with these effects, the space between the internal and external mitochondrial membrane, as well as in the intra-cristal space, was enlarged, reducing the performance of the mitochondria (Michl 2014). These observations are in agreement with reports, not necessarily associated with sex sorting, that ultrastructurally altered midpiece mitochondria, among other mitochondria with dilated intermembrane spaces, are associated with asthenozoospermia in humans (Pelliccione et al. 2011). The mitochondrial disturbances resulting from sex sorting of sperm are also reflected in their energy metabolism. Sander (2016) found that both the ATP production and mitochondrial membrane potential of bull sperm were reduced by the whole sorting process, whereas the fluorescent dye itself had no effect. Conversely, the combined exposure to Hoechst dye and laser increased mitochondrial condensation, and this may explain the loss of motility of bull sperm seen by Carvalho et al. (2010) under similar conditions.

4.6.2 Mechanical Forces

Sorting as such may not affect either viability or DNA defragmentation of sperm, but rather DNA damage may be caused by the mechanical shear forces associated with the procedure (Seidel and Garner 2002; De Ambrogi et al. 2006). This suggestion was supported by data from SCSA tests indicating sex-sorted sperm have less homogenous distribution of sperm chromatin than their unsorted counterparts (Boe-Hansen et al. 2005).

However, during the sorting procedure, mechanical forces hit sperm during their association with different sorter components. Firstly, sperm come into contact with shear forces in the nozzle assembly when core stream and sheath fluid with a set differential pressure focus the orientation of the sperm head as soon as it leaves the injection needle. Suh et al. (2005) thoroughly investigated the effects of the differential pressure on bovine and equine sperm motility and membrane integrity. They reported a significant improvement in both parameters at a relatively low pressure of 30 psi. However, as the sort resolution at high sorting speed also depends on the differential pressure, 40 psi has been recommended for routine work. Similar effects of differential pressure were found previously by Campos-Chillon and de la Torre (2003): in the bovine IVF system reported by these authors, higher cleavage and blastocyst formation rates were achieved from sperm sorted in a lower than a higher pressurised sorter. In stallion sperm, in addition to oxidative stress mitochondrial dysfunction was also attributed to high-pressure mechanical stress (da Silva et al. 2016a, b).

Secondly, the high-frequency impulses of the piezo crystal, with variable amplitude, induce forces on the assembly components as well as the fluidics and thereby on the sperm surface. The piezoelectric production of waves is necessary to form a discontinuous droplet stream, where ideally a separate droplet surrounds each sperm. The variable frequency is structurally related to the orifice diameter of the nozzle tip and the differential pressure of the system. Under conditions of high differential pressure, Suh et al. (2005) observed that sperm motility was better if the piezo was switched off, and no droplets were formed, than when it was switched on. Conversely, neither the mitochondrial membrane potential nor the mitochondrial ATP content, as measured by luminescence, changed in relation to the piezo amplitude (Sander 2016). Lowering the amplitude of the piezo crystal reduced the condensation of bovine mitochondria as observed by TEM in many but not all sperm (Michl 2014). Presumably, lower amplitude reduces the repeated change of the pressure pulses on the sperm surface and lowers the absolute cell membrane pressure. Consequently, sperm are more likely to withstand the stress of hydromechanical forces if the amplitude of the piezo crystal is reduced.

Thirdly, sorted sperm are pushed into the collection fluid with an approximate speed of 90Km/h. They arrive in a highly diluted state, surrounded by sheath medium that has washed away most of the membrane-protecting agents (decapacitation factors) of the seminal plasma, which at ejaculation are bound to the sperm surface (Maxwell and Johnson 1999). The beneficial components of boar seminal plasma, in the case of highly diluted boar sperm, have been isolated to the PSP-II subunit of the PSP-I/PSPII spermadhesin (Garcia et al. 2006). In the case of ram seminal plasma, the beneficial proteins may be RSVP-14/20 (Barrios et al. 2005) or ram spermadhesin (Bergeron et al. 2005). As these decapacitation factors, produced in the accessory sex glands, get lost during sorting, sperm membranes are destabilised and may pre-capacitate, shortening their fertilising lifespan. These changes may be reversible by the addition of seminal plasma fractions (Maxwell et al. 2007). Therefore, in addition to acting as a kind of mechanical cushion to break down the speed with which the sperm leave the flow sorter, collection tubes are preloaded with catch medium to provide a substitute for seminal plasma, which often incorporates decapacitation factors. Inclusion of seminal plasma in the staining extenders for boar and ram sperm, or in the collection medium for boar or bull sperm, has been shown to improve the viability and membrane integrity of sorted sperm (Maxwell and Johnson 1997, 1999; Centurion et al. 2003). Mostly, the collection medium is TES-TRIS-based with 2–20% egg yolk and 2% seminal plasma (Maxwell et al. 1998; Johnson 2000).

In the future, liposomes may replace the egg yolk component of the collection medium, which may have several advantages for commercial application as it avoids the use of animal-derived products and provides for a standardised commercial collection medium (Rath and Schmitz, unpublished data).

4.6.3   Electrical Charge

Charged droplets containing individual sperm are sorted in most standard flow cytometers by electrostatic deflection. The charge applied is related to the measured DNA content and is loaded on the last hanging droplet of the fluid stream. The charged droplets pass an electrostatic field of about 3000 V and, according to the polarisation of the charge, are deflected to either side of the centreline.

Repeated electric charging and electrostatic deflection have been proposed as another factor responsible for the reduced lifespan of sex-sorted sperm (Rath and Johnson 2008; Rath et al. 2009; Spinaci et al. 2006, 2010). The electrostatic voltage of a sorter is similar to that used for electroporation, which can induce the acrosome reaction (Tomkins and Houghton 1988). Therefore, the capacitation-like changes observed in the membranes of sex-sorted boar sperm by Maxwell and Johnson (1997) may have been an effect of the electric charge on the sperm membrane, rather than the result of mechanical forces. Such changes are similar to those observed in capacitated sperm, although actin cytoskeleton polymerisation and protein tyrosine phosphorylation seem to be less affected by sex sorting compared with normal capacitation in bull and boar sperm (Bucci et al. 2012). Furthermore, exposure of cells to an electrostatic field is known to induce the formation of ROS (Sauer et al. 2005), which damage sperm membranes (Leahy et al. 2010). A physiological level of ROS is necessary for hyperactivation, capacitation and the acrosome reaction in vitro (de Lamirande et al. 1997), but excessive ROS adversely affects the integrity of the bull sperm tail (Klinc et al. 2007; Klinc and Rath 2007). Moreover, the decreased mitochondrial membrane potential caused by ROS is correlated with decreasing motility of stallion (Baumber et al. 2000) and human sperm (Shi et al. 2012).

Whether or not the electrostatic field has a direct and independent effect on sperm mitochondria is difficult to interpret. While the electrostatic field is permanently present if switched on, unstained sperm that are not exposed to the laser are not deflected, as a discriminatory decision regarding the droplet charge cannot be made. In this case, therefore, they can only be recovered from the waste fluid, which has a high dilution effect and lacks the compensatory mechanisms normally applied by egg yolk or seminal plasma in the collection medium, and this alone may damage the sperm. Nevertheless, Spinaci et al. (2006) related the distribution change of HSP70 to the electrostatic field, and Michl (2014) found significantly higher percentages of condensed mitochondria in sex-sorted bovine sperm tails compared with controls, when passed through the sorter with deflection plates activated. There may also be indirect effects on mitochondrial activity through the action of ROS resulting from the electrical charge. However, De Ambrogi et al. (2006) reported that boar sperm membranes were damaged even if the electrostatic field was switched off and that the DNA fragmentation was still higher than in unsorted control sperm. Therefore, it remains controversial whether the effect of the electrostatic field has any direct relationship to the enrichment of ROS (Rath et al. 2013; Wang et al. 2013), which are known also to induce mitochondrial condensation in *Drosophila* (Walker and Benzer 2004).

4.6.4 Alternatives to Electrostatic Deflection

To avoid repeated charging and electrostatic stress during sex sorting, a recently developed method (Heisterkamp et al. 2015) replaces electrostatic deflection with laser irradiation of the sperm droplet flow. An acoustic-optical modulator (AOM) triggers the signal from a DPSS Er:YAG laser to the droplet stream. The laser does not kill sperm but rather deflects those droplets containing sperm with unwanted sex characteristics into the waste by a short surface-directed impulse. Most of the laser light is absorbed within the first micrometre of the droplet, and the emitted laser light generates recoil in the droplet by laser-based evaporation. Consequently, a steam jet formation produces an acceleration of the droplet. A high absorption coefficient of the liquid prevents sperm damage, and thermal interactions with the laser do not occur. This system allows sperm with desirable characteristics to pass through the sorter without being subjected to any deflection force. There were no differences in motility patterns or morphological integrity of bovine sperm after sorting with the laser-based deflection system compared with unsorted controls, whereas those sorted using electrostatic deflection lost 17% of their motility characteristics after 6 h post-sorting incubation (Rath et al. 2013).

4.7 Alternatives to Quantitative Flow Cytometry

4.7.1 Microfluidics

An alternative to high-speed flow cytometry, for more efficient sorting, might be the application of specially designed microfluidic chambers (for a review see Knowlton et al. 2015). The goal of these developments is not currently focussed on the separation of X and Y chromosome-bearing sperm but rather on quality characteristics, mainly to enrich intact human sperm populations for IVF or ICSI or for the culture of gametes and embryos (Suh et al. 2003). Based on the knowledge of flow cytometry, however, microfluidic sorting could provide a completely new approach to hydrodynamic cell orientation and chargeless deflection or impedance-related detection combined with a dielectrophoretic sorting (de Wagenaar et al. 2016). Schulte et al. (2007) reported a microfluidic sperm-sorting device for the selection of motile sperm with high DNA integrity. Similarly, to avoid centrifugation of the sperm sample, Li et al. (2016) used microfluidic chambers to select previously sex-sorted sperm for motility, in order to improve embryonic development after

IVF. Technologies based on microfluidics may replace traditional sex sorting by flow cytometry, sooner or later. However, the initial impact on sperm sexing will be in the parallel use of disposable microfluidic chambers as, for example, described in recent patent applications (Inguran 2013a, b, c).

4.7.2 Replacement of Quantitative Flow Cytometry by Qualitative Signal Creation Using Gold Nanoparticles

The current sex-sorting technique based on quantitative differences between the X and Y chromosome-bearing sperm populations has limitations in efficiency, mainly related to single-cell orientation in front of the laser beam. Numerous unsuccessful attempts have been made to sort sperm by physical methods, and qualitative surface markers have failed to find application on a wider scale (Cran and Johnson 1996). Nevertheless, the generation of a qualitatively different emission signal from sperm carrying one of the sex chromosomes would provide the possibility for a major improvement in sperm sexing. Haploid sperm differ in their DNA sequence on X and Y chromosomes, which are distinguishable in vitro by fluorescent in situ hybridisation (FISH). However, FISH requires disintegration of the sperm head as well as decondensation of chromatin (Kawarasaki et al. 1998), rendering the sperm non-functional.

Gold nanoparticles (AuNPs) have been assessed as suitable carriers for sequence-specific labelling of haploid mammalian sperm and for visualisation and tracing with an annealed DNA probe. Nanoparticles are known to have unique optical properties due to their strong surface plasmon resonance, and they can be made to function easily using thiol linkers. Compared with fluorochromes, AuNPs do not bleach and require only low energy because their quantum efficiency is very high (Taylor et al. 2014).

Different noble metals and metal alloys have been studied for nanotoxicity in various cells, tissues and organs, including sperm, oocytes and embryos (Tiedemann et al. 2014; Zhang et al. 2014; Feugang 2017). Toxicity of nanoparticles may be associated with their type, size and chemical characteristics (Taylor et al. 2012; Tiedemann et al. 2014) and is specifically associated with nanoparticle dosage. The toxicity affects the sperm membrane and leads to decreased motility and fertilisation potential of spermatozoa (Barchanski et al. 2015; Feugang et al. 2012; Taylor et al. 2015; Yoisungnern et al. 2015). An important precursor to their utilisation is the process by which the nanoparticles are produced from solid material. Pulsed laser ablation in liquids (PLAL) is a new method that synthesises totally ligand-free colloidal nanoparticles. PLAL has a significant advantage over chemical methods for nanoparticle production, as it does not leave any toxic chemical residues, but it does result in a broader distribution of particle sizes, possibly causing toxic side effects. This can be minimised if ablation and bio-conjugation are geometrically separated in a flow chamber where peptides are used to quench particles of unsuitable diameter or further supported by size quenching in the presence of electrolytes and pulsed laser melting of nanoparticles in liquids (Rehbock et al. 2014).

To use AuNPs as a qualitative marker to distinguish between X and Y sperm populations, three consecutive steps are necessary:

4.7.2.1 Internalisation of AuNPs Through the Sperm Membranes

So far, only capacitated sperm or those with a complete or partial acrosome reaction allow AuNP conjugates to enter through the plasma membrane. AuNPs have an external diameter of 10–100 nm, which is in the size range of those viruses against which sperm need protection. Internalisation of the nanoparticles by the sperm cell is also influenced by physical and chemical characteristics other than size, such as shape and electrochemical properties, the presence of ligands (Gao et al. 2005; Chithrani et al. 2006; Chithrani and Chan 2007; Jiang et al. 2008; Arvizo et al. 2010; Zhang et al. 2010), membrane fluidity, surface charge and functional molecules attached to the outer cell membrane. While ligand-free AuNPs enter cells by non-endosomal uptake (Salmaso et al. 2009; Taylor et al. 2010), particles with ordered arrangements of hydrophilic and hydrophobic functional groups are internalised by membrane wrapping (Verma et al. 2008).

In order to promote the internalisation of AuNPs into membrane-intact sperm heads, viral vectors (Everts et al. 2006), dendrimers (Shi et al. 2007) or supporting molecules have been tested and were attached together with a DNA probe to the AuNPs. Examples of cell-penetrating peptides (CPP) are TaT (transactivator of transcription) and penetratin. TaTs advance the internalisation of DNA (Tkachenko et al. 2004; Nativo et al. 2008; Mandal et al. 2009) and, when conjugated to AuNPs, support their endosomal transport into cells and cell nuclei (Petersen et al. 2011). Unfortunately, endosomal uptake does not occur in sperm.

The size of AuNPs used in the experiments of Taylor et al. (2009c, 2014) ranged around 50 nm. Further tests will be required to determine whether nanoclusters of from 3 to 5 nm will be able to pass through intact sperm membranes. Particles of larger size are not suitable for selective targeting in sperm heads due to diffusion limitations and a lack of plasmon coupling to the nanoclusters. This problem may be managed using small particles with distinctive optical properties, for example, fluorescent gold nanoclusters. Independently of internalisation, sperm cannot carry an unlimited number of AuNPs. A mass concentration dose of 10 mg/ml, equal to a total dose of 14,000 nanoparticles per sperm, leads to a decrease in their motility, presumably caused by the complexion of thiol or disulphide groups at the sperm surface. Ligand-free gold nanoparticles also impair sperm fertilising ability probably by agglomerate attachment to the sperm membrane, thereby mechanically interfering with sperm-oocyte interactions (Tiedemann et al. 2014; Taylor et al. 2015).

4.7.2.2 Non-invasive Coupling of a Specific DNA Probe with the Intact DNA Double Strand by Triplex Binding and Accumulation of Nanoparticles

To identify repeated Y chromosomal DNA sequences, a specific probe has to be generated and annealed to AuNPs. Furthermore, accumulation of the particles must neither affect the fertilising capacity and lifespan of sperm nor should it disintegrate the Y chromosome. Therefore, DNA hybridisation is performed using triplex forming oligonucleotides (TFOs) (Hoogsteen 1963). Triplex target sites and triplex hybridisation are very suitable for gold nano-targeting because they contain many poly-purine sequences interrupted by one or more base pair inversions. Target sequences very close to each other result in a better AuNP accumulation and therefore provide a better signal (Xodo et al. 2001). As a triplex hybridisation with DNA probes is a rather fragile connection, more stable binding can be established with DNA derivatives, such as locked nucleic acids (LNA) and peptide nucleic acids (Johnson and Fresco 1999; Buchini and Leumann 2003; Seidman and Glazer 2003). In vitro, AuNP-conjugated LNA probes form the most stable triplexes in solution (McKenzie et al. 2008).

4.7.2.3 Recognition of the Sex-Specific Signal Pattern to Sort the Sperm Population

The detection principle is based on the fact that AuNPs that are aggregated or accumulated, for example, due to binding of probes to highly repetitive DNA sequences, change their plasmon resonance peak. This can be measured as a spectral absorption (bathochromic) shift to the red as it changes the wavelength of the maximum light extinction, which can be used as a criterion to differentiate X from Y sperm (Jain 2007).

So far, the qualitative identification of Y chromosome-bearing sperm is still at the laboratory research stage. Many technical and functional aspects have been solved: repeated sequences have been used to identify the Y chromosome by triplex DNA formation; non-toxic functionalised AuNPs with DNA probes and transport promoting peptides have reached a high-throughput level; and bathochromic shifts of spectral absorption have been recognised. However, it has not been possible, so far, to internalise functionalised AuNPs of different diameter through the intact sperm membranes of different species. Further research is necessary, for example, to assess whether a temporal capacitation can be achieved and the change in membrane fluidity used to insert the nanoparticles. Once this last step has been elucidated, separation by qualitative signals may provide many advantages over the currently used quantitative sex-sorting method. For example, orientation of sperm in front of the laser will be unnecessary, all individually identified sperm would be sorted, and depending on the nanoparticle used, alternative methods to flow cytometry may be developed.

4.7.3 Promotion of Naturally Occurring Genome Variations by Gene Editing

As individual sperm sorting is highly inefficient in species that require large insemination doses, gene editing might be a valuable alternative, if ethical considerations allow its application. The methods available for gene editing are discussed elsewhere in this book. The purpose of gene editing, in the context of sex selection in farm animals, is to either modify or eliminate the coding from one of the sex chromosomes or its target autosome, so that fertile offspring of only one sex are born. Alternatively, gene editing, in order to discriminate specific functions like spermatogenesis and the related hormone production, can be used to modify offspring of the unwanted sex. Until recently, sperm DNA was unavailable for editing. However, genome editing has become a fast-growing research area since the development of sequence-specific nucleases and four different editing processes: meganuclease, zinc finger nuclease (ZFN), transcription activator-like effector nuclease (TALEN) and clustered regularly interspaced short palindromic repeats (CRISPR). The CRISPR-Cas9 system is distinctive, in that it has a relatively simple design and high efficiency. It induces a DNA double-strand break in the target sequence, which is repaired either by non-homologous end-joining (NHEJ), if no repair template is present, or by homology-directed repair (HDR). The NHEJ pathway produces small insertions and deletions (indels) with a relative high probability of forming coding regions to a frame shift and therefore a functional gene knockout. HDR is preferable for repairing mutations or adding DNA sequences. Daniel and Fahrenkrug (2016) provide a good overview of the possible applications of gene editing to sperm sexing (patent application EP3003021 2016). TALEN and CRISPR-Cas9 can be used, when applied in germ line (GS) cells, to study spermatogenesis and its genetic regulation. Sato et al. (2015) efficiently tested TALEN and double-nicking CRISPR-Cas9 on GS cells with two representative genes (Rosa26; Stra8) and recommended Rosa26-targeted cells as a means to differentiate competent sperm, whereas Stra8-targeted cells led to deficient initiation of meiosis. Both methods could be used during spermatogenesis to modify sperm or stem cell spermatogonia (Vassena et al. 2016; Wu et al. 2015).

Gene editing could be an important tool for pig reproduction in order to eliminate sexually mature boars and the boar taint in the carcass. Two different strategies are outlined below, from which either normal female piglets and phenotypical males without testicles – and the according hormone production – would be born or from which there would be no male offspring:

(a) The first strategy employs the CRISPR-Cas9 system, at the level of the germinal tissue, to induce a knockout of the SRY gene, thereby preventing fertile Y chromosome-bearing sperm from forming in the foetal genital ridge. Sertoli and Leydig cells would not be formed, and their hormonal products would not be secreted. In this default case, the Wolffian duct would not develop, and the Mullerian duct would not be supressed. The litters produced, in the case of multiparous species, would contain females and infertile, phenotypically, male offspring.

(b) For the second strategy, in which the production of male offspring is prevented, a multiple double-stranded tailoring with CRISPR-Cas9 would be required to modify the function of Y chromosome-bearing sperm. In consequence, only X sperm could participate in fertilisation. During spermatogenesis, the Y chromosome does not fulfil any functions that could adversely affect the animal in its development, so a modification of their genome would not be detrimental. The advantage of this approach is that only female offspring would be born.

4.8 Concluding Remarks

Farmers have long desired specific sex-directed reproduction in domestic animals. The technique of sex selection using flow cytometric sorting of sperm has gained a significant role in the food production chain that on the one hand has to fulfil the nutritive requirements of an exploding population and on the other needs to integrate sustainable agricultural concepts. Sex sorting is commercially available for several species but has been applied mostly in dairy cattle. The technique is challenging, as it has to identify, en masse, each individual sperm and maintain its fertilising capacity even after long-term storage. A good measure of the importance of sex sorting is the large number of international patent applications that have been made and patents granted, reflecting the intensive research and investment, that has been devoted to the technique. However, society should be aware that the significant networks generated by this intellectual property could create dependencies that might not be in consensus with marketing demands and may limit the developmental support for alternative sexing methods.

It is foreseeable that the existing quantitative sex-sorting technology, using flow cytometry, is only an intermediate method, as it is hampered by limited efficiency and high production costs. Alongside new qualitative sorting methods, identifying the specific sex-related characteristic of sperm, perhaps in combination with nanoparticles and with microfluidic systems replacing the traditional flow cytometer, will be new genetic techniques. These techniques will provide a simplified system to generate offspring not only of the required sex but also incorporating other important genetic characteristics. Gene editing based on meganucleases, ZFN, TALEN, CRISPR-Cas9 or future systems like CRISPR-AID are the first such techniques to appear on the scientific horizon and have quickly found their way into experimental reproduction. Whether or not gene editing will be applied in animal production will depend on ethical priorities rather than technical barriers. However, as the technology develops into the future, sex selection in livestock will continue to play a major role in providing sufficient food, of the highest quality and with sustainable protection of the environment, to meet the needs of a growing world human population.

Acknowledgements This article is dedicated to Dr. Lawrence A. Johnson, who contributed most to the development and introduction of sperm sexing in farm animal reproduction. The authors of this paper are very thankful for his constant support and friendship. Larry Johnson celebrated his 80th birthday on July 9th, 2016. We also gratefully acknowledge all the students, technicians and scientists, who contributed in the laboratories of both authors, and who made the research on sperm sexing such an interesting part of our lives. We honour the personal friendships created during these projects, including that between the authors, which has encompassed some 25 years of collaboration.

References

Abeydeera LR, Johnson LA, Welch GR, Wang WH, Boquest AC, Cantley TC, Rieke A, Day BN (1998) Birth of piglets preselected for gender following in vitro fertilization of in vitro matured pig oocytes by X and Y chromosome bearing spermatozoa sorted by high speed flow cytometry. Theriogenology 50(7):981–988

Ali JI, Eldridge FE, Koo GC, Schanbacher BD (1990) Enrichment of bovine X-chromosome and Y-chromosome bearing sperm with monoclonal H-Y antibody fluorescence-activated cell sorter. Arch Androl 24(3):235–245

Alkmin DV, Parrilla I, Tarantini T, del Olmo D, Vazquez JM, Martinez EA, Roca J (2016) Seminal plasma affects sperm sex sorting in boars. Reprod Fertil Dev 28(5):556–564

Alminana C, Caballero I, Heath PR, Maleki-Dizaji S, Parrilla I, Cuello C, Gil MA, Vazquez JL, Vazquez JM, Roca J, Martinez EA, Holt WV, Fazeli A (2014) The battle of the sexes starts in the oviduct: modulation of oviductal transcriptome by X and Y-bearing spermatozoa. BMC Genomics 15:293

Amann RP (1999) Issues affecting commercialization of sexed sperm. Theriogenology 52(8):1441–1457

Arvizo RR, Miranda OR, Thompson MA, Pabelick CM, Bhattacharya R, Robertson JD, Rotello VM, Prakash YS, Mukherjee P (2010) Effect of nanoparticle surface charge at the plasma membrane and beyond. Nano Lett 10(7):2543–2548

Barcelo-Fimbres M, Seidel GE (2004) Optimizing sperm concentration to maximize monospermy and minimize polyspermy with bovine in vitro fertilization. Poult Sci 83:371

Barcelo-Fimbres M, Campos-Chillon LF, Seidel GE (2011) In vitro fertilization using non-sexed and sexed bovine sperm: sperm concentration, sorter pressure, and bull effects. Reprod Domest Anim 46(3):495–502

Barchanski A, Taylor U, Sajti CL, Gamrad L, Kues WA, Rath D, Barcikowski S (2015) Bioconjugated gold nanoparticles penetrate into spermatozoa depending on plasma membrane status. J Biomed Nanotechnol 11(9):1597–1607

Barrios B, Fernández-Juan M, Muiño-Blanco T, Cebrián-Pérez JA (2005) Immunocytochemical localization and biochemical characterization of two seminal plasma proteins that protect ram spermatozoa against cold shock. J Androl 26:539–549

Baumber J, Ball BA, Gravance CG, Medina V, Davies-Morel MCG (2000) The effect of reactive oxygen species on equine sperm motility, viability, acrosomal integrity, mitochondrial membrane potential, and membrane lipid peroxidation. J Androl 21(6):895–902

Beal WE, White LM, Garner DL (1984) Sex ratio after insemination of bovine spermatozoa isolated using a bovine serum albumin gradient. J Anim Sci 58:1432–1436

Beernink FJ, Ericsson RJ (1982) Male sex preselection through sperm isolation. Fertil Steril 38(4):493–495

Beilby K, de Graaf S, Evans G, Maxwell WMC, Wilkening S, Wrenzycki C, Grupen C (2011) Quantitative mRNA expression in ovine blastocysts produced from X- and Y-chromosome bearing sperm, both in vitro and in vivo. Theriogenology 76(3):471–481

Bennett D, Boyse EA (1973) Sex-ratio progeny of mice inseminated with sperm treated with H-Y antiserum. Nature 246(5431):308–309

Bergeron A, Villemure M, Lazure C, Manjunath P (2005) Isolation and characterization of the major proteins of ram seminal plasma. Mol Reprod Dev 71:461–470

Bergmann A, Taylor U, Rath D (2012a) Flow-cytometric evaluation of lectin binding moieties on porcine uterine epithelial cells. Reprod Domest Anim 47:77

Bergmann A, Taylor U, Rath D (2012b) Interactions of spermatozoa and uterine epithelial cells in the pig: a cell culture study. Reprod Domest Anim 47:486

Bhattacharya BC (1962) Different sedimentation rates of X- and Y- sperm and the question of arbitrary sex determination. Zentralblatt für Wissenschaft & Zoologie 166:203–250

Bhattacharya BC, Bangham AD, Cro RJ, Keynes RD, Rowson LE (1966) An attempt to predetermine the sex of calves by artificial insemination with spermatozoa separated by sedimentation. N ature 211:863

Blecher SR, Howie R, Li S, Detmar J, Blahut LM (1999) A new approach to immunological sexing of sperm. Theriogenology 52(8):1309–1321

Blondin P, Beaulieu M, Fournier V, Morin N, Crawford L, Madan P, King WA (2009) Analysis of bovine sexed sperm for IVF from sorting to the embryo. Theriogenology 71(1):30–38

Blottner S, Bostedt H, Mewes K, Pitra C (1994) Enrichment of bovine X-spermatozoa and Y-spermatozoa by free-flow electrophoresis. Zentralbl Veterinarmed A 41(6):466–474

Boe-Hansen GB, Morris ID, Ersboll AK, Greve T, Christensen P (2005) DNA integrity in sexed bull sperm assessed by neutral Comet assay and sperm chromatin structure assay. Theriogenology 63(6):1789–1802

Boklage CE (2005) The epigenetic environment: secondary sex ratio depends on differential survival in embryogenesis. Hum Reprod 20(3):583–587

Bombardelli GD, Soares HF, Chebel RC (2016) Time of insemination relative to reaching activity threshold is associated with pregnancy risk when using sex-sorted semen for lactating Jersey cows. Theriogenology 85(3):533–539

Bourdon RM, Brinks JS (1987a) Simulated efficiency of range beef-production. 1. Growth and milk-production. J Anim Sci 65(4):943–955

Bourdon RM, Brinks JS (1987b) Simulated efficiency of range beef-production. 2. Fertility traits. J Anim Sci 65(4):956–962

Bourdon RM, Brinks JS (1987c) Simulated efficiency of range beef-production. 3. Culling strategies and nontraditional management-systems. J Anim Sci 65(4):963–969

Bucci D, Galeati G, Tamanini C, Vallorani C, Rodriguez-Gil JE, Spinaci M (2012) Effect of sex sorting on CTC staining, actin cytoskeleton and tyrosine phosphorylation in bull and boar spermatozoa. Theriogenology 77(6):1206–1216

Buchanan BR, Seidel GE, McCue PM, Schenk JL, Herickhoff LA, Squires EL (2000) Insemination of mares with low numbers of either unsexed or sexed spermatozoa. Theriogenology 53(6):1333–1344

Buchini S, Leumann CJ (2003) New nucleoside analogues for the recognition of pyrimidine-purine inversion sites. Nucleosides Nucleotides Nucleic Acids 22(5–8):1199–1201

Byskov AG (1986) Differentiation of mammalian embryonic gonad. Physiol Rev 66(1):71–117

Campos-Chillon LF, de la Torre JF (2003) Effect of concentration of sexed bovine sperm sorted at 40 and 50 psi on developmental capacity of in vitro produced embryos. Theriogenology 59:506 (abstract)

Carvalho JO, Sartori R, Machado GM, Mourao GB, Dode MAN (2010) Quality assessment of bovine cryopreserved sperm after sexing by flow cytometry and their use in in vitro embryo production. Theriogenology 74(9):1521–1530

Carvalho JO, Michalczechen-Lacerda VA, Sartori R, Rodrigues FC, Bravim O, Franco MM, Dode MAN (2012) The methylation patterns of the IGF2 and IGF2R genes in bovine spermatozoa are not affected by flow-cytometric sex sorting. Mol Reprod Dev 79(2):77–84

Catt JW (1996) Intracytoplasmic sperm injection (ICSI) and related technology. Anim Reprod Sci 42(1–4):239–250

Catt SL, Sakkas D, Bizzaro D, Bianchi PG, Maxwell WMC, Evans G (1997) Hoechst staining and exposure to UV laser during flow cytometric sorting does not affect the frequency of detected endogenous DNA nicks in abnormal and normal human spermatozoa. Mol Hum Reprod 3(9):821–825

Centurion F, Vazquez JM, Calvete JJ, Roca J, Sanz L, Parilla I, Garcia EM, Martinez E (2003) Influence of porcine spermadhesins on the susceptibility of boar spermatozoa to high dilution. Biol Reprod 69:640–646

Chang LB, Chou C-J, Shiu J-S, Tu P-A, Gao S-X, Peng S-Y, Wu S-C (2017) Artificial insemination of Holstein heifers with sex-sorted semen during the hot season in a subtropical region. Trop Anim Health Prod 49(6):1157–1162

Chithrani BD, Chan WCW (2007) Elucidating the mechanism of cellular uptake and removal of protein-coated gold nanoparticles of different sizes and shapes. Nano Lett 7(6):1542–1550

Chithrani BD, Ghazani AA, Chan WCW (2006) Determining the size and shape dependence of gold nanoparticle uptake into mammalian cells. Nano Lett 6(4):662–668

Clulow JR, Buss H, Evans G, Sieme H, Rath D, Morris LHA, Maxwell WMC (2012) Effect of staining and freezing media on sortability of stallion spermatozoa and their post-thaw viability after sex-sorting and cryopreservation. Reprod Domest Anim 47(1):1–7

Clulow JR, Buss H, Sieme H, Rodger JA, Cawdell-Smith AJ, Evans G, Rath D, Morris LHA, Maxwell WMC (2008) Field fertility of sex-sorted and non-sorted frozen-thawed stallion spermatozoa. Anim Reprod Sci 108(3–4):287–297

Cran DG (1997) Production of lambs by low dose intrauterine insemination with flow cytometrically sorted and unsorted semen. Theriogenology 47:267

Cran DG, Johnson LA (1996) The predetermination of embryonic sex using flow cytometrically separated X and Y spermatozoa. Hum Reprod Update 2(4):355–363

Cran DG, Johnson LA, Polge C (1995) Sex preselection in cattle - a field trial. Vet Rec 136(19):495–496

Cran DG, Johnson LA, Miller NGA, Cochrane D, Polge C (1993) Production of bovine calves following separation of X-chromosome and Y-chromosome bearing sperm and in vitro fertilization. Vet Rec 132(2):40–41

Daniel C, Fahrenkrug S (2016) Genetic techniques for making animals with sortable sperm. Patent application "EP3003021"

da Silva CMB, Ortega-Ferrusola C, Morrell JM, Martinez HR, Pena FJ (2016a) Flow cytometric chromosomal sex sorting of stallion spermatozoa induces oxidative stress on mitochondria and genomic DNA. Reprod Domest Anim 51(1):18–25

da Silva CMB, Ferrusola CO, Bolanos JMG, Davila MP, Martin-Munoz P, Morrell JM, Martinez HR, Pena FJ (2014) Effect of overnight staining on the quality of flow cytometric sorted stallion sperm: comparison with traditional protocols. Reprod Domest Anim 49(6):1021–1027

da Silva CMB, Ortega-Ferrusola C, Rodriguez AM, Bolanos JMG, Davila MP, Morrell JM, Martinez HR, Tapia JA, Aparicio IM, Pena FJ (2013) Sex sorting increases the permeability of the membrane of stallion spermatozoa. Anim Reprod Sci 138(3–4):241–251

da Silva CMB, Ortega-Ferrusola C, Morrell JM, Rodriguez Martinez H, Pena FJ (2016b) Flow cytometric chromosomal sex sorting of stallion spermatozoa induces oxidative stress on mitochondria and genomic DNA. Reprod Domest Anim 51(1):18–25

De Ambrogi M, Spinaci M, Galeati G, Tamanini C (2006) Viability and DNA fragmentation in differently sorted boar spermatozoa. Theriogenology 66(8):1994–2000

De Cecco M, Spinaci M, Zannoni A, Bernardini C, Seren E, Forni M, Bacci ML (2010) Coupling sperm mediated gene transfer and sperm sorting techniques: a new perspective for swine transgenesis. Theriogenology 74(5):856–862

de Graaf SP, Evans G, Maxwell WMC, O'Brien JK (2006) In vitro characteristics of fresh and frozen-thawed ram spermatozoa after sex sorting and re-freezing. Reprod Fertil Dev 18(8):867–874

de Graaf SP, Beilby KH, Underwood SL, Evans G, Maxwell WMC (2009) Sperm sexing in sheep and cattle: the exception and the rule. Theriogenology 71(1):89–97

de Graaf SP, Beilby K, O'Brien JK, Osborn D, Downing JA, Maxwell WMC, Evans G (2007a) Embryo production from superovulated sheep inseminated with sex-sorted ram spermatozoa. Theriogenology 67(1):550–555

de Graaf SP, Evans G, Gillan L, Guerra MMP, Maxwell WMC, O'Brien JK (2007b) The influence of antioxidant, cholesterol and seminal plasma on the in vitro quality of sorted and non-sorted ram spermatozoa. Theriogenology 67(2):217–227

de Graaf SP, Evans G, Maxwell WMC, Downing JA, O'Brien JK (2007c) Successful low dose insemination of flow cytometrically sorted ram spermatozoa in sheep. Reprod Domest Anim 42:648–653

de Wagenaar B, Dekker S, de Boer HL, Bomer JG, Olthuis W, van den Berg A, Segerink LI (2016) Towards microfluidic sperm refinement: impedance-based analysis and sorting of sperm cells. Lab Chip 16(8):1514–1522

Dean PN, Pinkel D, Mendelsohn ML (1978) Hydrodynamic orientation of sperm heads for flow cytometry. Biophys J 23:7–13

del Olmo D, Parrilla I, Sanchez-Osorio J, Gomis J, Angel MA, Tarantini T, Gil MA, Cuello C, Vazquez JL, Roca J, Vaquez JM, Martinez EA (2014) Successful laparoscopic insemination with a very low number of flow cytometrically sorted boar sperm in field conditions. Theriogenology 81(2):315–320

de Lamirande E, Leclerc P, Gagnon C (1997) Capacitation as a regulatory event that primes spermatozoa for the acrosome reaction and fertilization. Mol Hum Reprod 3(3):175–194

Dmowski WP, Gaynor L, Rao R, Lawrence M, Scommegna A (1979) Use of albumin gradients for X-sperm and Y-sperm separation and clinical experience with male sex preselection. Fertil Steril 31(1):52–57

Downing TW, Garner DL, Ericsson SA, Redelman D (1991) Metabolic toxicity of fluorescent stains on thawed cryopreserved bovine sperm cells. J Histochem Cytochem 39(4):485–489

Durand RE, Olive PL (1982) Cytotoxicity, mutagenicity and DNA damage by Hoechst 33342. J Histochem Cytochem 30:111–116

Eggers S, Sinclair A (2012) Mammalian sex determination-insights from humans and mice. Chromosom Res 20(1):215–238

Eggers S, Ohnesorg T, Sinclair A (2014) Genetic regulation of mammalian gonad development. Nat Rev Endocrinol 10(11):673–683

Engelmann U, Krassnigg F, Schatz H, Schill W-B (1988) Separation of human X and Y spermatozoa by free-flow electrophoresis. Gamete Res 19(2):151–160

Erickson RP, Lewis SE, Butley M (1981) Is haploid gene-expression possible for sperm antigens. J Reprod Immunol 3(4):195–217

Ericsson RJ, Langevin CN, Nishino M (1973) Isolation of fractions rich in human Y sperm. Nature 246(5433):421–424

Ettema JF (2007) Economic opportunities for sexed semen on commercial dairies. Western Dairy News 7(3):67–68

Everts M, Saini V, Leddon JL, Kok RJ, Stoff-Khalili M, Preuss MA, Millican CL, Perkins G, Brown JM, Bagaria H, Nikles DE, Johnson DT, Zharov VP, Curiel DT (2006) Covalently linked au nanoparticles to a viral vector: potential for combined photothermal and gene cancer therapy. Nano Lett 6(4):587–591

Flaherty SP, Michalowska J, Swann NJ, Dmowski WP, Matthews CD, Aitken RJ (1997) Albumin gradients do not enrich Y-bearing human spermatozoa. Hum Reprod 12(5):938–942

Feugang JM, Youngblood RC, Greene JM, Fahad AS, Monroe WA, Willard ST, Ryan PL (2012) Application of quantum dot nanoparticles for potential non-invasive bio-imaging of mammalian spermatozoa. J Nanobiotechnol 10:45

Feugang JM (2017) Novel agents for sperm purification, sorting, and imaging. Mol Reprod Dev 84(9):832–841

Frijters ACJ, Mullaart E, Roelof RMG, van Hoorne RP, Moreno JF, Moreno O, Merton JS (2009) What affects fertility of sexed bull semen more, low sperm dosage or the sorting process? Theriogenology 71(1):64–67

Fulwyler MJ (1977) Hydrodynamic orientation of cells. J Histochem Cytochem 25(7):781–783

Gao HJ, Shi WD, Freund LB (2005) Mechanics of receptor-mediated endocytosis. Proc Natl Acad Sci U S A 102(27):9469–9474

Garcia EM, Vázquez JM, Calvete JJ, Sanz L, Caballero I, Parilla I, Gil MA, Roca J, Martinez EA (2006) Dissecting the protective effect of the seminal plasma sperm adhesin PSP-I/PSP-II on boar sperm functionality. J Androl 27:434–442

Garner DL (2006) Flow cytometric sexing of mammalian sperm. Theriogenology 65(5):943–957

Garner DL (2009) Hoechst 33342: the dye that enabled differentiation of living X-and Y-chromosome bearing mammalian sperm. Theriogenology 71(1):11–21

Garner DL, Suh TK (2002) Effect of Hoechst 33342 staining and Laser illumination on the viability of sex-sorted bovine sperm. Theriogenology 57:746 abstract

Garner DL, Seidel GE (2008) History of commercializing sexed semen for cattle. Theriogenology 69(7):886–895

Garner DL, Gledhill BL, Pinkel D, Lake S, Stephenson D, Vandilla MA, Johnson LA (1983) Quantification of the X-chromosome-bearing and Y-chromosome-bearing spermatozoa of domestic-animals by flow-cytometry. Biol Reprod 28(2):312–321

Gibb Z, Lambourne SR, Aitken RJ (2012) Do spermatozoa from fertile thoroughbred stallions live fast and die young? Reprod Domest Anim 47:587–588

Gibb Z, Butler TJ, Morris LHA, Maxwell WMC, Grupen CG (2013) Quercetin improves the post-thaw characteristics of cryopreserved sex-sorted and non sorted stallion sperm. Theriogenology 79(6):1001–1009

Gledhill BL, Lake S, Steinmetz LL, Gray JW, Crawford JR, Dean PN, Vandilla MA (1976) Flow microfluorometric analysis of sperm DNA content - effect of cell-shape on fluorescence distribution. J Cell Physiol 87(3):367–375

Grossfeld R, Klinc P, Sieg B, Rath D (2005) Production of piglets with sexed semen employing a non-surgical insemination technique. Theriogenology 63(8):2269–2277

Guthrie HD, Johnson LA, Garrett WM, Welch GR, Dobrinsky JR (2002) Flow cytometric sperm sorting: effects of varying laser power on embryo development in swine. Mol Reprod Dev 61(1):87–92

Hancock RJT, Duncan D, Carey S, Cockett ATK, May A (1983) Anti-sperm antibodies, Hla anti-gens, and semen analysis. Lancet 2(8354):847–848

Hanley NA, Hagan DM, Clement-Jones M, Ball SG, Strachan T, Salas-Cortés L, McElreavey K, Lindsay S, Robson S, Bullen P, Ostrer H, Wilson DI (2000) SRY, SOX9, and DAX1 expression patterns during human sex determination and gonadal development. Mech Dev 91(1–2):403–407

Heer P (2007) Anpassung der Konservierungsprozesse für Hengstsperma an die Beltsville Sperm Sexing Technology. (Adaptation of cryo-preservation of stallion semen to the Beltsville Sperm Sexing Technology) Dissertation, Veterinary University, Hannover, Germany

Heisterkamp A, Lorbeer R, Masterrind GmbH, Meyer H, Rath D (2015) Apparatus and method for selecting particles. Pat.Appl.: US000009034260

Hendriksen PJM, Tieman M, Vanderlende T, Johnson LA (1993) Binding of anti-H-Y monoclonal-antibodies to separated X-chromosome and Y-chromosome bearing porcine and bovine sperm. Mol Reprod Dev 35(2):189–196

Hohenboken WD (1999) Applications of sexed semen in cattle production. Theriogenology 52(8):1421–1433

Hollinshead FK, O'Brien JK, Maxwell WMC, Evans G (2002) Production of lambs of predeter-mined sex after the insemination of ewes with low numbers of frozen-thawed sorted X- or Y-chromosome-bearing spermatozoa. Reprod Fertil Dev 14(8):503–508

Hoogsteen K (1963) Crystal and molecular structure of a hydrogen-bonded complex between 1-methylthymine and 9-methyladenine. Acta Crystallogr 16(9):907–916

Inaba Y, Abe R, Geshi M, Matoba S, Nagai T, Somfai T (2016) Sex-sorting of spermatozoa affects developmental competence of in vitro fertilized oocytes in a bull-dependent manner. J Reprod Dev 62(5):451–456

Inguran LLC, US (2013a) Device for high throughput sperm sorting. Pat. Appl.: US020140273192

Inguran LLC, US (2013b) Device for high throughput sperm sorting. Pat. Appl.: US020140273179

Inguran LLC, US (2013c) Methods for high throughput sperm sorting. Pat. Appl.: US020140273059

Jacobs PA, Ross A (1966) Structural abnormalities of the Y chromosome in man. Nature 210(5034):352–354

Jain KK (2007) Applications of nanobiotechnology in clinical diagnostics. Clin Chem 53(11):2002–2009

Jiang W, Kim BYS, Rutka JT, Chan WCW (2008) Nanoparticle-mediated cellular response is size-dependent. Nat Nanotechnol 3(3):145–150

Johnson LA (1988) Flow cytometric determination of sperm sex-ratio in semen purportedly enriched for X-bearing or Y-bearing sperm. Theriogenology 29(1):265–265

Johnson LA (1991) Sex preselection in swine - altered sex-ratios in offspring following surgi-cal insemination of flow sorted X-bearing and Y-bearing sperm. Reprod Domest Anim 26(6):309–314

Johnson LA (1997) Advances in gender preselection in swine. J Reprod Fertil Suppl 52:255–266

Johnson LA (2000) Sexing mammalian sperm for production of offspring: the state-of-the-art. Anim Reprod Sci 61:93–107

Johnson LA, Pinkel D (1986) Modification of a laser-based flow cytometer for high-resolution DNA analysis of mammalian spermatozoa. Cytometry 7(3):268–273

Johnson LA, Clarke RN (1988) Flow sorting of X-chromosome-bearing and Y-chromosome-bearing mammalian sperm - activation and pronuclear development of sorted bull, boar, and ram sperm microinjected into hamster oocytes. Gamete Res 21(4):335–343

Johnson LA, Welch GR (1999) Sex preselection: high-speed flow cytometric sorting of X and Y sperm for maximum efficiency. Theriogenology 52(8):1323–1341

Johnson LA, Flook JP, Look MV (1987) Flow cytometry of X and Y chromosome-bearing sperm for DNA using an improved preparation method and staining with Hoechst 33342. Gamete Res 17(3):203–212

Johnson LA, Flook JP, Hawk HW (1989) Sex preselection in rabbits - live births from X-sperm and Y-sperm separated by DNA and cell sorting. Biol Reprod 41(2):199–203

Johnson LA, Rath D, Vazquez JM, Maxwell WMC, Dobrinsky JR (2005) Preselection of sex of off-spring in swine for production: current status of the process and its application. Theriogenology 63(2):615–624

Johnson MD, Fresco JR (1999) Third-strand in situ hybridization (TISH) to non-denatured meta-phase spreads and interphase nuclei. Chromosoma 108(3):181–189

Junge S, Taylor U, Schuberth HJ, Baulain U, Rath D (2010) Influence of inseminate components on the presence of leukocytes and spermatozoa in the porcine uterus 2 hours after artificial insemination (AI). Reprod Domest Anim 45:66

Junge S, Taylor U, Schuberth HJ, Guenther J, Baulain U, Rath D (2011) Seminal plasma and spermatozoa modulate gene expression in the porcine uterus. Reprod Domest Anim 46:105

Junge S, Taylor U, Bergmann A, Schuberth HJ, Guenther J, Baulein U, Rath D (2012) Modulated gene expression in the porcine uterus after contact with seminal plasma and spermatozoa - results of a microarray study. Reprod Domest Anim 47:29

Kaneko S, Yamaguchi J, Kobayashi T, Iizuka R (1983) Separation of human X-bearing and Y-bearing sperm using percoll density gradient centrifugation. Fertil Steril 40(5):661–665

Kawarasaki T, Welch GR, Long CR, Yoshida M, Johnson LA (1998) Verification of flow cytometrically-sorted X- and Y-bearing porcine spermatozoa and reanalysis of spermatozoa for DNA content using the fluorescence in situ hybridization (FISH) technique. Theriogenology 50(4):625–635

Klinc P, Rath D (2005) State of the art and perspectives of application of sorted sperm cells in farm animals. Züchtungskunde 77(2–3):218–229

Klinc P, Rath D (2007) Reduction of oxidative stress in bovine spermatozoa during flow cytometric sorting. Reprod Domest Anim 42(1):63–67

Klinc P, Frese D, Osmers H, Rath D (2007) Insemination with sex sorted fresh bovine spermatozoa processed in the presence of antioxidative substances. Reprod Domest Anim 42(1):58–62

Knop K, Hoffmann N, Rath D, Sieme H (2005) Effects of cushioned centrifugation technique on sperm recovery and sperm quality in stallions with good and poor semen freezability. Anim Reprod Sci 89(1–4):294–297

Knowlton SM, Sadasivam M, Tasoglu S (2015) Microfluidics for sperm research. Trends Biotechnol 33(4):221–229

Kobayashi J, Oguro H, Uchida H, Kohsaka T, Sasada H, Sato E (2004) Assessment of bovine X- and Y-bearing spermatozoa in fractions by discontinuous Percoll gradients with rapid fluo-rescence in situ hybridization. J Reprod Dev 50(4):463–469

Koundouros S, Verma P (2012) Significant enrichment of Y-bearing chromosome human sperma-tozoa using a modified centrifugation technique. Int J Androl 35(6):880–886

Krausz C, Casamonti E (2017) Spermatogenic failure and the Y chromosome. Hum Genet 136(5):637–655

Krisher RL, Wheeler MB (2010) Towards the use of microfluidics for individual embryo culture. Reprod Fertil Dev 22(1):32–39

Krueger C, Rath D, Johnson LA (1999) Low dose insemination in synchronized gilts. Theriogenology 52(8):1363–1373

Kurykin J, Hallap T, Jalakas M, Padrik P, Kaart T, Johannisson A, Jaakma U (2016) Effect of insemination-related factors on pregnancy rate using sexed semen in Holstein heifers. Czeh J Anim Sci 61(12):568–577

Leahy T, Celi P, Bathgate R, Evans G, Maxwell WMC, Marti JI (2010) Flow-sorted ram sperma-tozoa are highly susceptible to hydrogen peroxide damage but are protected by seminal plasma and catalase. Reprod Fertil Dev 22(7):1131–1140

Li JC, Zhu SB, He XJ, Sun R, He QY, Gan Y, Liu SJ, Funahashi H, Li YB (2016) Application of a microfluidic sperm sorter to in vitro production of dairy cattle sex-sorted embryos. Theriogenology 85(7):1211–1218

Libbus BL, Perreault SD, Johnson LA, Pinkel D (1987) Incidence of chromosome-aberrations in mammalian sperm stained with Hoechst-33342 and UV-laser irradiated during flow sorting. Mutat Res 182(5):265–274

Lindsey AC, Varner DD, Seidel GE, Bruemmer JE, Squires EL (2005) Hysteroscopic or rectally guided, deep-uterine insemination of mares with spermatozoa stored 18 h at either 5 degrees C or 15 degrees C prior to flow-cytometric sorting. Anim Reprod Sci 85(1–2):125–130

Lopez SR, de Souza JC, Gonzalez JZ, Sanchez AD, Romero-Aguirregomezcorta J, de Carvalho RR, Rath D (2013) Use of sex-sorted and unsorted frozen/thawed sperm and in vitro fertilization events in bovine oocytes derived from ultrasound-guided aspiration. Revista Brasileira De Zootecnia 42(10):721–727

Lu KH, Seidel GE (2004) Effects of heparin and sperm concentration on cleavage and blastocyst development rates of bovine oocytes inseminated with flow cytometrically-sorted sperm. Theriogenology 62(5):819–830

Lu KH, Cran DG, Seidel GE (1999) In vitro fertilization with flow-cytometrically-sorted bovine sperm. Theriogenology 52(8):1393–1405

Mancini R, Sieg B, Rath D (2013) Bull sperm motility and molecular kinetic of Hoechst dye are effected by the buffer system of extenders. Reprod Domest Anim 48:82

Mandal D, Maran A, Yaszemski MJ, Bolander ME, Sarkar G (2009) Cellular uptake of gold nanoparticles directly cross-linked with carrier peptides by osteosarcoma cells. J Materials Sci Mater Med 20(1):347–350

Manger M, Bostedt H, Schill WB, Mileham AJ (1997) Effect of sperm motility on separation of bovine X- and Y-bearing spermatozoa by means of free-flow electrophoresis. Andrologia 29(1):9–15

Mari G, Bucci D, Love CC, Mislei B, Rizzato G, Giaretta E, Merlo B, Spinaci M (2015) Effect of cushioned or single layer semen centrifugation before sex sorting on frozen stallion semen quality. Theriogenology 83(6):953–958

Martinez EA, Vazquez JM, Roca J, Lucas X, Gil MA, Parrilla I, Vazquez JL, Day BN (2001) Successful non-surgical deep intrauterine insemination with small numbers of spermatozoa in sows. Reproduction 122(2):289–296

Martinez EA, Vazquez JM, Parrilla I, Cuello C, Gil MA, Rodriguez-Martinez H, Roca J, Vazquez JL (2006) Incidence of unilateral fertilizations after low dose deep intrauterine insemination in spontaneously ovulating sows under field conditions. Reprod Domest Anim 41(1):41–47

Matthijs A, Engel B, Woelders H (2003) Neutrophil recruitment and phagocytosis of boar spermatozoa after artificial insemination of sows, and the effects of inseminate volume, sperm dose and specific additives in the extender. Reproduction 125(3):357–367

Matthijs A, Harkema W, Engel B, Woelders H (2000) In vitro phagocytosis of boar spermatozoa by neutrophils from peripheral blood of sows. J Reprod Fertil 120(2):265–273

Maxwell WMC, Johnson LA (1997) Chlortetracycline analysis of boar spermatozoa after incubation, flow cytometric sorting, cooling, or cryopreservation. Mol Reprod Dev 46(3):408–418

Maxwell WMC, Johnson LA (1999) Physiology of spermatozoa at high dilution rates: the influence of seminal plasma. Theriogenology 52:1353–1362

Maxwell WMC, Long CR, Johnson LA, Dobrinsky JR, Welch GR (1998) The relationship between membrane status and fertility of boar spermatozoa after flow cytometric sorting in the presence or absence of seminal plasma. Reprod Fertil Dev 10:433–440

Maxwell WMC, Evans G, Hollinshead FK, Bathgate R, de Graaf SP, Eriksson BM, Gillan L, Morton KM, O'Brien JK (2004) Integration of sperm sexing technology into the ART toolbox. Anim Reprod Sci 82-3:79–95

Maxwell WMC, de Graaf SP, El-Hajj Ghaoui R, Evans G (2007) Seminal plasma effects on the sperm handling and female fertility. In: Juengel JL, Murray JF, Smith MF (eds) Reproduction in domestic ruminants VI. Nottingham University Press, Nottingham, pp 13–37

McKenzie F, Faulds K, Graham D (2008) LNA functionalized gold nanoparticles as probes for double stranded DNA through triplex formation. Chem Commun 20:2367–2369

McNutt TL, Johnson LA (1996) Flow cytometric sorting of sperm: influence on fertilization and embryo fetal development in the rabbit. Mol Reprod Dev 43(2):261–267

Michl J (2014) Ultrastrukturelle Charakterisierung geschlechtsspezifisch sortierter Spermien. (Ultrastructural characterization of sex sorted sperm). Dissertation, University of Goettingen, Germany

Mittwoch U (2013) Sex determination. EMBO Rep 14(7):588–592

Moce E, Graham JK, Schenk JL (2006) Effect of sex-sorting on the ability of fresh and cryopreserved bull sperm to undergo an acrosome reaction. Theriogenology 66(4):929–936

Morris LHA, Hunter RHF, Allen WR (2000) Hysteroscopic insemination of small numbers of spermatozoa at the uterotubal junction of preovulatory mares. J Reprod Fertil 118(1):95–100

Morton KM, Catt SL, Hollinshead FK, Maxwell WMC, Evans G (2004) Production of lambs after the transfer of fresh and cryopreserved in vitro produced embryos from prepubertal lamb oocytes and unsorted and sex-sorted frozen-thawed spermatozoa. Reprod Domest Anim 39(6):454–461

Morton KM, Herrmann D, Sieg B, Struckmann C, Maxwell WMC, Rath D, Evans G, Lucas-Hahn A, Niemann H, Wrenzycki C (2007) Altered mRNA expression patterns in bovine blastocysts after fertilisation in vitro using flow-cytometrically sex-sorted sperm. Mol Reprod Dev 74(8):931–940

Moruzzi JF (1979) Selecting a mammalian-species for the separation of X-chromosome-bearing and Y-chromosome-bearing spermatozoa. J Reprod Fertil 57(2):319–323

Nativo P, Prior IA, Brust M (2008) Uptake and intracellular fate of surface-modified gold nanoparticles. ACS Nano 2(8):1639–1644

Niemann H, Kuhla B, Flachowsky G (2011) Perspectives for feed-efficient animal production. J Anim Sci 89(12):4344–4363

Palma GA, Olivier NS, Neumuller C, Sinowatz F (2008) Effects of sex-sorted spermatozoa on the efficiency of in vitro fertilization and ultrastructure of in vitro produced bovine blastocysts. Anat Histol Embryol 37(1):67–73

Panarace M, Pellegrini RO, Basualdo MO, Bele M, Ursino DA, Cisterna R, Desimone G, Rodriguez E, Medina MJ (2014) First field results on the use of stallion sex-sorted semen in a large-scale embryo transfer program. Theriogenology 81(4):520–525

Parrilla I, Vazquez JM, Cuello C, Gil MA, Roca J, Di Berardino D, Martinez EA (2004) Hoechst 33342 stain and U.V. Laser exposure do not induce genotoxic effects in flow-sorted boar spermatozoa. Reproduction 128(5):615–621

Pelliccione F, Micillo A, Cordeschi G, D'Angeli A, Necozione S, Gandini L, Lenzi A, Francavilla F, Francavilla S (2011) Altered ultrastructure of mitochondrial membranes is strongly associated with unexplained asthenozoospermia. Fertil Steril 95(2):641–646

Petersen S, Barchanski A, Taylor U, Klein S, Rath D, Barcikowski S (2011) Penetratin-conjugated gold nanoparticles design of cell penetrating nanomarkers by femtosecond laser ablation. J Phys Chem C 115(12):5152–5159

Pinkel D, Garner DL, Gledhill BL, Lake S, Stephenson D, Johnson LA (1985) Flow cytometric determination of the proportions of X-chromosome-bearing and Y-chromosome-bearing sperm in samples of purportedly separated bull sperm. J Anim Sci 60(5):1303–1307

Pinkel D, Lake S, Gledhill BL, Vandilla MA, Stephenson D, Watchmaker G (1982) High-resolution DNA content measurements of mammalian sperm. Cytometry 3(1):1–9

Probst S, Rath D (2003) Production of piglets using intracytoplasmic sperm injection (ICSI) with flowcytometrically sorted boar semen and artificially activated oocytes. Theriogenology 59(3–4):961–973

Pyrzak R (1994) Separation of X-bearing and Y-bearing human spermatozoa using albumin gradients. Hum Reprod 9(10):1788–1790

Polge C, Salamon S, Wilmut I (1970) Fertilizing capacity of frozen boar semen following surgical insemination. Vet Rec 87:424–428

Quesnel FN, Wilcox CJ, Simerl NA, Sharma AK, Thatcher WW (1995) Effects of fetal sex and sire and other factors on periparturient and postpartum performance of dairy cattle. Braz J Genet 18(4):541–545

Rath D, Johnson LA (2008) Application and commercialization of flow cytometrically sex-sorted semen. Reprod Domest Anim 43:338–346

Rath D, Sieme H (2003) Sexing of stallion semen. Pferdeheilkunde 19(6):675–676

Rath D, Niemann H, Johnson LA (1994a) Gamete intrafallopian transfer (Gift), an alternative to in-vitro fertilization procedures for special applications. Reprod Domest Anim 29(5):349–351

Rath D, Johnson LA, Welch GR (1993) In vitro culture of porcine embryos: development to blastocysts after in vitro fertilization (IVF) with flow cytometrically sorted and unsorted semen. Theriogenology 39:293

Rath D, Johnson LA, Welch GR, Niemann H (1994b) Successful gamete intrafallopian transfer (Gift) in the porcine. Theriogenology 41(5):1173–1179

Rath D, Moench-Tegeder G, Taylor U, Johnson LA (2009) Improved quality of sex-sorted sperm: a prerequisite for wider commercial application. Theriogenology 71(1):22–29

Rath D, Johnson LA, Dobrinsky JR, Welch GR, Niemann H (1997) Production of piglets preselected for sex following in vitro fertilization with X and Y chromosome-bearing spermatozoa sorted by flow cytometry. Theriogenology 47(4):795–800

Rath D, Long CR, Dobrinsky JR, Welch GR, Schreier LL, Johnson LA (1999) In vitro production of sexed embryos for gender preselection: high-speed sorting of X-chromosome-bearing sperm to produce pigs after embryo transfer. J Anim Sci 77(12):3346–3352

Rath D, Tiedemann D, Gamrad L, Johnson LA, Klein S, Kues W, Mancini R, Rehbock C, Taylor U, Barcikowski S (2015) Sex-sorted boar sperm - an update on related production methods. Reprod Domest Anim 50:56–60

Rath D, Barcikowski S, de Graaf S, Garrels W, Grossfeld R, Klein S, Knabe W, Knorr C, Kues W, Meyer H, Michl J, Moench-Tegeder G, Rehbock C, Taylor U, Washausen S (2013) Sex selection of sperm in farm animals: status report and developmental prospects. Reproduction 145(1):R15–R30

Rehbock C, Jakobi J, Gamrad L, van der Meer S, Tiedemann D, Taylor U, Kues W, Rath D, Barcikowski S (2014) Current state of laser synthesis of metal and alloy nanoparticles as ligand-free reference materials for nano-toxicological assays. Beilstein J Nanotechnol 5:1523–1541

Rens W, Welch GR, Johnson LA (1998) A novel nozzle for more efficient sperm orientation to improve sorting efficiency of X and Y chromosome-bearing sperm. Cytometry 33(4):476–481

Roca J, Carvajal G, Lucas X, Vazquez JM, Martinez EA (2003) Fertility of weaned sows after deep intrauterine insemination with a reduced number of frozen-thawed spermatozoa. Theriogenology 60(1):77–87

Roca J, Vazquez JM, Gil MA, Cuello C, Parrilla I, Martinez EA (2006) Challenges in pig artificial insemination. Reprod Domest Anim 41:43–53

Roca J, Parrilla I, Rodriguez-Martinez H, Gil MA, Cuello C, Vazquez JM, Martinez EA (2011) Approaches towards efficient use of boar semen in the pig industry. Reprod Domest Anim 46:79–83

Roelofs JB, Bouwman EB, Pedersen HG, Rasmussen ZR, Soede NM, Thomsen PD, Kemp B (2006) Effect of time of artificial insemination on embryo sex ratio in dairy cattle. Anim Reprod Sci 93(3–4):366–371

Rohde W, Porstmann T, Prehn S, Dorner G (1975) Gravitational pattern of Y-bearing human spermatozoa in density gradient centrifugation. J Reprod Fertil 42(3):587–591

Rorie RW (1999) Effect of timing of artificial insemination on sex ratio. Theriogenology 52(8):1273–1280

Rorie RW, Lester TD, Lindsey BR, McNew RW (1999) Effect of timing of artificial insemination on gender ratio in beef cattle. Theriogenology 52(6):1035–1041

Ross A, Robinson JA, Evans HJ (1975) Failure to confirm separation of X-bearing and Y-bearing human sperm using Bsa gradients. Nature 253(5490):354–355

Salamon S, Visser D (1973) Fertility after surgical insemination with frozen boar semen. Aust J Biol Sci 27(5):499–504

Salmaso S, Caliceti P, Amendola V, Meneghetti M, Magnusson JP, Pasparakis G, Alexander C (2009) Cell up-take control of gold nanoparticles functionalized with a thermoresponsive polymer. J Mater Chem 19(11):1608–1615

Samper JC, Morris L, Pena FJ, Plough TA (2012) Commercial breeding with sexed stallion semen: reality or fiction? J Equine Vet 32(8):471–474

Sander S (2016) Luminometrische Verlaufskontrolle des ATP-Gehaltes von flowzytometrisch geschlechtsdifferenzierten bovinen Spermien. (Luminometrical control of ATP content in flow cytometrically sex sorted bovine sperm). Dissertation, Veterinary University Hannover, Germany

Sato T, Sakuma T, Yokonishi T, Katagiri K, Kamimura S, Ogonuki N, Ogura A, Yamamoto T, Ogawa T (2015) Genome editing in mouse spermatogonial stem cell lines using TALEN and double-nicking CRISPR/Cas9. Stem Cell Rep 5(1):75–82

Sauer H, Bekhite MM, Hescheler J, Wartenberg M (2005) Redox control of angiogenic factors and CD31-positive vessel-like structures in mouse embryonic stem cells after direct current electrical field stimulation. Exp Cell Res 304(2):380–390

Schenk JL, Seidel GE (2007) Pregnancy rates in cattle with cryopreserved sexed spermatozoa; effects of laser intensity, staining conditions and catalase. Reprod Domes Ruminants VI Soc Reprod Fertil 64:165–167

Schenk JL, Suh TK, Seidel GE (2006) Embryo production from superovulated cattle following insemination of sexed sperm. Theriogenology 65(2):299–307

Schilling E, Thormaehlen D (1977) Enrichment of human X-chromosome and Y-chromosome bearing spermatozoa by density gradient centrifugation. Andrologia 9(1):106–110

Schulte RT, Chung YK, Ohl DA, Takayama S, Smith GD (2007) Microfluidic sperm sorting device provides a novel method for selecting motile sperm with higher DNA integrity. Fertil Steril 88:S76–S76

Seidel GE, Garner DL (2002) Current status of sexing mammalian spermatozoa. Reproduction 124(6):733–743

Seidel GE (2003a) Sexing mammalian sperm--intertwining of commerce, technology, and biology. Anim Reprod Sci 79(3–4):145–156

Seidel GE (2003b) Economics of selecting for sex: the most important genetic trait. Theriogenology 59(2):585–598

Seidel GE, Whittier JC (2015) BEEF SPECIES SYMPOSIUM: beef production without mature cows. J Anim Sci 93(9):4244–4251

Seidel GE, Schenk JL, Herickhoff LA, Doyle SP, Brink Z, Green RD, Cran DG (1999) Insemination of heifers with sexed sperm. Theriogenology 52(8):1407–1420

Seidel GE, Allen CH, Johnson LA, Holland MD, Brink Z, Welch GR, Graham JK, Cattell MB (1997) Uterine horn insemination of heifers with very low numbers of nonfrozen and sexed spermatozoa. Theriogenology 48(8):1255–1264

Seidman MM, Glazer PM (2003) The potential for gene repair via triple helix formation. J Clin Investig 112(4):487–494

Sevinc A (1968) Experiments on sex control by electrophoretic separation of spermatozoa in rabbit. J Reprod Fertil 16(1):7

Sharpe JC, Evans KM (2009) Advances in flow cytometry for sperm sexing. Theriogenology 71(1):4–10

Shastry PR, Hegde UC, Rao SS (1977) Use of ficoll-sodium metrizoate density gradient to separate human X-bearing and Y-bearing spermatozoa. Nature 269(5623):58–60

Shi XG, Wang SH, Meshinchi S, Van Antwerp ME, Bi XD, Lee IH, Baker JR (2007) Dendrimer-entrapped gold nanoparticles as a platform for cancer-cell targeting and imaging. Small 3(7):1245–1252

Shi TY, Chen G, Huang X, Yuan Y, Wu X, Wu B, Li Z, Shun F, Chen H, Shi H (2012) Effects of reactive oxygen species from activated leucocytes on human sperm motility, viability and morphology. Andrologia 44:696–703

Shirai M, Matsuda S (1974) Galvanic separation of X-bearing and Y-bearing human spermatozoa. Jpn J Urol 65(5):297–302

Shirai M, Matsuda S, Mitsukaw S (1974) Electrophoretic separation of X- and Y-chromosome-bearing sperm in human semen. Tohoku J Exp Med 113(3):273–281

Shishito S, Shirai M, Sasaki K (1975) Galvanic separation of X-bearing and Y-bearing human spermatozoa. Int J Fertil 20(1):13–16

Sills ES, Kirman I, Colombero LT, Hariprashad J, Rosenwaks Z, Palermo GD (1998) H-Y antigen expression patterns in human X- and Y-chromosome-bearing spermatozoa. Am J Reprod Immunol 40(1):43–47

Sinclair AH, Berta P, Palmer MS, Hawkins JR, Griffiths BL, Smith MJ, Foster JW, Frischauf AM, Lovellbadge R, Goodfellow PN (1990) A gene from the human sex-determining region encodes a protein with homology to a conserved DNA-binding motif. Nature 346(6281):240–244

Sinha RP, Hader DP (2002) UV-induced damage and repair: a review. Photochem Photobiol 1:225–236

Spinaci M, De Ambrogi M, Volpe S, Galeati G, Tamanini C, Seren E (2005) Effect of staining and sorting on boar sperm membrane integrity, mitochondrial activity and in vitro blastocyst development. Theriogenology 64(1):191–201

Spinaci M, Volpe S, Bernardint C, De Ambrogi M, Tamanini C, Seren E, Galeati G (2006) Sperm sorting procedure induces a redistribution of Hsp70 but not Hsp60 and Hsp90 in boar spermatozoa. J Androl 27(6):899–907

Spinaci M, Vallorani C, Bucci D, Bernardini C, Tamanini C, Seren E, Galeati G (2010) Effect of liquid storage on sorted boar spermatozoa. Theriogenology 74(5):741–748

Spinaci M, Perteghella S, Chlapanidas T, Galeati G, Vigo D, Tamanini C, Bucci D (2016) Storage of sexed boar spermatozoa: limits and perspectives. Theriogenology 85(1):65–73

Steverink DWB, Soede NM, Bouwman EG, Kemp B (1997) Influence of insemination-ovulation interval and sperm cell dose on fertilization in sows. J Reprod Fertil 111(2):165–171

Steverink DWB, Soede NM, Bouwman EG, Kemp B (1998) Semen backflow after insemination and its effect on fertilisation results in sows. Anim Reprod Sci 54(2):109–119

Stovel RT, Sweet RG, Herzenberg LA (1978) Means for orienting flat cells in flow systems. Biophys J 23(1):1–5

Suh TK, Schenk JL, Seidel GE (2005) High pressure flow cytometric sorting damages sperm. Theriogenology 64(5):1035–1048

Suh RS, Phadke N, Ohl DA, Takayama S, Smith GD (2003) Rethinking gamete/embryo isolation and culture with microfluidics. Hum Reprod Update 9(5):451–461

Taylor U, Barchanski A, Kues W, Barcikowski S, Rath D (2012) Impact of metal nanoparticles on germ cell viability and functionality. Reprod Domest Anim 47:359–368

Taylor U, Rath D, Zerbe H, Schuberth HJ (2008) Interaction of intact porcine spermatozoa with epithetial cells and neutrophilic granulocytes during uterine passage. Reprod Domest Anim 43(2):166–175

Taylor U, Petersen S, Barcikowski S, Rath D, Klein S (2009c) Verification of gold nanoparticle uptake by bovine immortalised cells using laser scanning confocal microscopy. Cytometry Part A 75a(8):714

Taylor U, Rehbock C, Streich C, Rath D, Barcikowski S (2014) Rational design of gold nanoparticle toxicology assays: a question of exposure scenario, dose and experimental setup. Nanomedicine 9(13):1971–1989

Taylor U, Klein S, Petersen S, Kues W, Barcikowski S, Rath D (2010) Nonendosomal cellular uptake of ligand-free, positively charged gold nanoparticles. Cytometry Part A 77a(5):439–446

Taylor U, Tiedemann D, Rehbock C, Kues WA, Barcikowski S, Rath D (2015) Influence of gold, silver and gold-silver alloy nanoparticles on germ cell function and embryo development. Beilstein J Nanotechnology 6:651–664

Taylor U, Schuberth HJ, Rath D, Michelmann HW, Sauter-Louis C, Zerbe H (2009a) Influence of inseminate components on porcine leucocyte migration in vitro and in vivo after pre- and post-ovulatory insemination. Reprod Domest Anim 44(2):180–188

Taylor U, Zerbe H, Seyfert HM, Rath D, Baulain U, Langner KFA, Schuberth HJ (2009b) Porcine spermatozoa inhibit post-breeding cytokine induction in uterine epithelial cells in vivo. Anim Reprod Sci 115(1–4):279–289

Tiedemann D, Taylor U, Rehbock C, Jakobi J, Klein S, Kues WA, Rath D (2014) Reprotoxicity of gold, silver, and gold-silver alloy nanoparticles on mammalian gametes. Analyst 139(5):931–942

Tkachenko AG, Xie H, Liu YL, Coleman D, Ryan J, Glomm WR, Shipton MK, Franzen S, Feldheim DL (2004) Cellular trajectories of peptide-modified gold particle complexes: comparison of nuclear localization signals and peptide transduction domains. Bioconjug Chem 15(3):482–490

Tomkins PT, Houghton JA (1988) The rapid induction of the acrosome reaction of human-spermatozoa by electropermeabilization. Fertil Steril 50(2):329–336

Trigal B, Gomez E, Caamano JN, Munoz M, Moreno J, Carrocera S, Martin D, Diez C (2012) In vitro and in vivo quality of bovine embryos in vitro produced with sex-sorted sperm. Theriogenology 78(7):1465–1475

Uwland J, Willems CM (1975) Results of semen separation using a modified electromagneto-chemical method. Tijdschr Diergeneeskd 100:369–374

van Munster EB (2002) Interferometry in flow to sort unstained X- and Y-chromosome-bearing bull spermatozoa. Cytometry 47(3):192–199

Vassena R, Heindryckx B, Peco R, Pennings G, Raya A, Sermon K, Veiga A (2016) Genome engineering through CRISPR/Cas9 technology in the human germline and pluripotent stem cells. Hum Reprod Update 22(4):411–419

Vazquez JM, Martinez EA, Parrilla I, Gil MA, Lucas X, Roca J (2002) Motility characteristics and fertilizing capacity of boar spermatozoa stained with Hoechst 33342. Reprod Domest Anim 37(6):369–374

Vazquez JM, Martinez EA, Parrilla I, Roca J, Gil MA, Vazquez JL (2003) Birth of piglets after deep intrauterine insemination with flow cytometrically sorted boar spermatozoa. Theriogenology 59(7):1605–1614

Vazquez JM, Roca J, Gil MA, Cuello C, Parrilla I, Vazquez JL, Martinez EA (2008b) New developments in low-dose insemination technology. Theriogenology 70(8):1216–1224

Vazquez JM, Roca J, Gil MA, Cuello C, Parrilla I, Caballero I, Vazquez JL, Martinez EA (2008a) Low-dose insemination in pigs: problems and possibilities. Reprod Domest Anim 43:347–354

Vazquez JM, Martinez EA, Roca J, Gil MA, Parrilla I, Cuello C, Carvajal G, Lucas X, Vazquez JL (2005) Improving the efficiency of sperm technologies in pigs: the value of deep intrauterine insemination. Theriogenology 63(2):536–547

Verma A, Uzun O, Hu YH, Hu Y, Han HS, Watson N, Chen SL, Irvine DJ, Stellacci F (2008) Surface-structure-regulated cell-membrane penetration by monolayer-protected nanoparticles. Nat Mater 7(7):588–595

Vidal F, Moragas M, Catala V, Torello MJ, Santalo J, Calderon G, Gimenez C, Barri PN, Egozcue J, Veiga A (1993) Sephadex filtration and human serum-albumin gradients do not select spermatozoa by sex-chromosome – a fluorescent in-situ hybridization study. Hum Reprod 8(10):1740–1743

Vidament A (2005) French field results (1985–2005) on factors affecting fertility of frozen stallion semen. Anim Reprod Sci 89(1–4):115–136

Viring S, Einarsson S (1981) Sperm distribution within the genital-tract of naturally inseminated gilts. Nord Vet Med 33(3):145–149

Waberski D, Meding S, Dirksen G, Weitze KF, Leiding C, Hahn R (1994) Fertility of long-term-stored boar semen – influence of extender (Androhep and Kiev), storage time and plasma droplets in the semen. Anim Reprod Sci 36(1–2):145–151

Walker DW, Benzer S (2004) Mitochondrial "swirls" induced by oxygen stress and in the Drosophila mutant hyperswirl. Proc Natl Acad Sci U S A 101(28):10290–10295

Wang HX, Flaherty SP, Swann NJ, Matthews CD (1994a) Assessment of the separation of X-bearing and Y-bearing sperm on albumin gradients using double-label fluorescence in-situ hybridization. Fertil Steril 61(4):720–726

Wang HX, Flaherty SP, Swann NJ, Matthews CD (1994b) Discontinuous percoll gradients enrich X-bearing human spermatozoa – a study using double-label fluorescence in-situ hybridization. Hum Reprod 9(7):1265–1270

Wang XH, Fang HQ, Huang ZL, Shang W, Hou TT, Cheng AW, Cheng HP (2013) Imaging ROS signaling in cells and animals. J Mol Med 91(8):917–927

Watkins AM, Chan PJ, Kalugdan TH, Patton WC, Jacobson JD, King A (1996) Analysis of the flow cytometer stain Hoechst 33342 on human spermatozoa. Mol Hum Reprod 2(9):709–712

Welch GR, Johnson LA (1999) Sex preselection: laboratory validation of the sperm sex ratio of flow sorted X- and Y-sperm by sort reanalysis for DNA. Theriogenology 52(8):1343–1352

Windsor DP, Evans G, White IG (1993) Sex predetermination by separation of X and Y chromosome-bearing sperm: a review. Reprod Fertil Dev 5(1):155–171

Wu YX, Zhou H, Fan XY, Zhang Y, Zhang M, Wang YH, Xie ZF, Bai MZ, Yin Q, Liang D, Tang W, Liao JY, Zhou CK, Liu WJ, Zhu P, Guo HS, Pan H, Wu CL, Shi HJ, Wu LG, Tang FC, Li JS (2015) Correction of a genetic disease by CRISPR-Cas9-mediated gene editing in mouse spermatogonial stem cells. Cell Res 25(1):67–79

Xodo LE, Rathinavelan T, Quadrifoglio F, Manzini G, Yathindra N (2001) Targeting neighbouring poly(purine center dot pyrimidine) sequences located in the human bcr promoter by triplex-forming oligonucleotides. Eur J Biochem 268(3):656–664

Xu J, Guo Z, Su L, Nedambale TL, Zhang J, Schenk J, Moreno JF, Dinnyes A, Ji W, Tian XC, Yang X, Du F (2006) Developmental potential of vitrified Holstein cattle embryos fertilized in vitro with sex-sorted sperm. J Dairy Sci 89(7):2510–2518

Xu KP, Yadav BR, King WA, Betteridge KJ (1992) Sex-related differences in developmental rates of bovine embryos produced and cultured in vitro. Mol Reprod Dev 31(4):249–252

Yoisungnern T, Choi Y-J, Han JW, Kang M-H, Das J, Gurunathan S, Chang WK (2015) Internalization of silver nanoparticles into mouse spermatozoa results in poor fertilization and compromised embryo development. Sci Rep 5:11170

Zhang XD, Wu HY, Wu D, Wang YY, Chang JH, Zhai ZB, Meng AM, Liu PX, Zhang LA, Fan FY (2010) Toxicologic effects of gold nanoparticles in vivo by different administration routes. Int J Nanomedicine 5:771–781

Zhang Y, Bai Y, Jia J, Gao N, Li Y, Zhang R, Yan B (2014) Perturbation of physiological systems by nanoparticles. Chem Soc Rev 43(10):3762–3809

Zobel R, Gereš D, Pipal I, Buić V, Gračner D, Tkalcic S (2011) Influence of the semen deposition site on the calves' sex ratio in simmental dairy cattle. Reprod Domest Anim 46(4):595–601

Embryo Transfer Technology in Cattle

5

Gabriel A. Bó and Reuben J. Mapletoft

Abstract

Although the first mammalian embryo transfers were done more than 100 years ago, commercial bovine embryo transfer came into being in the early 1970s with the importation of European breeds of cattle into North America. Since that time commercial bovine embryo transfer has grown throughout the world, and in 2016, approximately one million bovine embryos were transferred, and several thousands of embryos were transported internationally. Because in vivo-derived bovine embryos can be made specified pathogen-free by washing procedures, they provide the ideal means of moving animal genetics around the world. Embryo transfer techniques have improved over the years so that new methods of controlling ovarian function facilitate superstimulation of donors and synchronization of recipients and nonsurgical procedures facilitate on-farm embryo transfer.

5.1 Introduction

The commercial bovine embryo transfer industry arose in the early 1970s in North America (Betteridge 2003; Seidel Jr 1981). European breeds of cattle that had been imported into Canada were valuable and scarce, and embryo transfer offered a means by which their numbers could be multiplied rapidly. Private veterinary practitioners and small embryo transfer companies adopted a research technology for

G. A. Bó
Instituto de Reproducción Animal Córdoba (IRAC), Córdoba, Argentina

R. J. Mapletoft (✉)
Western College of Veterinary Medicine, University of Saskatchewan, Saskatoon, SK, Canada
e-mail: Reuben.mapletoft@usask.ca

© Springer International Publishing AG, part of Springer Nature 2018
H. Niemann, C. Wrenzycki (eds.), *Animal Biotechnology 1*,
https://doi.org/10.1007/978-3-319-92327-7_5

commercial use. They also founded the International Embryo Technology Society (IETS) in 1974 to facilitate sharing of ideas and technical information which was considered necessary for progress to be made (Carmichael 1980; Schultz 1980). The IETS became the main forum for scientific and regulatory exchange and discussion of embryo transfer and associated technologies. In particular, the Import/Export Committee of the IETS (now referred to as the Health and Safety Advisory Committee; HASAC) has been instrumental in gathering and disseminating scientific information on the potential for disease control with bovine embryo transfer. The *Manual of the International Embryo Technology Society: A Procedural Guide and General Information for the Use of Embryo Transfer Technology Emphasizing Sanitary Procedures* has become the reference source for sanitary procedures used in embryo export protocols (IETS Manual 4th Edition 2010).

In 2016, practitioners from around the world collected 632,638 in vivo-derived embryos from 93,815 donors; 195,563 were transferred into recipients immediately after collection (fresh), while the remainder were cryopreserved for transfer at a later date (Perry 2017). North America accounted for 52.5% of in vivo-derived bovine embryos, while Europe accounted for 20.4% and South America 7.5%. In addition, 666,215 in vitro-produced bovine embryos were produced in 2016, 56.8% of which were in South America, 39.1% in North America, and 2.8% in Europe. In vitro embryo production has increased very rapidly in both South and North America in the last two decades.

Very briefly, bovine embryo transfer involves the selection, management, and treatment of donors and recipients, and the collection and transfer of embryos within a narrow window of time 6–8 days after estrus. This technology has been incorporated into dairy and beef cattle operations and often involves the participation of herd veterinarians. The following chapter draws heavily on material contained in prior reviews and extensive literature of primary research on the topic, including reviews (Mapletoft 1985; Mapletoft and Hasler 2005) and research reports from the authors' laboratories. Some of the important uses of bovine embryo transfer follow. A more detailed review of bovine embryo transfer is available online (Mapletoft and Bó 2016).

5.2   Applications

5.2.1   Planned Matings

The most common use of bovine embryo transfer has been the proliferation of so-called desirable phenotypes. As AI has permitted the widespread dissemination of a male's genetic potential, embryo transfer has provided the opportunity to disseminate the genetics of elite females. Embryo transfer has also been used to expand a limited gene pool rapidly, e.g., the dramatic rise of the embryo transfer industry in North America in the early 1970s. The production of AI bulls through embryo transfer is currently a common application of planned matings (Lohuis 1995; Teepker and Keller 1989).

5.2.2 Genetic Improvement

Smith (1988a) introduced the concept of MOET (multiple ovulation and embryo transfer). He showed that MOET programs could result in increased selection intensity and reduced generation intervals, resulting in increased genetic gains. The establishment of nucleus herds and "juvenile MOET" in heifer offspring was shown to result in genetic gains near twice that achieved with traditional progeny test schemes (Smith 1988b). Genomic techniques are now being used to select embryo donors, and genomic analysis has become essential for the selection of bull dams used in embryo transfer (Ponsart et al. 2014; Seidel Jr 2010). The commercial cattle industry has benefited greatly from the use of bulls produced through MOET programs (Christensen 1991).

5.2.3 Disease Control

For an infectious agent to be transmitted by embryo transfer, the pathogen must be present within the cells of the embryo (true embryonic infection), in association with the zona pellucida, or in the medium bathing the embryo. Infectious agents have not been shown to pass through the zona pellucida of the bovine embryo, but some tend to adhere to the outer surface of the zona pellucida. Similarly, embryos with damaged or compromised zonae pellucidae could allow an infectious agent to invade the embryo itself. Thus, procedures for decontaminating the zona pellucida feature strongly in the sanitary handling protocols advocated in the IETS Manual 4th Edition (2010). These protocols include inspection of the embryo microscopically at a magnification of at least 50 times to insure that the zona pellucida is intact and free of adherent material that might trap infectious agents. In addition, the embryo must be washed at least ten times with fresh medium at a 100-fold dilution and a new sterile pipette for each wash. Enveloped viruses may stick to the zona pellucida, but they can be removed/inactivated by trypsin treatments. Thus, two trypsin treatments between washes 5 and 6 are recommended when viruses adhering to the zona pellucida are of concern (Singh 1985; Stringfellow 2010).

It is now clear that in vivo-derived bovine embryos do not transmit infectious diseases providing they are handled correctly between collection and transfer. This includes inspection of the zona pellucida at >50× magnification and washing/trypsin treatment procedures. The IETS has categorized disease agents based on the risk of transmission by in vivo-derived bovine embryos (IETS Manual 4th Edition; Stringfellow and Givens 2010). Category 1 includes disease agents for which sufficient evidence has accrued to show that the risk of transmission is negligible. This category includes enzootic bovine leukosis, foot-and-mouth disease (cattle), bluetongue (cattle), *Brucella abortus* (cattle), infectious bovine rhinotracheitis, pseudorabies in swine, and bovine spongiform encephalopathy. Category 2, 3, and 4 diseases are those for which less research information has been generated, but it is noteworthy that there is no evidence that an infectious agent has been transmitted by in vivo-derived bovine embryos. Consequently, embryo transfer procedures are recommended for the salvage of genetics in the face of a disease outbreak (Wrathall et al. 2004).

5.2.4 Embryo Import-Export

The ability to utilize in vivo-derived bovine embryos to prevent the transmission of infectious disease makes them ideal for the international movement of animal germplasm. Benefits of embryos also include reduced transportation and quarantine costs, a wider genetic base from which to select, the retention of the original genetics within the exporting country, and adaptation. Although handling procedures recommended by the IETS make it possible to safely export in vivo-derived embryos, the zona pellucida of in vitro-produced bovine embryos differs, and pathogens are more difficult to remove by washing/trypsin treatment procedures (Stringfellow and Givens 2000). Thus, specific export protocols which include donor, herd, and media testing are usually required for the international movement of in vitro-produced embryos (Stringfellow et al. 2004).

5.3 General Procedural Steps

Donors are usually superstimulated with gonadotrophins to increase the numbers of retrievable embryos. Methods of controlling ovarian function have resulted in increased embryo production per unit time. Donors are now being superstimulated as often as every 30 days, and more embryos are being produced per year with no change in the actual superstimulation protocol (Bó and Mapletoft 2014). The donor may be inseminated naturally or artificially, and embryos are normally collected nonsurgically 6–8 days after estrus. Following collection, embryos must be identified, evaluated, and maintained in a physiological medium prior to transfer. They may also be subjected to manipulations or cryopreservation (Hasler 2003). The following sections outline the control of ovarian function in donors and recipients.

5.4 Superovulation

The objective of superstimulation treatments is to obtain the maximum number of transferable embryos with a high probability of producing pregnancies. Wide ranges in superovulatory response and embryo yield have been detailed in several reviews of commercial embryo transfer records (Hasler et al. 1983; Looney 1986). These reports demonstrate a high degree of unpredictability that affects the efficiency and profitability of bovine embryo transfer.

5.4.1   Gonadotrophins and Superovulation

Two different types of gonadotrophins have been used to induce superovulation in cattle: equine chorionic gonadotrophin (eCG) and pituitary extracts containing follicle-stimulating hormone (FSH; Kelly et al. 1997; Murphy et al. 1984). Equine chorionic gonadotrophin is a complex glycoprotein with both FSH and luteinizing hormone (LH) activity and has been shown to have a half-life of approximately 40 h in the cow (Schams et al. 1978); thus, eCG is normally administered once to induce superovulation (Murphy and Martinuk 1991). Recommended doses range from 1500 to 3000 IU/animal with 2500 IU by intramuscular injection commonly used. The long half-life of eCG also causes protracted ovarian stimulation, abnormal endocrine profiles, large follicles, and reduced embryo quality (Mikel-Jenson et al. 1982; Saumande et al. 1978). These problems have been overcome by the intravenous administration of antibodies to eCG at the time of the first insemination (Dieleman et al. 1993; Gonzalez et al. 1994). However, antibodies to eCG are not available commercially, and so eCG is seldom used to superstimulate cattle.

Pituitary extracts are most commonly used to superstimulate cattle (Armstrong 1993). As the biological half-life of pituitary FSH in the cow has been estimated to be 5 h or less (Laster 1972), it must be injected twice daily to induce superovulation (Monniaux et al. 1983; Walsh et al. 1993). The usual regimen is 4 or 5 days of twice daily intramuscular treatments with FSH. Forty-eight to 72 h after initiation of treatment, prostaglandin F2α (PGF) is administered to induce luteolysis. Estrus (and preovulatory LH release) occurs in 36–48 h, with ovulation 24–36 h later (Reviewed in Bó and Mapletoft 2014). Purified pituitary extracts with LH removed are now available; Folltropin-V (Vetoquinol) is a porcine pituitary extract with approximately 84% of the LH removed (Gonzalez-Reyna et al. 1990). It has been used successfully in constant or decreasing dose schedules with PGF given either 48 or 72 h after initiating treatment (Mapletoft et al. 2002).

Recombinant bovine FSH (rbFSH) has also been used to induce superovulation in cattle. Wilson et al. (1993) reported high superovulatory responses following twice daily administration of rbFSH, and more recently, Carvalho et al. (2004b) reported the successful superstimulation of Holstein heifers with a single administration of a long-acting rbFSH. Although there are no products available for cattle currently, rFSH is used in human medicine suggesting that recombinant FSH is likely to be used in cattle, provided it gains registration and is affordable.

5.4.2   Follicular Wave Dynamics and Superovulation

We have demonstrated a greater superovulatory response when gonadotrophin treatments are initiated on Day 9 of the estrous cycle (Day 8 post-ovulation) as compared to Days 3, 6, or 12 (Lindsell et al. 1986). Ultrasonography has now shown that the second follicle wave emerges 8.5–10.5 days after ovulation (Adams 1994; Ginther et al. 1989; Pierson and Ginther 1987). We have also shown that superovulatory response is greater when FSH treatments are initiated at the time of follicle wave emergence (Nasser et al. 1993). While initiation of FSH treatments in the presence of a dominant follicle resulted in a 40–50% decrease in superovulatory response (Bungartz and Niemann 1994; Guilbault et al. 1991; Kim et al. 2001; Shaw and Good 2000), the presence of a large number of follicles 3–6 mm in diameter 8–10 days after ovulation, in the presence of a large follicle, provides evidence for dominant follicle regression and emergence of a new follicle wave (Adams et al. 2008; Singh et al. 2004).

5.4.3   Manipulation of the Follicular Wave for Superstimulation

The conventional protocol of initiating ovarian superstimulation during mid-cycle (8–12 days after estrus) has now been supported by ultrasonographic evidence indicating that mid-cycle is the approximate time of emergence of the second follicular wave. However, the day of emergence of the second follicular wave differs among individuals within wave type and is 1 or 2 days later in two- than three-wave cycles (Adams et al. 2008). In addition, the necessity of waiting until mid-cycle to initiate superstimulatory treatments implies monitoring estrus and an obligatory delay making it difficult to superstimulate large numbers of donors at the same time. An alternative approach is to initiate superstimulation treatments subsequent to the synchronization of follicular wave emergence. There are three methods of synchronizing follicle wave emergence for superstimulation.

5.4.3.1 Follicle Ablation
The most efficacious approach to the synchronization of follicle wave emergence involves transvaginal ultrasound-guided ablation of all follicles ≥5 mm, regardless of stage of the estrous cycle (Bergfelt et al. 1994; Garcia and Salaheddine 1998). This removes the suppressive effects of follicular products (estradiol and inhibin) on FSH release, resulting in an FSH surge and emergence of a new follicular wave 1 day later (Adams et al. 1992a). Superstimulatory treatments are then administered, beginning 1 or 2 days after ablation (Bergfelt et al. 1997). The timing of estrus was more synchronous when a progestin device was inserted for the period of superstimulation and two injections of PGF were administered on the day of device removal. Transvaginal ultrasound-guided ablation of only the dominant follicle (Bungartz and Niemann 1994; Shaw and Good 2000) during mid-diestrus, followed in 2 days by superstimulation, also resulted in a higher superovulatory response than when the dominant follicle was not ablated. We have also shown that ablation of the two largest follicles at random stages of the estrous cycle was efficacious in synchronizing follicular wave emergence for superstimulation (Baracaldo et al. 2000). Unfortunately, follicle ablation is difficult to utilize under field conditions. The recommended protocol for superstimulation utilizing follicle ablation is illustrated in Fig. 5.1.

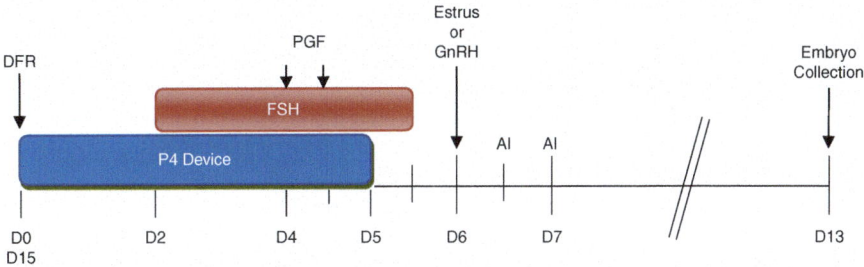

Fig. 5.1 Treatment schedule for superstimulating donors after the removal of the dominant folli-cle (DFR). On Day 0, donors receive a progesterone-releasing (P4) device, and all follicles ≥5 mm in diameter, or the two largest follicles, are ablated using ultrasound-guided follicle aspiration. Superstimulatory treatments are initiated 1.5 or 2 days later, with twice daily intramuscular doses of FSH over 4 or 5 days. PGF is administered with the fifth and sixth (or seventh and eighth) FSH injections, and P4 devices are removed 24 h after the first PGF. Donors in estrus or receiving GnRH 24 h after P4 device removal are inseminated 12 and 24 h later. Embryos are collected 7 days after estrus or GnRH treatment

5.4.3.2 Estradiol and Progesterone

We have shown that treatment of progestin-treated cattle with estradiol results in synchronous emergence of a new follicle wave (Bó et al. 1995, 1996). The mecha-nism apparently involves suppression of FSH, and possibly LH, which results in regression of FSH- and LH-dependent follicles. Once follicle regression begins and the estradiol is metabolized, FSH surges, and a new follicle wave emerges, 1 day later (Adams et al. 1992a). The use of estradiol-17β in progestin-treated cattle was followed by the emergence of a new wave 3–5 days later, regardless of the stage of follicular growth at the time of treatment. Estradiol-17β is normally injected with 50–100 mg of progesterone at placement of a progestin device to prevent estrogen-induced LH release in animals without a functional corpus luteum (CL). Data from experimental (Bó et al. 1996) and commercial (Bó et al. 2002) embryo transfer records show that the superovulatory response of donors given estradiol-17β and progesterone at unknown stages of the estrous cycle was comparable to those super-stimulated 8–12 days after estrus.

Unfortunately, estradiol-17β is not available commercially in many coun-tries (Lane et al. 2008), and so we investigated the use of estrogen esters. Treatment with 2.5 mg estradiol benzoate plus 50 mg progesterone at the time of progestin device insertion resulted in emergence of a new follicular wave 3–4 days later. Superstimulatory treatments initiated 4 days after treatment resulted in responses comparable to the use of estradiol-17β plus progesterone or superstimulation initiated 8–12 days after estrus (Bó et al. 2002). The rec-ommended protocol for superstimulating cattle utilizing estradiol and proges-terone is shown in Fig. 5.2.

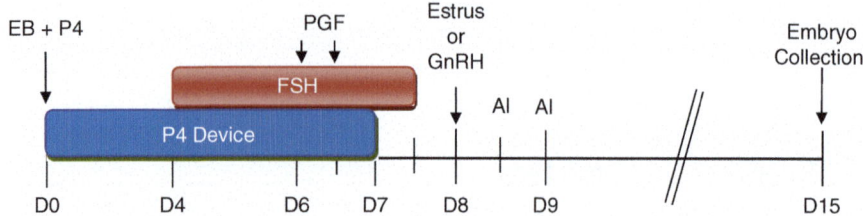

Fig. 5.2 Treatment schedule for superstimulating donors after synchronizing follicle wave emergence with estradiol and progesterone. On Day 0, donors receive a progesterone-releasing (P4) device and 2 or 2.5 mg estradiol benzoate (EB) and 50 or 100 mg of P4 intramuscularly. Superstimulatory treatments are initiated on Day 4 with twice daily intramuscular doses of FSH over 4 or 5 days. PGF is administered with the fifth and sixth (or seventh and eighth) FSH injections, and P4 devices are removed 24 h after the first PGF. Donors in estrus or receiving GnRH 24 h after P4 device removal are inseminated 12 and 24 h later. Embryos are collected 7 days after estrus or GnRH

5.4.3.3 Gonadotrophin-Releasing Hormone (GnRH)

The administration of GnRH or porcine LH (pLH) has been shown to induce ovulation of a dominant follicle present at the time of treatment followed by emergence of a new follicle wave in 2 days (Martinez et al. 1999; Pursley et al. 1995; Thatcher et al. 1993). However, neither GnRH nor pLH always induces ovulation, and if ovulation does not occur, follicle wave emergence will not be synchronized (Martinez et al. 1999). The reported asynchrony of follicular wave emergence (range, 3 days before treatment to 5 days after treatment) suggested that GnRH-based approaches may not be feasible for superstimulation. However, three reports revealed no differences in the numbers of transferable embryos when donors were superstimulated 2 days after treatment with GnRH as compared to treatment with estradiol (Hinshaw 1999; Steel and Hasler 2009; Wock et al. 2008). It is noteworthy that in these studies, GnRH was administered 2–3 days after insertion of a progestin device which may have resulted in an unovulated dominant follicle which would be more responsive to GnRH treatment.

Bó et al. (2008) reported on another protocol for superstimulation following the administration of GnRH. It was based on a study in which a persistent follicle was induced by the administration of PGF at the time of insertion of a progestin device 7–10 days before GnRH (Small et al. 2009); ovulation and follicle wave emergence occurred 1–2 days after the administration of GnRH in >90% of cows, indicating that this approach could be used in groups of randomly cycling donors. As superovulatory responses following administration of GnRH 2 vs 7 days after insertion of a progestin device were not significantly different (Hinshaw et al. 2015), either approach would appear to be efficacious. The recommended protocols for superstimulating cattle following the use of GnRH to synchronize follicle wave emergence are shown in Fig. 5.3.

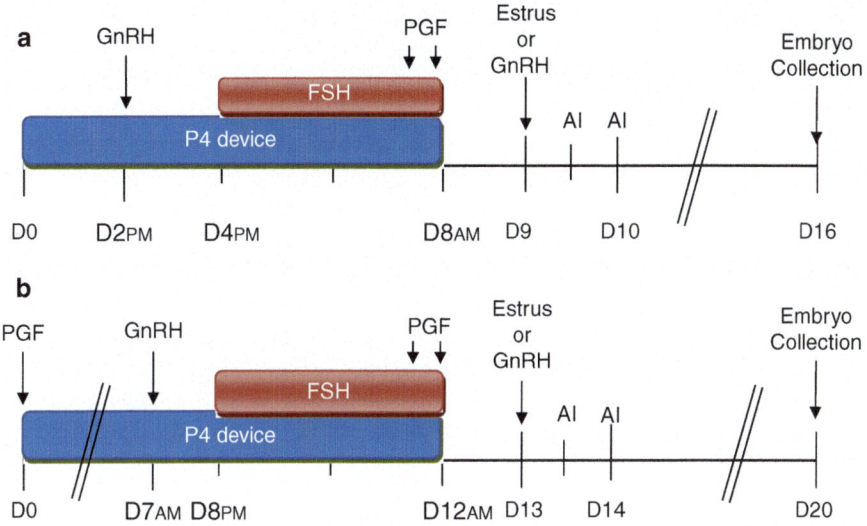

Fig. 5.3 Treatment schedules for superstimulating donors after synchronization of follicle wave emergence with GnRH. Donors receive a progesterone-releasing (P4) device alone (**a**) or along with a dose of PGF (**b**). GnRH is administered 2 or 7 days later, and superstimulatory treatments are initiated 36–48 h after GnRH with twice daily intramuscular doses of FSH over 4 or 5 days. PGF is administered with the fifth and sixth (or seventh and eighth) FSH injections, and P4 devices are removed 24 h after the first PGF. Donors in estrus or receiving GnRH 24 h after P4 device removal are inseminated 12 and 24 h later. Embryos are collected 7 days after estrus or GnRH

5.4.4   Superstimulation of Donors with Abnormal Ovarian Function

Cows with abnormal ovarian function are difficult to superstimulate because they usually do not have a functional CL, they show estrus at unpredictable times, and stage of follicle development is difficult to predict. It was the need to superstimulate cows with abnormal ovarian function that led to the use of estradiol prior to the administration of FSH. In a retrospective study, embryo production did not differ between 190 cows with abnormal ovarian function which were superstimulated 7 days after receiving a norgestomet implant and an injection of norgestomet and estradiol valerate and 260 control cows superstimulated between Days 8 and 12 of the estrous cycle. Subsequently, it was shown that estradiol valerate treatment resulted in emergence of a new follicle wave (reviewed in Mapletoft and Bó 2004).

5.4.5 Follicle Numbers and Superovulation

The numbers of antral follicles in the ovary as determined by ultrasonography have been shown to vary, and superstimulatory response has been shown to be correlated with the numbers of small antral follicles at the time of initiating FSH treatments (Ireland et al. 2007; Singh et al. 2004). In humans, circulating antimullerian hormone (AMH) concentrations have been found to be an informative serum marker for ovarian follicle reserve (Toner and Seifer 2013), and information is accumulating that circulating AMH concentrations may be a reliable marker for predicting antral follicle numbers in cattle (Batista et al. 2014; Ireland et al. 2011; Monniaux et al. 2013). There was high repeatability across different phases of the estrous cycle, days in milk, levels of milk production, and parities making AMH determinations particularly useful to select potential donors or to predict superovulatory response in selected donors (Souza et al. 2014).

5.4.6 Reducing the Need for Multiple Treatments with FSH

Because the half-life of pituitary FSH is short in the cow (Laster 1972), traditional superstimulatory treatment protocols have consisted of twice daily intramuscular injections over 4 or 5 days (Bó et al. 1994). This requires constant attention and increases the possibility of failures due to non-compliance. Twice daily treatments may also cause stress in donors with a subsequent decreased superovulatory response and/or altered preovulatory LH surge (Edwards et al. 1987; Stoebel and Moberg 1982). Therefore, simplified protocols may be expected to reduce donor handling and improve response.

A single subcutaneous administration of FSH has been shown to induce a superovulatory response equivalent to the traditional twice daily treatment protocol in beef cows in high body condition, i.e., body condition score of >3 out of 5 (Bó et al. 1994; Hiraizumi et al. 2015), but results were not repeatable in Holsteins which presumably had less adipose tissue. However, superovulatory responses were improved in Holsteins when the FSH dose was split into two; 75% administered subcutaneously on the first day of treatment, and the remaining 25% administered 48 h later when PGF is normally administered (Lovie et al. 1994).

An alternative in inducing superovulation with a single administration of FSH is to utilize agents that cause FSH to be released over several days. These are commonly referred to as polymers which are biodegradable and nonreactive in tissues facilitating use in animals (Sutherland 1991). In a series of experiments, FSH was diluted in a 2% hyaluronan solution and administered as a single intramuscular injection (to avoid the effects of body condition); a similar number of ova/embryos were produced as with the twice-daily FSH protocol (Tríbulo et al. 2011). However, 2% hyaluronan was viscous and difficult to mix with FSH. More dilute preparations (1% or 0.5% hyaluronan) were easier to mix with FSH but were less efficacious in a single administration protocol. Their use was improved by splitting the total dose of FSH into two injections administered 48 h apart (Tríbulo et al. 2012). When compared to the twice daily treatments, the number of transferable embryos with the two-injection protocol did not differ. A report derived from commercial embryo transfer data confirmed these results in beef cattle in North America (Hasler and Hockley 2012). However, the single- or two-injection protocol of FSH in hyaluronan is not recommended for lactating dairy cattle where results have been inconsistent and generally unsatisfactory.

5.4.7 Fixed-Time AI of Superstimulated Donors

Bó et al. (2006) developed a protocol for fixed-time AI in *Bos taurus* beef donors, without the need for estrus detection, by monitoring the timings of ovulations ultrasonically. Basically, the time of progestin device removal was delayed to prevent early ovulations and allow late developing follicles to "catch-up" and ovulation was induced with GnRH or pLH. In this protocol, follicular wave emergence was synchronized with estradiol and a progestin device on random days of the estrous cycle (Day 0) and FSH treatments were initiated on Day 4. On Day 6, PGF was administered in the AM and PM, and the progestin device was removed on Day 7 AM (24 h after the first PGF). On Day 8 AM (24 h after the removal of the progestin device), GnRH or pLH was administered and fixed-time AI were done 12 and 24 h later. Delaying the removal of the progestin device from Day 6 PM to Day 7 AM resulted in a higher number of ova/embryos and fertilized ova. From a practical perspective, fixed-time AI of donors has been shown to be useful in eliminating estrus detection for busy embryo transfer practitioners with no adverse effect on embryo production (Larkin et al. 2006).

Studies in high-producing Holstein cows (*Bos taurus*) in Brazil have indicated that it is preferable to allow an additional 12 h before removing the progestin device (i.e., Day 7 PM) followed by GnRH or pLH 24 h later, i.e., Day 8 PM (Martins et al. 2012). Baruselli et al. (2006) also reported that it is preferable to remove the progestin device on Day 7 PM in *Bos indicus* beef breeds, followed by GnRH 12 h later (i.e., Day 8 AM). Baruselli et al. (2006) also showed it is possible to use a single insemination with high-quality semen 16 h after pLH. This protocol has also been used successfully with sex-selected semen, except that inseminations were delayed by an additional 6 h i.e., 18 and 30 h after GnRH (Soares et al. 2011).

5.4.8 Semen and Semen Quality

Superstimulated donors are normally inseminated 12 and 24 h after onset of estrus (around 60 and 72 h after injection of PGF; Schiewe et al. 1987). However, superstimulation places extraordinary pressure on the capacity of frozen/thawed semen to fertilize multiple oocytes. As ovulation rate increases, the number of accessory sperm decreases, and unfertilized oocytes from superovulated cattle seldom have sperm attached to the zona pellucida (DeJarnette et al. 1992). However, viability may also be compromised; Saacke et al. (1988) showed that the number of viable sperm in the lower isthmus of the oviduct is less for a shorter period of time in superstimulated cattle. Thus the two inseminations would appear to be warranted.

In 1988, Hawk et al. (1988) reported that insemination of superstimulated cattle with 4.4 billion fresh sperm resulted in a greater number of fertilized ova and higher fertilization rates than 70 million frozen-thawed sperm. To investigate this observation, we selected three bulls with normal spermiograms and cryopreserved their semen in insemination dosages of 20, 50, or 100 million sperm in 0.25 ml or 0.5 ml straws. When used in a single insemination at 12 and 24 h after onset of estrus in superstimulated heifers, there were no differences in fertilization rates (Garcia et al. 1994). We concluded that the key was normal spermiogram when dosages of at least 10 million motile sperm (20 million pre-freeze) were used.

Semen used in superstimulated cattle should exceed the minimum standards established by the American Society for Theriogenology for frozen/thawed semen (Barth 1993). Briefly, these are a minimum of 70% morphologically normal sperm and, immediately after thawing, 25% directional motility with a rate of 3/5 and a minimum of 60% intact acrosomes. After 2 h, directional motility must exceed 15% (rate 2), and percentage of intact acrosomes must exceed 40%.

5.4.9   Superstimulation for Ovum Pickup (OPU)

Although *Bos indicus* cattle have high antral follicle counts and are not normally superstimulated prior to OPU, most *Bos taurus* breeds are treated with a half dose of FSH prior to oocyte aspiration. The common approach is to synchronize follicle wave emergence and administer four or six intramuscular injections of FSH over 2 or 3 days. Following a "coasting" period (with no FSH treatments) of approximately 40 h, oocytes for in vitro maturation and fertilization are recovered from antral follicles by ultrasound-guided oocyte aspiration (Blondin et al. 2002). Superstimulation prior to OPU has resulted in a significant increase in blastocyst production in Holstein donors (Vieira et al. 2014), and dilution of FSH in 0.5% hyaluronan prior to a single intramuscular administration has been shown to be equally efficacious (Vieira et al. 2015).

5.5   Estrus Synchronization in Recipients

High pregnancy rates are partially dependent upon the onset of estrus in recipients being within 24 h of synchrony with that of the embryo donor (Hasler et al. 1987). Recipients may be selected for embryo transfer by estrus detection of untreated animals or after drug-induced estrus synchronization. Regardless of the method used, timing and critical attention to estrus detection are important. Recipients synchronized with PGF must be treated 12–24 h before donors because PGF-induced estrus occurs in 60–72 h in single-ovulating cattle (Kastelic et al. 1990) and in 36–48 h in superstimulated donors (Bó et al. 2002, 2006). The success of estrus synchronization programs is dependent on an understanding of estrous cycle physiology, pharmacological agents and their effects on the estrous cycle, and herd management factors that reduce anestrus and increase conception rates. Treatment alternatives are discussed below.

5.5.1 Prostaglandin (PGF)

Prostaglandin has become the most common treatment for estrus synchronization in cattle (Larson and Ball 1992; Odde 1990), but PGF is not effective in inducing luteolysis in the first 5 days of the cycle, and when luteolysis is effectively induced, the ensuing estrus is distributed over a 6-day period (Kastelic et al. 1990). This is due to the status of the dominant follicle at the time of treatment. In a two-dose PGF protocol, an interval of 10 or 11 days between treatments has been used because all animals should have a responsive CL at the time of the second PGF. However, a 14-day interval is usually preferred for AI (Folman et al. 1990).

5.5.2 Progestins

Various progestins (progesterone and progesterone-like compounds) have been utilized for estrus synchronization (Mapletoft et al. 2003). Progesterone prevents ovulation in cattle and suppresses LH pulse frequency, which causes suppression of the growth of LH-dependent follicles (i.e., dominant follicle), but it does not suppress FSH secretion (Adams et al. 1992b). Thus, follicle waves continue to emerge in the presence of a functional CL. Progestins given for longer than the CL life-span (i.e., for 14 days or more) result in synchronous estrus upon withdrawal, but fertility is low (Revah and Butler 1996). Progestins used to control the estrous cycle in cattle have relatively less suppressive effects on LH secretion than the CL and are associated with the development of "persistent" follicles, which contain aged oocytes with low fertility. Although an early study indicated no effect of a CL resulting from a persistent follicle on pregnancy rates in recipients (Wehrman et al. 1997), Mantovani et al. (2005) reported reduced pregnancy rates.

To synchronize estrus, progestin devices are normally placed in the vagina for 7 days; PGF is given 24 h before device removal, and estrus detection begins 48 h later. Because of the short period of progestin treatment (7 days), the incidence of persistent follicles is reduced. Progesterone-releasing vaginal devices are also well suited to protocols used to synchronize follicular development and ovulation (Mapletoft et al. 2003).

5.5.3 Fixed-Time Embryo Transfer (FTET)

In recipients, the need for estrus detection can be eliminated by utilizing protocols that have been developed for fixed-time AI in cattle (Mapletoft et al. 2003). Basically,

two approaches have been used: the so-called Ovsynch or Cosynch protocols utilizing GnRH (Pursley et al. 1995; Wiltbank 1997) or pLH (Martinez et al. 1999), with
or without a progestin device (Lamb et al. 2001; Martinez et al. 2002), or estradiol
and progesterone to synchronize follicle wave emergence and ovulation in progestin-
treated animals (Bó et al. 2012; Mapletoft et al. 2003).

5.5.3.1 GnRH

If treatment of cattle with GnRH induces ovulation of a growing dominant follicle
(Thatcher et al. 1993), emergence of a new follicular wave occurs approximately
2 days later (Martinez et al. 1999). Treatment with PGF 7 days after GnRH results
in luteal regression and ovulation of the new dominant follicle, especially when a
second GnRH injection is given 36–56 h later (referred to as the Ovsynch protocol;
Pursley et al. 1995). However, the Ovsynch protocol has been more efficacious in
lactating dairy cows than in heifers. The cause for this variability is not known, but
ovulation to the first GnRH occurred in a higher percentage of cows than heifers
(Martinez et al. 1999; Pursley et al. 1995), and Wiltbank (1997) reported that 19%
of heifers showed estrus before the injection of PGF making fixed-time AI difficult.
However, the addition of a CIDR to a 7-day GnRH-based protocol improved pregnancy rates after fixed-time AI in heifers and improved pregnancy rates in non-
cycling, lactating beef cows (Lamb et al. 2001).

GnRH-based protocols have been used to synchronize ovulation in recipients
that received in vivo-derived (Baruselli et al. 2000, 2010; Hinshaw 1999) or
in vitro-produced (Ambrose et al. 1999) embryos. In these studies, more recipients
received embryos than when estrus detection was used because GnRH-based protocols do not depend on estrus detection; thus, pregnancy rates are higher than in
controls. Prevention of early ovulations by addition of a progestin-releasing device
to a 7-day GnRH-based protocol is usually used for FTET; Hinshaw (1999) treated
1637 recipients with GnRH plus a progestin-releasing device and transferred
in vivo-derived embryos, without estrus detection, with an overall pregnancy rate
of 59.9%.

Recent studies have shown that reducing the period of follicle dominance (by
removing the progestin device 5 days after insertion) and increasing the time from
progestin device removal to GnRH improve pregnancy per AI in GnRH-based protocols (Bridges et al. 2008; Lima et al. 2011; Santos et al. 2010). However, due to a
shorter interval between the first GnRH and induction of luteolysis in the 5-day
protocol, two injections of PGF are necessary to induce complete regression of the
GnRH-induced CL. However, Colazo and Ambrose (2011) showed that when the
first GnRH in the 5-day Cosynch protocol was not given in heifers and a single PGF
was administered, pregnancy rate to FTAI was not affected. We have preliminary
evidence indicating that this modification of the 5-day GnRH-based protocol results
in a comparable proportion of recipients receiving an embryo and becoming pregnant per embryo transfer as with other FTET protocols (Bó et al. 2012) or estrus
detection (Sala et al. 2016). The two recommended protocols for FTET in bovine
recipients using GnRH are shown in Fig. 5.4.

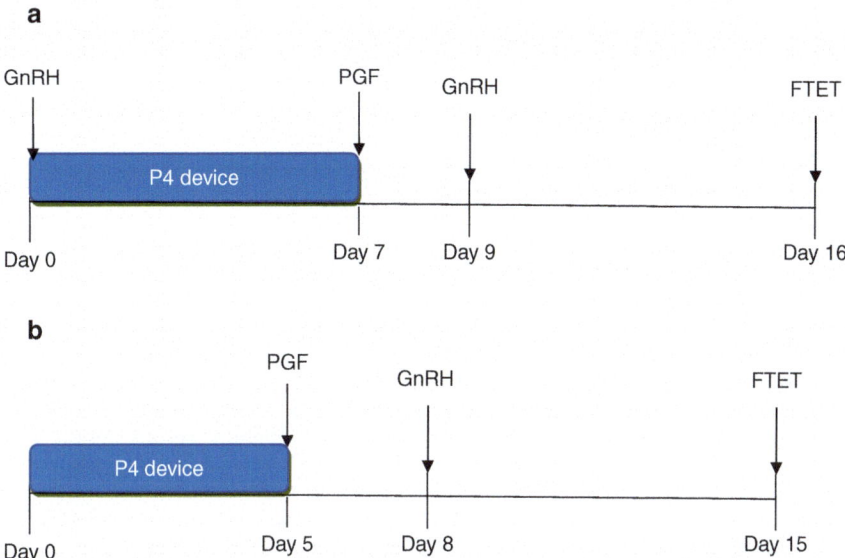

Fig. 5.4 Two protocols for FTET in bovine recipients using GnRH. (**a**) Traditional 7-day GnRH + P4 device protocol (GnRH plus P4 device on Day 0 and GnRH on Day 9). (**b**) Modified 5-day GnRH + P4 device protocol (P4 device on Day 0 and GnRH on Day 8). Recipients with a CL detected by palpation or ultrasonography receive an embryo at a fixed-time (FTET) 7 days after GnRH. An injection of 400 IU of eCG may also be given to *Bos indicus* recipients or suckled *Bos taurus* recipients at the time of P4 device removal

5.5.3.2 Estradiol and Progesterone

As indicated earlier, treatment with estradiol and progestins has been used to synchronize estrus in cattle, but Bó et al. (1995) demonstrated that estradiol treatment also synchronizes follicle development. In fixed-time AI protocols, a second, lower dose of estradiol is usually given 24 h after progestin device removal to induce LH release, which occurs approximately 16–18 h later, and ovulation in approximately 24 h (Mapletoft et al. 2003). Estradiol treatment protocols are the most commonly used treatment to synchronize follicle wave emergence and ovulation in beef and dairy recipients in South America (Baruselli et al. 2010, 2011). The progestin device is usually removed on Day 8, and ovulation is induced by the administration of 0.5 or 1 mg of estradiol cypionate at that time, or 1 mg of estradiol benzoate 24 h after progestin removal, or administration of GnRH or pLH 48 to 54 h after progestin removal (reviewed in Bó et al. 2002, 2012; Baruselli et al. 2010, 2011). As estrus detection is usually not performed, Day 9 is considered to be the day of estrus. When estrus detection is performed, all the recipients not in estrus by 48 h after progestin device removal receive GnRH. All recipients with a functional CL on Day 17 receive an embryo; conception rates were comparable to embryo transfer 7 days after observed estrus. The recommended protocols for FTET in recipients using estradiol and progestin are shown in Fig. 5.5.

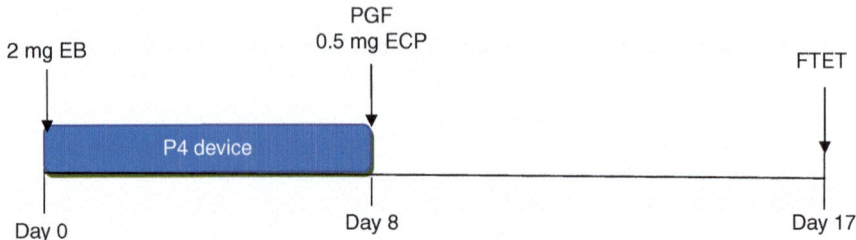

Fig. 5.5 Simplified estradiol-based synchronization protocol for FTET in bovine recipients. On Day 0, recipients receive a P4 device and estradiol benzoate (EB). On Day 8, P4 devices are removed, and PGF is administered. Ovulation is induced with estradiol cypionate on Day 8 (simplified) or EB on Day 9. Recipients with a CL detected by palpation or ultrasonography receive an embryo at a fixed-time (FTET) on Day 17. An injection of 400 IU of eCG may also be given to *Bos indicus* recipients or suckled *Bos taurus* recipients at the time of P4 device removal. If some estrus detection is implemented with tail patches or tail paint, recipients not showing estrus by 48 h after P4 device removal receive GnRH at that time

5.5.3.3 Use of eCG to Improve Pregnancy Rates in Recipients

A common strategy to increase pregnancy rates in pasture-managed beef recipients in South America is the addition of 400 IU of eCG on either Day 5 or Day 8 of the estradiol/progestin treatment protocol. Overall, 75–85% of the recipients treated with eCG receive an embryo (compared to 50% or less with simple PGF synchronization), progesterone concentrations at the time of embryo transfer are high, and conception rates following transfer exceed 50% (reviewed in Baruselli et al. 2010, 2011; Bó et al. 2012). The efficacy of the estradiol benzoate, progestin, and eCG treatment protocol for FTET has been confirmed in several different parts of the world in more than 15,000 recipients (Argentina, Bó et al. 2005; Brazil, Nasser et al. 2011; China, Remillard et al. 2006; Mexico, Looney et al. 2010). In each of these studies, treatment with eCG increased the number of recipients receiving an embryo resulting in higher pregnancy rates.

The use of eCG in GnRH-based protocols has also been evaluated. In a Canadian study designed to evaluate the potential use of eCG in beef cattle recipients synchronized with GnRH/progestin for FTET (Small et al. 2007), recipient selection rates did not differ, but in a Colombian study (Mayor et al. 2008), eCG significantly increased pregnancy rates following FTET in recipients treated with the GnRH/progestin protocol.

In summary, the addition of eCG to estradiol- or GnRH-based protocols which included the use of progestin devices resulted in increased pregnancy rates depending on the type, body condition, and cyclicity of the recipients. However, treatment with eCG may not improve pregnancy rates in *Bos taurus* recipients managed under more optimal conditions.

5.5.3.4 Other Treatments to Increase Pregnancy Rates in Recipients

Several studies have investigated the relationship between circulating progesterone concentrations and pregnancy rates in recipients (reviewed in Baruselli et al. 2010;

Carter et al. 2008). However, the use of supplementary progesterone has resulted in inconsistent effects on pregnancy rates. Generally, the beneficial effects of increasing circulating concentrations of progesterone seem to be more evident when pregnancy rates in untreated recipients were lower than expected. An alternative strategy is to create an accessory CL by induction of ovulation of the first-wave dominant follicle, around the time of embryo transfer (reviewed in Thatcher et al. 2001). Again, results have been inconsistent; in *Bos indicus* recipients, treatment with human chorionic gonadotrophin (hCG) on Day 7 increased progesterone concentrations (Marques et al. 2002), and treatment with GnRH (Rodrigues et al. 2003) or GnRH, hCG, and pLH or a progestin device (Marques et al. 2003) at the time of embryo transfer resulted in increased pregnancy rates. However, in another study involving *Bos indicus*-cross recipients synchronized with the progestin/estradiol plus eCG protocol, pregnancy rates were not affected by treatment with hCG or GnRH at the time of FTET (Tribulo et al. 2005). Small et al. [2004] were also unable to improve pregnancy rates in *Bos taurus* recipients treated with GnRH or pLH on Days 5 or 7 after estrus. In a more recent study, administration of 1000 IU hCG at the time of embryo transfer (Wallace et al. 2011) resulted in higher serum progesterone concentrations in recipients with lower body condition scores but not with higher body condition scores. The authors concluded that giving hCG at the time of embryo transfer increased the incidence of accessory CL and higher serum progesterone concentrations which resulted in higher pregnancy rates in recipients with lower body condition scores. Lower embryonic losses were also observed in recipients that received GnRH 2 days prior to receiving in vitro-produced embryos (García Guerra et al. 2016).

5.5.4 Management Factors

The two management factors that determine the success or failure of an estrus synchronization program are nutrition (body condition score) and postpartum interval. If cows lose weight during pregnancy, the onset of estrous cycles after calving will be delayed, while cows that are fed adequately during pregnancy but fail to gain weight between calving and breeding will cycle but have reduced conception rates (Carvalho et al. 2014b) and may also have reduced pregnancy rates after receiving a viable embryo by embryo transfer (Bó et al. 2005). In a field study, pregnancy rates were significantly higher in beef recipients scoring 3 and 4 than in those scoring 1, 2 (thin), or 5 (obese; reviewed in Mapletoft 1986). Therefore, the nutritional status of recipients must be evaluated before using them for embryo transfer.

5.6 Embryo Recovery

In early commercial bovine embryo transfer programs, embryos were collected surgically 4 days after estrus (reviewed by Betteridge 2003). However, three methods of nonsurgical embryo recovery were described in 1976 (Drost et al. 1976; Elsden et al. 1976; Rowe et al. 1976). Nonsurgical techniques are preferred as they are not damaging to the reproductive tract, are repeatable, and can be performed on the farm. Briefly, the donor is restrained, and the rectum is evacuated of feces and air. The number of CL is usually estimated at this time or just prior to ova/embryo recovery. The perineal region and vulvar labia are washed thoroughly and dried, and the tail is tied out of the way. Embryo recovery is not attempted until a satisfactory epidural anesthetic is completed.

Nonsurgical embryo collections involve the passage of a cuffed catheter through the cervix and into one of the uterine horns on Days 6–8 after estrus. Once the catheter is in place, the cuff is inflated with collection medium. Although original reports involved the use of two-way and three-way Foley catheters (Rowe et al. 1976), the two-way Rusch catheter has been preferred by many practitioners (Schneider and Hahn 1979). The catheter is stiffened for passage through the cervix by a stainless steel stilette, which locks into the Luer-Lok fittings. The Rusch catheter is long enough for large cows and is stiff enough that it can be easily threaded down the uterine lumen. Several other catheters are now available from embryo transfer suppliers, but they are really modifications of the Rusch catheter.

Basically, there are two methods of embryo collection (Mapletoft 1986): the closed-circuit continuous or interrupted flow system and the interrupted-syringe technique. However, any combination of these two techniques is possible. It must be recognized that each system has advantages and disadvantages relative to the other. With the closed system, it is easier to maintain sterility, and there is less chance of losing medium and consequently, embryos. However, it is cumbersome and the extra tubing provides extra potential for contamination by either microbes or chemicals. With the interrupted-syringe method, it is possible to use fully disposable equipment and to search for embryos while the collection is in progress.

The embryo recovery medium is normally prepared in advance. Dulbecco's phosphate-buffered saline (PBS) or other simple salt solutions are normally prepared in 500–1000 ml quantities for embryo collection. In addition, quantities of heat-inactivated fetal calf serum (FCS) and antibiotic/antimycotic solution for each volume of PBS may be kept frozen. The holding medium, containing a higher concentration of serum, is normally held in a "plastic on plastic" syringe for use (an antioxidant on the rubber plunger in plastic syringes has been shown to be toxic to embryos; Hasler 2003). Holding medium is normally passed through a disposable 0.22 μm Millipore filter prior to use, but the first 4–5 ml should be discarded as it may also affect embryo survival. Ready-made embryo collection and holding media are now available commercially; they have been filtered and are ready for use. However, if media contain animal products, e.g., serum or BSA, they must be refrigerated. Recently, collection and holding media that do not contain animal products have become available (Hasler 2010).

Temperature does not seem to be critical for embryo survival, provided extremes are avoided; room temperature seems satisfactory. Similarly, sterility is not possible, but one should attempt to be clean. Sterilization with chemicals is more likely to kill embryos than microbial contaminants. As a routine, embryos should be passed through ten washes of fresh, sterile medium prior to transfer or freezing to remove all contaminants. Certain infectious agents such *Bovine herpesvirus* have been shown to "stick" to the zona pellucida, so two trypsin treatments after wash number five is often recommended, followed by the final five washes (Singh 1985).

5.7   Embryo Handling

Embryos are normally held in the same or a similar medium to that in which they were collected. Media must be buffered to maintain a pH of 7.2–7.6 and have an osmolarity of around 300 mos/L. Dulbecco's PBS or more complex media with the HEPES buffer and enriched with FCS or BSA and antibiotics are normally used in the field. Embryos are located with a stereoscopic dissecting microscope at 10× magnification after passing the collection medium through a filter with pores that are approximately 50–70 μm in diameter (Mapletoft 1986). Although embryos are usually transferred as soon as possible after collection, it is possible to maintain embryos in holding medium for several hours at room temperature or to cool bovine embryos in holding medium and maintain them in the refrigerator for 2 or 3 days. As a final alternative, embryos may be cryopreserved. Embryo collection, holding, and freezing media that are free of animal products not only eliminates the need for refrigeration but also increase biosecurity (Hasler 2010).

5.7.1   Embryo Evaluation

Evaluation of bovine embryos must be done at 50–100× magnification, with the embryo in a small culture dish. The IETS Manual describes a numerical system for classification of embryo developmental stage, ranging from 1 (single-cell zygote) to 8 (hatched blastocyst), and quality, from 1 (good and excellent) to 4 (dead). It is important to be able to recognize the various stages of development and to compare these with the developmental stage that the embryo should be based on the days from estrus. Often a decision as to whether an embryo is worthy of transfer will depend on the availability of a recipient. Fair-quality embryos should be transferred fresh, while good and excellent quality embryos have a high probability of surviving cryopreservation. The IETS considers the export of poor- and fair-quality embryos to be improper.

The overall diameter of the bovine embryo is 150–190 μm, including a zona pellucida thickness of 12–15 mm. The overall diameter of the embryo remains virtually unchanged from the one-cell stage to blastocyst stage. The best predictor of an embryo's viability is its stage of development relative to what it should be on a given day after ovulation. An ideal embryo is compact and spherical. The blastomeres should be of similar size with a homogenous color and texture. The cytoplasm should not be granular or vesiculated, and the perivitelline space should be clear and contain no cellular debris. The zona pellucida should be uniform, should neither be cracked nor misshapen, and should not contain debris on its surface. Embryos of good and excellent quality (IETS quality code 1) and at the developmental stages of late morula to blastocyst yield the highest pregnancy rates, fresh or frozen/thawed. The IETS Manual has a complete library of embryo photographs.

5.8   Embryo Transfer

Transfer of high-quality embryos in cattle will result in high pregnancy rates providing that estrus in the donor and recipient occurred within 24 h of each other (Hasler et al. 1987). Alternately, recipients must be synchronous with the stage of development of embryos that had been previously cryopreserved. Bovine embryo transfers were initially done surgically, while most are done today using nonsurgical methods (Rowe et al. 1980; Wright 1981). Nonsurgical embryo transfer techniques involve the use of specialized embryo transfer pipettes. After confirming synchrony of estrus, the recipient is restrained, and the rectum is evacuated of feces and air. At that time, the presence and side of a functional CL is confirmed. An epidural anesthetic is administered, and the vulva is washed with water (no soap) and dried with a paper towel. The embryo is placed in 0.25 ml straw between at least two air bubbles and two columns of medium, and the straw is placed in the embryo transfer pipette. Care is taken to insure that the straw engages the sheath tightly so as to avoid leakage. The sheath is coated with sterile, nontoxic obstetrical lubricant, and the sheathed pipette is passed through the vulvar labia while avoiding contamination. The embryo is placed in the uterine horn adjacent to the ovary bearing the CL by passing the pipette through the cervix, very similar to AI. However, an attempt is usually made to pass the pipette at least halfway down the uterine horn. Care must be taken to prevent trauma to the endometrium. The embryo is deposited slowly and firmly, while the tip of the transfer pipette is withdrawn slightly. Practice and dexterity seem to improve one's ability to achieve high pregnancy rates suggesting that trauma to the endometrium may be a limiting factor. Stimulation of the cervix or inadvertent introduction of bacterial contaminants does not seem to affect pregnancy rates.

5.9   Summary and Conclusions

Commercial embryo transfer in cattle has become a well-established industry. Although a very small number of offspring are produced on an annual basis, its impact is large because of the quality of animals being produced. Embryo transfer is now being used for genetic improvement, especially in the dairy industry, and most semen used today comes from bulls that have been produced by embryo transfer. An even greater benefit of bovine embryo transfer may be that in vivo-derived embryos can be made specified pathogen-free by washing procedures, making them ideal for disease control programs or in the international movement of animal genetics. Techniques have improved over the past 40 years so that frozen-thawed embryos can be transferred to suitable recipients as easily and simply as AI. A combination of embryo transfer with proven cows inseminated with semen from proven bulls and industry-wide AI is a common commercial application of bovine embryo transfer.

References

Adams GP (1994) Control of ovarian follicular wave dynamics in cattle; implications for synchronization and superstimulation. Theriogenology 4:19–24

Adams GP, Matteri RL, Kastelic JP et al (1992a) Association between surges of follicle stimulating hormone and the emergence of follicular waves in heifers. J Reprod Fertil 94:177–188

Adams GP, Matteri RL, Ginther OJ (1992b) The effect of progesterone on growth of ovarian follicles, emergence of follicular waves and circulating FSH in heifers. J Reprod Fertil 95: 627–640

Adams GP, Jaiswal R, Singh J et al (2008) Progress in understanding ovarian follicular dynamics in cattle. Theriogenology 69:72–80

Ambrose JD, Drost RL, Monson RL et al (1999) Efficacy of timed embryo transfer with fresh and frozen in vitro-produced embryos to increase pregnancy rates in heat-stressed dairy cattle. J Dairy Sci 82:2369–2376

Armstrong D (1993) Recent advances in superovulation of cattle. Theriogenology 39:7–24

Baracaldo MI, Martinez M, Adams GP et al (2000) Superovulatory response following transvaginal follicle ablation in cattle. Theriogenology 53:1239–1250

Barth AD (1993) Evaluation of frozen bovine semen by the veterinary practitioner. Reviewers: WG Parker, EG Robertson, RG Saacke, WH Cardwell, JR Mitchell, GW McKay. In: Society for Theriogenology Handbook B-9. Society for Theriogenology, Pike Road, AL, USA

Baruselli PS, Marques MO, Carvalho NAT et al (2000) Ovsynch protocol with fixed-time embryo transfer increasing pregnancy rates in bovine recipients. Arq Fac Vet UFRGS, Porto Alegre, Brazil 28:205 (Abstract)

Baruselli PS, Sá Fhilo M, Martins CM et al (2006) Superovulation and embryo transfer in *Bos indicus* cattle. Theriogenology 65:77–88

Baruselli PS, Ferreira RM, Sá Filho MF et al (2010) Bovine embryo transfer recipient synchronization and management in tropical environments. Reprod Fertil Dev 22:67–74

Baruselli PS, Ferreira RM, Sales JNS et al (2011) Timed embryo transfer programs for the management of donor and recipient cows. Theriogenology 76:1583–1593

Batista EOS, Macedo GG, Sala RV et al (2014) Plasma anti-Mullerian hormone as a predictor of ovarian antral follicular population in *Bos indicus* (Nelore) and *Bos taurus* (Holstein) heifers. Reprod Domest Anim 49:448–452

Bergfelt DR, Lightfoot KC, Adams GP (1994) Ovarian dynamics following ultrasound-guided transvaginal follicle ablation in heifers. Theriogenology 42:895–907

Bergfelt DR, Bó GA, Mapletoft RJ et al (1997) Superovulatory response following ablation-induced follicular wave emergence at random stages of the oestrous cycle in cattle. Anim Reprod Sci 49:1–12

Betteridge KJ (2003) A history of farm animal embryo transfer and some associated techniques. Anim Reprod Sci 79:203–244

Blondin P, Bousquet D, Twagiramungu H et al (2002) Manipulation of follicular development to produce developmentally competent bovine oocytes. Biol Reprod 66:38–43

Bó GA, Mapletoft RJ (2014) Historical perspectives and recent research on superovulation in cattle. Theriogenology 81:38–48

Bó GA, Hockley DK, Nasser LF et al (1994) Superovulatory response to a single subcutaneous injection of Folltropin-V in beef cattle. Theriogenology 42:963–975

Bó GA, Adams GP, Pierson RA et al (1995) Exogenous control of follicular wave emergence in cattle. Theriogenology 43:31–40

Bó GA, Adams GP, Pierson RA et al (1996) Effect of progestogen plus E-17β treatment on superovulatory response in beef cattle. Theriogenology 45:897–910

Bó GA, Baruselli PS, Moreno D et al (2002) The control of follicular wave development for self-appointed embryo transfer programs in cattle. Theriogenology 57:53–72

Bó GA, Cutaia L, Chesta P et al (2005) Application of fixed-time artificial insemination and embryo transfer programs in beef cattle operations. In: Proceedings of the joint meeting of the American Embryo Transfer Association & Canadian Embryo Transfer Association, Minneapolis, MN, pp.37–59

Bó GA, Baruselli PS, Chesta P et al (2006) The timing of ovulation and insemination schedules in superstimulated cattle. Theriogenology 65:89–101

Bó GA, Guerrero DC, Adams GP (2008) Alternative approaches to setting up donor cows for superstimulation. Theriogenology 69:81–87

Bó GA, Coelho Peres L, Cutaia LE et al (2012) Treatments for the synchronisation of bovine recipients for fixed-time embryo transfer and improvement of pregnancy rates. Reprod Fertil Dev 24:272–277

Bridges GA, Helser LA, Grum DE et al (2008) Decreasing the interval between GnRH and PGF2α from 7 to 5 days and lengthening proestrus increases timed-AI pregnancy rates in beef cows. Theriogenology 69:843–851

Bungartz L, Niemann H (1994) Assessment of the presence of a dominant follicle and selection of dairy cows suitable for superovulation by a single ultrasound examination. J Reprod Fertil 101:583–591

Carmichael RA (1980) History of the international embryo transfer society – Part I. Theriogenology 13:3–6

Carter F, Forde N, Duffy P et al (2008) Effect of increasing progesterone concentration from Day 3 of pregnancy on subsequent embryo survival and development in beef heifers. Reprod Fertil Dev 20:368–375

Carvalho PD, Souza AH, Amundson MC et al (2014a) Relationships between fertility and postpartum changes in body condition and body weight in lactating dairy cows. J Dairy Sci 97:1–18

Carvalho PD, Hackbart KS, Bender RW et al (2014b) Use of a single injection of long-acting recombinant bovine FSH to superovulate Holstein heifers: a preliminary study. Theriogenology 82:481–489

Christensen LG (1991) Use of embryo transfer in future cattle breeding schemes. Theriogenology 35:141–156

Colazo MG, Ambrose DJ (2011) Neither duration of progesterone insert nor initial GnRH treatment affected pregnancy per timed-insemination in dairy heifers subjected to a Co-synch protocol. Theriogenology 76:578–588

DeJarnette JM, Saacke RG, Bame J et al (1992) Accessory sperm: their importance to fertility and embryo quality, and attempts to alter their numbers in artificially inseminated cattle. J Anim Sci 70:484–491

Dieleman S, Bevers M, Vos P et al (1993) PMSG/anti-PMSG in cattle: a simple and efficient superovulatory treatment. Theriogenology 39:25–42

Drost M, Brand A, Aaarts MH (1976) A device for nonsurgical recovery of bovine embryos. Theriogenology 6:503–508

Edwards L, Rahe C, Griffin J et al (1987) Effect of transportation stress on ovarian function in superovulated Hereford heifers. Theriogenology 28:291–299

Elsden RP, Hasler JF, Seidel GE Jr (1976) Non-surgical recovery of bovine eggs. Theriogenology 6:523–532

Folman Y, Kaim M, Herz Z et al (1990) Comparison of methods for the synchronization of estrous cycles in dairy cows. 2. Effects of progesterone and parity on conception. J Dairy Sci 73:2817–2825

García Guerra A, Sala RV, Baez GM et al (2016) Treatment with GnRH on Day 5 reduces pregnancy loss in heifers receiving in vitro-produced expanded blastocysts. Reprod Fertil Dev 28:185 (Abstract)

Garcia A, Salaheddine M (1998) Effects of repeated ultrasound-guided transvaginal follicular aspiration on bovine oocyte recovery and subsequent follicular development. Theriogenology 50:575–585

Garcia A, Mapletoft RJ, Kennedy R (1994) Effect of semen dose on fertilization and embryo quality in superovulated cows. Theriogenology 41:202 (Abstract)

Ginther OJ, Knopf L, Kastelic JP (1989) Temporal associations among ovarian events in cattle during oestrous cycles with two or three follicular waves. J Reprod Fertil 87:223–230

Gonzalez A, Wang H, Carruthers TD et al (1994) Increased ovulation rates in PMSG – stimulated beef heifers treated with a monoclonal PMSG antibody. Theriogenology 41:1631–1642

Gonzalez-Reyna A, Lussier JG, Carruthers TD et al (1990) Superovulation of beef heifers with Folltropin: a new FSH preparation containing reduced LH activity. Theriogenology 33:519–529

Guilbault LA, Grasso F, Lussier JG et al (1991) Decreased superovulatory responses in heifers superovulated in the presence of a dominant follicle. J Reprod Fertil 91:81–89

Hasler JF (2003) The current status and future of commercial embryo transfer in cattle. Anim Reprod Sci 79:245–264

Hasler JF (2010) Synthetic media for culture, freezing and vitrification of bovine embryos. Reprod Fertil Dev 22:119–125

Hasler J, Hockley D (2012) Efficacy of hyaluronan as a diluent for a two injection FSH superovulation protocol in Bos taurus beef cows. Reprod Domest Anim 47:459 (Abstract)

Hasler JF, McCauley AD, Schermerhorn EC et al (1983) Superovulatory responses of Holstein cows. Theriogenology 20:1983–1999

Hasler JF, McCauley AD, Lathrop WF et al (1987) Effect of donor-embryo-recipient interactions on pregnancy rate in a large-scale bovine embryo transfer program. Theriogenology 27:139–168

Hawk HW, Conley HH, Wall RJ et al (1988) Fertilization rates in superovulating cows after deposition of semen on the infundibulum, near the uterotubal junction or after insemination with high numbers of sperm. Theriogenology 29:1131–1142

Hinshaw RH (1999) Formulating ET contracts. In: Proceedings of the society for theriogenology, Nashville, TN, USA, pp 399–404

Hinshaw RH, Switzer ML, Mapletoft RJ et al (2015) A comparison of two approaches for the use of GnRH to synchronize follicle wave emergence for superovulation. Reprod Fertil Dev 27:263 (Abstract)

Hiraizumi S, Nishinomiya H, Oikawa T et al (2015) Superovulatory response in Japanese Black cows receiving a single subcutaneous porcine follicle–stimulating hormone treatment or six intramuscular treatments over three days. Theriogenology 83:466–473

Ireland JJ, Ward F, Jimenez-Krassel F et al (2007) Follicle numbers are highly repeatable within individual animals but are inversely correlated with FSH concentrations and the proportion of good-quality embryos after ovarian stimulation in cattle. Hum Reprod 22:1687–1695

Ireland JJ, Smith GW, Scheetz D et al (2011) Does size matter in females? An overview of the impact of the high variation in the ovarian reserve on ovarian function and fertility, utility of anti-Mullerian hormone as a diagnostic marker for fertility and causes of variation in the ovarian reserve in cattle. Reprod Fertil Dev 23(1):14

Kastelic JP, Knopf L, Ginther OJ (1990) Effect of day of prostaglandin F treatment on selection and development of the ovulatory follicle in heifers. Anim Reprod Sci 23:169–180

Kelly P, Duffy P, Roche JF et al (1997) Superovulation in cattle: effect of FSH type and method of administration on follicular growth, ovulatory response and endocrine patterns. Anim Reprod Sci 46:1–14

Kim HI, Son DS, Yeon H et al (2001) Effect of dominant follicle removal before superstimulation on follicular growth, ovulation and embryo production in Holstein cows. Theriogenology 55:937–945

Lamb GC, Stevenson JS, Kesler DJ et al (2001) Inclusion of an intravaginal progesterone insert plus GnRH and prostaglandin F2α for ovulation control in postpartum suckled beef cows. J Anim Sci 79:2253–2259

Lane EA, Austin EJ, Crowe MA (2008) Estrus synchronisation in cattle-current options following the EU regulations restricting use of estrogenic compounds in food-producing animals: a review. Anim Reprod Sci 109:1–16

Larkin S, Chesta P, Looney C et al (2006) Distribution of ovulation and subsequent embryo production using Lutropin and estradiol-17β for timed AI of superstimulated beef females. Reprod Fertil Dev 18:289 (Abstract)

Larson LL, Ball PJH (1992) Regulation of estrous cycles in dairy cattle: a review. Theriogenology 38:255–267

Laster DB (1972) Disappearance of and uptake of ^{125}I FSH in the rat, rabbit, ewe and cow. J Reprod Fertil 30:407–415

Lima FS, Ayres H, Favoreto MG et al (2011) Effects of gonadotropin releasing hormone at initiation of the 5-d timed artificial insemination (AI) program and timing of induction of ovulation relative to AI on ovarian dynamics and fertility of dairy heifers. J Dairy Sci 94:4997–5004

Lindsell CE, Murphy BD, Mapletoft RJ (1986) Superovulatory and endocrine responses in heifers treated with FSH-P at different stages of the estrous cycle. Theriogenology 26:209–219

Lohuis MM (1995) Potential benefits of bovine embryo-manipulation technologies to genetic improvement programs. Theriogenology 43:51–60

Looney CR (1986) Superovulation in beef females. In: Proceedings of the annual meeting of the American Embryo Transfer Association, Fort Worth, TX, pp 16–29

Looney CR, Stutts KJ, Novicke AK et al (2010) Advancements in estrus synchronization of Brahman-influenced embryo transfer recipient females. In: Proceedings of the joint meeting of the American Embryo Transfer Association & Canadian Embryo Transfer Association, Charlotte, NC, pp 17–22

Lovie M, Garcia A, Hackett A et al (1994) The effect of dose schedule and route of administration on superovulatory response to Folltropin in Holstein cows. Theriogenology 41:241 (Abstract)

Mantovani AP, Reis EL, Gacek F et al (2005) Prolonged use of a progesterone-releasing intravaginal device (CIDR®) for induction of persistent follicles in bovine embryo recipients. Anim Reprod 2:272–277

Mapletoft RJ (1985) Embryo transfer in the cow: general procedures. Rev Sci Tech Off Int Epiz 4:843–858

Mapletoft RJ (1986) Bovine embryo transfer. In: Morrow DA (ed) Current therapy in theriogenology II. WB Saunders Co, Philadelphia, PA, pp 54–63

Mapletoft RJ, Bó GA (2004) The control of ovarian function for embryo transfer: superstimulation of cows with normal or abnormal ovarian function. In: Proceedings of 23 world buiatrics congress, Quebec City, QC, 34(1 and 2), pp 67–68

Mapletoft RJ, Bó GA (2016) Bovine embryo transfer. In: I.V.I.S. (Ed.), IVIS reviews in veterinary medicine. International Veterinary Information Service (www.ivis.org), Ithaca. Document No. R0104.1106S

Mapletoft RJ, Hasler JF (2005) Assisted reproductive technologies in cattle: a review. Rev Sci Tech Off Int Epiz 24:393–403

Mapletoft RJ, Steward KB, Adams GP (2002) Recent advances in the superovulation of cattle. Reprod Nutr Dev 42:1–11

Mapletoft RJ, Martinez MF, Colazo MG et al (2003) The use of controlled internal drug release devices for the regulation of bovine reproduction. J Anim Sci 1(E. Suppl 2):E28–E36

Marques MO, Madureira EH, Bó GA et al (2002) Ovarian ultrasonography and plasma progesterone concentration Bos taurus x Bos indicus heifers administered different treatments on Day 7 of the estrous cycle. Theriogenology 57:548 (Abstract)

Marques MO, Nasser LF, Silva RCP et al (2003) Increased pregnancy rates in Bos taurus x Bos indicus embryo recipients with treatments that increase plasma progesterone concentrations. Theriogenology 59:369 (Abstract)

Martinez MF, Adams GP, Bergfelt D et al (1999) Effect of LH or GnRH on the dominant follicle of the first follicular wave in heifers. Anim Reprod Sci 57:23–33

Martinez MF, Kastelic JP, Adams GP et al (2002) The use of a progesterone-releasing device (CIDR) or melengestrol acetate with GnRH, LH or estradiol benzoate for fixed-time AI in beef heifers. J Anim Sci 80:1746–1751

Martins CM, Rodrigues CA, Vieira LM et al (2012) The effect of timing of the induction of ovulation on embryo production in superstimulated lactating Holstein cows undergoing fixed-time artificial insemination. Theriogenology 78:974–980

Mayor JC, Tribulo HE, Bó GA (2008) Pregnancy rates following fixed-time embryo transfer in Bos indicus recipients synchronized with progestin devices and estradiol or GnRH and treated with eCG. Reprod Domest Anim 43(Suppl 3):180 (Abstract)

Mikel-Jenson A, Greve T, Madej A et al (1982) Endocrine profiles and embryo quality in the PMSG-PGF$_{2\alpha}$-treated cow. Theriogenology 18:33–34

Monniaux D, Chupin D, Saumande J (1983) Superovulatory responses of cattle. Theriogenology 19:55–82

Monniaux D, Drouilhet L, Rico C et al (2013) Regulation of anti-Mullerian hormone production in domestic animals. Reprod Fertil Dev 25(1):16

Murphy B, Martinuk S (1991) Equine chorionic gonadotropin. Endocr Rev 12:27–44

Murphy BD, Mapletoft RJ, Manns J et al (1984) Variability in gonadotrophin preparations as a factor in the superovulatory response. Theriogenology 21:117–125

Nasser LF, Adams GP, Bó GA et al (1993) Ovarian superstimulatory response relative to follicular wave emergence in heifers. Theriogenology 40:713–724

Nasser LFT, Penteado L, Rezende CR et al (2011) Fixed time artificial insemination and embryo transfer programs in Brazil. Acta Sci Vet 39(Suppl 1):s15–s22

Odde KG (1990) A review of synchronization of estrus in postpartum cattle. J Anim Sci 68:817–830

Perry G (2017) 2016 statistics of embryo collection and transfer in domestic farm animals. Embryo Technology Newsletter 35(4):8–23

Pierson RA, Ginther OJ (1987) Follicular populations during the estrous cycle in heifers: I. Influence of day. Anim Reprod Sci 14:165–176

Ponsart C, Le Bourhis D, Knijn H et al (2014) Reproductive technologies and genomic selection in dairy cattle. Reprod Fertil Dev 26:12–21

Pursley JR, Mee MO, Wiltbank MC (1995) Synchronization of ovulation in dairy cows using PGF2α and GnRH. Theriogenology 44:915–923

Remillard R, Martínez MF, Bó GA et al (2006) The use of fixed-time techniques and eCG to synchronize recipients for frozen-thawed bovine IVF embryos. Reprod Fertil Dev 18:204 (Abstract)

Revah I, Butler WR (1996) Prolonged dominance of follicles and reduced viability of bovine oocytes. J Reprod Fertil 106:39–47

Rodrigues CA, Mancilha RF, Dalalio M et al (2003) Increase of conception rates in IVF embryo recipients treated with GnRH at embryo transfer moment. Acta Sci Vet 33:550–551 (Abstract)

Rowe RF, Del Campo MR, Eilts CL et al (1976) A single cannula technique for nonsurgical collection of ova from cattle. Theriogenology 6:471–484

Rowe RF, Del Campo MR, Critser JK et al (1980) Embryo transfer in cattle: nonsurgical transfer. Am J Vet Res 41:1024–1028

Saacke RG, Nebel RL, Karabius DS et al (1988) Sperm transport and accessory sperm evaluation. In: Proceedings of 12th NAAB technical conference on artificial insemination, pp 7–14

Sala LC, Sala RV, Fosado M et al (2016) Factors that influence fertility in an IVF embryo transfer program in dairy heifers. Reprod Fertil Dev 28:183 (Abstract)

Santos JEP, Narciso CD, Rivera F et al (2010) Effect of reducing the period of follicle dominance in a timed AI protocol on reproduction of dairy cows. J Dairy Sci 93:2976–2988

Saumande J, Chupin D, Mariana J et al (1978) Factors affecting the variability of ovulation rates after PMSG stimulation. In: Sreenan JM (ed) Control of reproduction in the cow. Martinus Nijhoff, The Hague, pp 195–224

Schams D, Menzer D, Schalenberger E et al (1978) Some studies of the pregnant mare serum gonadotrophin (PMSG) and on endocrine responses after application for superovulation in cattle. In: Sreenan JM (ed) Control of reproduction in the cow. Martinus Nijhoff, The Hague, pp 122–142

Schiewe MC, Looney CR, Johnson CA et al (1987) Transferable embryo recovery rates following different insemination schedules in superovulated beef cattle. Theriogenology 28:395–406

Schneider U, Hahn J (1979) Embryo transfer in Germany. Theriogenology 11:63–80

Schultz RH (1980) History of the international embryo transfer society – Part II. Theriogenology 13:7–12

Seidel GE Jr (1981) Superovulation and embryo transfer in cattle. Science 211:351–358

Seidel GE Jr (2010) Brief introduction to whole genome selection in cattle using single nucleotide polymorphisms. Reprod Fertil Dev 22:138–144

Shaw DW, Good TE (2000) Recovery rates and embryo quality following dominant follicle ablation in superovulated cattle. Theriogenology 53:1521–1528

Singh EL (1985) Disease control: procedures for handling embryos. Rev Sci Tech Off Int Epiz 4:867–872

Singh J, Dominguez M, Jaiswal R et al (2004) A simple ultrasound test to predict superstimulatory response in cattle. Theriogenology 62:227–243

Small J, Colazo M, Ambrose D et al (2004) Pregnancy rate following transfer of in vitro- and in vivo-produced bovine embryos to LH-treated recipients. Reprod Fertil Dev 16:213 (Abstract)

Small JA, Colazo MG, Kastelic JP et al (2007) The effects of CIDR and eCG treatment in a GnRH-based protocol for timed-AI or embryo transfer on pregnancy rates in lactating beef cows. Reprod Fertil Dev 19:127–128 (Abstract)

Small JA, Colazo MG, Kastelic JP et al (2009) Effects of progesterone pre-synchronization and eCG on pregnancy rates to GnRH-based, timed-AI in beef cattle. Theriogenology 71:698–706

Smith C (1988a) Applications of embryo transfer in animal breeding. Theriogenology 29:203–212

Smith C (1988b) Genetic improvement of livestock using nucleus breeding units. World Anim Rev 65:2–10

Soares JG, Martins CM, Carvalho NAT et al (2011) Timing of insemination using sex-sorted sperm in embryo production with *Bos indicus* and *Bos taurus* superovulated donors. Anim Reprod Sci 127:148–153

Souza AH, Rozner A, Carvalho PD et al (2014) Relationship between circulating AMH (anti-Mullerian hormone) and embryo production in superovulated high producing donor cows. In: Proceedings of the joint meeting of the American and Canadian Embryo Transfer Associations, Madison, WI, pp 12–16

Steel R, Hasler J (2009) Comparison of three different protocols for superstimulation of dairy cattle. Reprod Fertil Dev 21:246 (Abstract)

Stoebel D, Moberg G (1982) Repeated acute stress during the follicular phase and luteinizing hormone surge of dairy heifers. J Dairy Sci 65:92–96

Stringfellow DA (2010) Recommendations for the sanitary handling of in-vivo-derived embryos. In: Stringfellow DA, Givens MD (eds) Manual of the international embryo transfer society. 4th edn. Savoy, IL, pp 65–68

Stringfellow DA, Givens MD (2000) Epidemiologic concerns relative to in vivo and in vitro production of livestock embryos. Anim Prod Sci 60–61:629–642

Stringfellow DA, Givens MD (eds) (2010) Manual of the international embryo transfer society, 4th edn. Savoy, IL

Stringfellow DA, Givens MD, Waldrop JG (2004) Biosecurity issues associated with current and emerging embryo technologies. Reprod Fertil Dev 16:93–102

Sutherland W (1991) Biomaterials – novel material from biological sources. In: Byrom D (ed) Stockton Press, New York, NY, pp 307–333

Teepker G, Keller DS (1989) Selection of sires originating from a nucleus breeding unit for use in a commercial dairy population. Can J Anim Sci 69:595–604

Thatcher WW, Drost M, Savio JD et al (1993) New clinical uses of GnRH and its analogues in cattle. Anim Reprod Sci 33:27–49

Thatcher WW, Moreira F, Santos JEP et al (2001) Effects of hormonal treatments on reproductive performance and embryo production. Theriogenology 55:75–89

Toner JP, Seifer DB (2013) Why we may abandon basal follicle-stimulating hormone testing: a sea change in determining ovarian reserve using antimullerian hormone. Fertil Steril 99:1825–1830

Tribulo R, Balla E, Cutaia L et al (2005) Effect of treatment with GnRH or hCG at the time of embryo transfer on pregnancy rates in cows synchronized with progesterone vaginal devices, estradiol benzoate and eCG. Reprod Fertil Dev 17:234 (Abstract)

Tríbulo A, Rogan D, Tribulo H et al (2011) Superstimulation of ovarian follicular development in beef cattle with a single intramuscular injection of Folltropin-V. Anim Reprod Sci 129:7–13

Tríbulo A, Rogan D, Tríbulo H et al (2012) Superovulation of beef cattle with a split-single intramuscular administration of Folltropin-V in two concentrations of hyaluronan. Theriogenology 77:1679–1685

Vieira LM, Rodrigues CA, Castro Netto A et al (2014) Superstimulation prior to the ovum pick-up to improve in vitro embryo production in lactating and non-lactating Holstein cows. Theriogenology 82:318–324

Vieira LM, Rodrigues CA, Castro Netto A et al (2015) Efficacy of a single intramuscular injection of porcine FSH in hyaluronan prior to ovum pick-up in Holstein cattle. Theriogenology 84:1–10

Wallace LD, Breiner CA, Breiner RA et al (2011) Administration of human chorionic gonadotropin at embryo transfer induced ovulation of a first wave dominant follicle, and increased progesterone and transfer pregnancy rates. Theriogenology 75:1506–1515

Walsh JH, Mantovani R, Duby RT et al (1993) The effects of once or twice daily injections of p-FSH on superovulatory response in heifers. Theriogenology 40:313–321

Wehrman ME, Fike KE, Melvin EJ et al (1997) Development of a persistent ovarian follicle and associated elevated concentrations of 17β-estradiol preceding ovulation does not alter the pregnancy rate after embryo transfer in cattle. Theriogenology 47:1413–1421

Wilson JW, Jones AL, Moore K et al (1993) Superovulation of cattle with a recombinant-DNA bovine follicle stimulating hormone. Anim Reprod Sci 33:71–82

Wiltbank MC (1997) How information of hormonal regulation of the ovary has improved understanding of timed breeding programs. In: Proceedings of the annual meeting society for theriogenology, Montreal, QC, pp 83–97

Wock J, Lyle L, Hockett M (2008) Effect of gonadotropin-releasing hormone compared with estradiol-17β at the beginning of a superstimulation protocol on superovulatory response and embryo quality. Reprod Fertil Dev 20:228 (Abstract)

Wrathall AE, Simmons HA, Bowles DJ et al (2004) Biosecurity strategies for conserving valuable livestock genetic resources. Reprod Fertil Dev 16:103–112

Wright JM (1981) Non-surgical embryo transfer in cattle embryo-recipient interactions. Theriogenology 15:43–56

ET-Technologies in Small Ruminants

6

Sergio Ledda and Antonio Gonzalez-Bulnes

Abstract

In the last decades small ruminants have become increasingly important, and nowadays sheep and goat are continuously increasing in the number of breeds and their geographic distribution. An important feature of small ruminants is that they can live and produce on land that is unfavorable for other forms of agriculture. The increase in small ruminant breeding has been supported more recently by the development and improvement of assisted reproductive technologies (ARTs). However, while some ARTs have reached widespread application, including estrus induction, estrus synchronization, and artificial insemination, other ARTs, such as superovulation and embryo transfer, in vitro embryo production, and embryo cryopreservation, are only rarely used. Multiple ovulation and embryo transfer (MOET) programs in small ruminants are usually restricted to few countries and still remain experimental. The success of this technique is unpredictable due to many limiting factors that contribute to the overall results, such as the reproductive seasonality with a long, naturally occurring anestrus period, high variability of the superovulatory response, fertilization failures, and the need of surgery for collection and transfer of gametes and embryos. However recent progress in better understanding of the follicular wave patterns, the elucidation of follicular dominance, and the integration of this information into superovulation treatments are instrumental in predicting good responders and reducing variability. Protocols that control follicular dominance have been developed to allow the initiation of precise hyperstimulation protocols which are designed to recruit and stimulate a homogeneous pool of small follicles that are gonadotrophin respon-

S. Ledda (✉)
Dipartimento di Medicina Veterinaria, Sezione Ostetricia e Ginecologia, Sassari, Italy
e-mail: giodi@uniss.it

A. Gonzalez-Bulnes (✉)
Comparative Physiology Group-RA SGIT-INIA, Madrid, Spain
e-mail: bulnes@inia.es

© Springer International Publishing AG, part of Springer Nature 2018   135
H. Niemann, C. Wrenzycki (eds.), *Animal Biotechnology 1*,
https://doi.org/10.1007/978-3-319-92327-7_6

sive, thereby enhancing superovulatory response and embryo yields. Significant improvements in the development of nonsurgical techniques are paving the way to reducing stress and costs of donors and recipient management, indicating the possible repeated use of individual donors. In addition, the progress with IVP embryos generated from adult and juvenile animals, combined with the genomic analysis of economically productive tracts, is opening new perspectives and could be instrumental for improving MOET programs in small ruminants.

6.1 Introduction

In the last decades small ruminants have become increasingly important, and nowadays sheep and goat breeding plays a crucial, economic and social role, as shown by the continuous increase in the number of breeds and their geographic distribution. An important feature of small ruminants is that they can live and produce on land that is unfavorable for other forms of agriculture.

According to FAO (faostat.fao.org, 2013), the number of sheep and goats in the world was 1169 and 996 million, respectively. Sheep and goats are shown to have a global distribution with emphasis in Africa and America with 848 and 929 million, respectively, whereas in Asia, Europe, and Oceania, 173 million sheep and 67 million goats were held. The global economic value of sheep and goat milk was 5.6 and 6.4 billion USD and for meat it was 37 and 25 billion, respectively. International sheep meat trade is limited (around 7% of the total production), and the bulk of this trade consists of export from the southern hemisphere (New Zealand has 47% and Australia has 36% of the total) to the European Union, North Asia, the Middle East, and North America. In many parts of the world, particularly in temperate regions, meat from sheep and goats is the most consumed product, and its importance as source of high-quality protein is steadily increasing.

The increase in small ruminant breeding has been supported in the last decades by the development and improvement of assisted reproductive technologies (Armstrong and Evans 1983; Loi et al. 1998). The control of reproduction and its modulation is an efficient tool for achieving genetic progress in these productive species.

While some assisted reproductive technologies (ARTs) have reached widespread application, including estrus induction, estrus synchronization, and artificial insemination, other ARTs, such as superovulation and embryo transfer, in vitro embryo production, and embryo cryopreservation, are only rarely used compared to cattle. Multiple ovulation and embryo transfer (MOET) programs in small ruminants are usually restricted to few countries and still remain experimental, even if this technique is considered an efficient and provides a low-cost option to exporting genetic material across international boundaries. However, the success of this technique is rather unpredictable due to many factors that contribute to the overall results, and many practitioners consider MOET one of the most frustrating ART in small ruminants.

The main limitation to field application in small ruminants is the reproductive seasonality with a long, naturally occurring anestrus period, high variability of the superovulatory response, fertilization failures, and the need of surgery for collection and transfer of gametes and embryos (reviewed by Cognié 1999; Cognié et al. 2003). This unpredictability combined with high costs of the pharmacological stimulation treatments have prevented large-scale use of MOET in sheep and goats, and up to now this technique is considered as being not enough robust to be applicable in large-scale breeding systems.

New prospects offered by in vitro embryo production (IVP) and repeated ovum pick-up from live adult and juvenile female donors are suggesting that IVP technology can be used as an alternative system to MOET programs, thus moving this technology from the research status in the laboratory to the field (Cognié et al. 2004; Paramio and Izquierdo 2014). Recent improvements of embryo production and freezing technologies could allow a wider propagation of valuable genetics in small ruminant populations and could also be used for establishing flocks without risk of disease transmission. In addition, they can make a substantial contribution to the preservation of endangered species or breeds.

The aim of this review is to provide an overview of some recent developments in MOET programs in small ruminants, updating recent information regarding estrus synchronization methods, follicular wave synchronization, and/or ovulation induction techniques during superovulatory treatments in ewes, as well as embryo collection and transfer techniques. The possibility offered by the generation of in vitro-produced embryos obtained from selected adult and juvenile donors will be also discussed with regard to the possibility offered by these new techniques to accelerate genetic progression of highly selected valuable animals.

6.2   Management of Reproductive Activity and Control of the Ovarian Cycle in Donor and Recipient Females

Sheep and goats are characterized by seasonal cycles of reproduction, consisting of a breeding season (which usually begins in late summer or early autumn in response to decreasing day length and ends in the late winter or early spring in response to increasing day length) and an anovulatory period (which covers the late spring to midsummer), which are separated by transition periods.

The breeding season is composed of a succession of sexual cycles (named estrous cycles since they are characterized by sexual receptivity (named estrus from the Latin word *estruus*), in the period preceding ovulation. The estrous cycles in small ruminants average 17 days in sheep and 21 days in goats and include the follicular phase and the luteal phase. The objective of the ovarian cycle is the development of a follicle able to ovulate and release an oocyte competent to be fertilized and able to develop in a viable embryo and, afterward, the maintenance of a corpus luteum competent for maintaining pregnancy.

Hence, the adequate management of reproductive activity and the ovarian cycle is indispensable in both donor and recipient females involved in MOET programs. The main objective is to render the ovarian follicular population responsive to the gonado-trophin treatments in a healthy and large number and, as in the case of in vivo embryo production, to control the timing of ovulation in donor females. The precise control of the timing of ovulation and the availability of corpora lutea competent for maintaining pregnancy are the main objectives in recipient females. Several methods have been proposed to regulate seasonality and control ovarian activity in small ruminants.

6.2.1 Administration of Progesterone and Analogues (Progestagens)

The most widely used methods for synchronization of estrous cycle and ovulation are based on the administration of progesterone or its analogues (progestagens; the most common being fluorogestone acetate and medroxyprogesterone acetate). These treatments simulate the action of natural progesterone produced by the corpus luteum during the luteal phase of the cycle and allow control of LH secretion from the pitu-itary gland and thus prevent occurrence of ovulation. Removal of the substances leads to the appearance of a follicular phase with the growth of a preovulatory follicle and the occurrence of estrus and ovulation. The first successful protocol was devel-oped in the early 1960s for sheep and consisted of intravaginally inserted sponges impregnated with progestagens (Robinson et al. 1967). The treatment was found to be equally effective for inducing ovulation in both the breeding season and the anovulatory period, with a high degree of synchronization in females treated at the same time. Thereafter, the method was found to be useful also for estrus synchroni-zation in goats (Ritar et al. 1984). An alternative to intravaginal sponges is the con-trolled internal drug release (CIDR) dispenser, which is made with an inert silicone elastomer usually impregnated with natural progesterone (Welch et al. 1984).

The use of either progesterone, fluorogestone acetate, or medroxyprogesterone acetate seems not to affect superovulatory yields (Bartlewski et al. 2015); con-versely, the protocol of administration seems to have a determinant effect (Gonzalez-Bulnes et al. 2004b).

Protocols for the administration of progesterone and progestagens aim to exceed the life span of the corpus luteum in the ovary and last for 14–16 days in sheep and goats, respectively. However, plasma concentrations rise during the first 48 h after insertion (Robinson et al. 1967), and, at the end of the treatment, the levels may even be too low for suppressing LH secretion effectively (Kojima et al. 1992) which in turn may lead to inadequate follicular growth with the appearance of persistent large estrogenic follicles (Johnson et al. 1996; Leyva et al. 1998; Viñoles et al. 1999). In superovulatory treatments, the appearance of persistent large follicles has a dramatic negative effects on oocyte and embryo yields (Gonzalez-Bulnes et al. 2004b), and low plasma levels of progesterone/progestagens during the superovula-tory treatment could be avoided by the use of two CIDRs/sponges from early onward (Thompson et al. 1990; Dingwall et al. 1994).

However, the use of long-term treatments and high doses has been associated with alterations in final follicle growth (Gonzalez-Bulnes et al. 2005), in patterns of the LH release (Scaramuzzi et al. 1988; Gordon 1975; Menchaca and Rubianes 2004), in the quality of ovulations (Killian et al. 1985; Gonzalez-Bulnes et al. 2005; Viñoles et al. 2001), and/or in sperm transport and survival in the female reproductive tract (Hawk and Conley 1971). An alternative would be the use of short-term treatments (6-day length), which would avoid the abovementioned shortcomings caused by the use of long-term and high-dose treatments (Ungerfeld and Rubianes 1999; Menchaca and Rubianes 2004; Letelier et al. 2009). However, a 6-day treatment period is shorter than the half-life of a possible corpus luteum in the ovaries; thus it is necessary to apply a single dose of prostaglandin $F_{2\alpha}$ ($PGF_{2\alpha}$) or its analogues for inducing regression of the corpus luteum.

6.2.2 Administration of Prostaglandins and Analogues (Prostanoids)

The objective of the administration of $PGF_{2\alpha}$ or its analogue (prostanoids) is to eliminate the corpus luteum and, in consequence, to induce growth of a follicular phase with ovulation (Abecia et al. 2012). Treatments with prostaglandin are therefore only effective in cycling animals with a functional corpus luteum. In association with the short-term treatment with progesterone/progestagens, the goal is to remove the corpus luteum, to allow the appearance of the follicular phase. Treatment with $PGF_{2\alpha}$ alone for estrus synchronization in a group of females requires two injections 9–10 days apart, thereby assuring that nearly all animals will be in mid-luteal phase at second $PGF_{2\alpha}$ dose and thus will respond with estrus behavior and ovulation. This treatment is effective in synchronizing estrus, but its practical application has been limited by reduction in fertility when compared to progestagen sponges (Killian et al. 1985; Scaramuzzi et al. 1988). However, most of the animals treated at 9–10 days intervals are in the midluteal phase of the estrous cycle, which coincides with a follicular wave with reduced fertility. Treatments during the early luteal phase (achieved by two doses of $PGF_{2\alpha}$ 5–6 days apart) may be an adequate alternative for synchronizing estrus (Contreras-Solis et al. 2009a, b).

$PGF_{2\alpha}$-based treatments were implemented in MOET protocols by Mayorga et al. (2011) who showed that it is possible to produce high enough numbers of transferable embryos during natural estrus induced by $PGF_{2\alpha}$ without the use of progestagen sponges.

6.2.3 Use of Melatonin

The discovery of the melatonin function in photoperiod-dependent breeding animals opened up new ways to control reproduction in these species, by inducing changes in the function of the photoperiod and the annual pattern of reproduction. Administration of melatonin could simulate the females during the reproductive season, but the effectivity of the synchronization obtained by this treatment and efficacy in increasing the superovulation response is still under debate. In fact, it has been shown (McEvoy et al. 1998) that a melatonin treatment of embryo donor and recipient ewes during anestrus affects their endocrine status, but not the ovulation rate, embryo survival, or pregnancy. On the other hand, Zhang et al. (2013) reported that the number of corpora lutea in ewes with subcutaneous 40 or 80 mg melatonin implants was significantly higher than that in the control group ($p < 0.05$). Similarly, the number of recovered embryos from ewes having received subcutaneous 40 or 80 mg melatonin implants was higher than in the control group ($p < 0.05$). After transfer of embryos collected from 40 to 80 mg melatonin-treated donors, pregnancy and birth rates were significantly increased compared to control ewes.

Melatonin implants inserted 3 months prior to the superovulatory treatment in aged high-prolificacy Rasa Aragonesa ewes (Forcada et al. 2006) did not improve the superovulation rate but was associated with the recovery of embryos with a better viability compared to controls and an increase in the number of blastocysts. These blastocysts were also more viable after cryopreservation. Moreover, a melatonin treatment reduced the number of nonviable (degenerate and retarded) embryos.

6.3   Induction of Superovulation by Exogenous Gonadotrophin Treatments

The superovulatory gonadotrophin treatment aims to increase the number of follicles growing to the preovulatory stage and ultimately yields a higher ovulation rate. Administration of gonadotrophins is concurrent with the last days of a progestative treatment to avoid premature ovulations and to synchronize ovulations. The first gonadotrophin protocols consisted of a single high dose of equine chorionic gonadotrophin (eCG). However, such protocol was associated with high variability and a high inconsistency of the ovulatory response in successive treatments (Cognié 1999).

Superovulatory protocols have mainly been based on multiple doses of FSH, administered twice daily, due to the short half-life of the hormone. A superovulatory treatment induces the growth of a high number of follicles, but the supply of large amounts of exogenous gonadotrophins necessary to achieve a superovulatory response may be associated with detrimental effects. There is evidence showing that the response to superovulatory treatments is associated with alterations in follicular development, oocyte maturation, and/or ovulation failures similar to other ruminants (Rubianes et al. 1997). Thus the number of transferable embryos obtained after a superovulatory treatment can sometimes be disappointingly low (Cognié 1999). The main causes of the decrease in viability of embryos collected from superovulated ewes can be related to alterations in follicular-oocyte competence, changes in the periovulatory and preimplantation endocrine patterns, and decreased intrinsic developmental capacity of the embryos and/or negative effects from the uterine environment (Gonzalez-Bulnes et al. 2004b). Some of these alterations are common to all superovulatory treatments, but some factors like the source of the gonadotrophin preparation, its purity, and the way of administration affect the final outcome.

Source and purity of the gonadotrophin preparations have been identified as main factors affecting the ovulatory response mainly due to the variable LH contents

(Lindsell et al. 1986). Some researchers have employed a recombinant follicle-stimulating hormone agonist (Rutigliano et al. 2014) for avoiding the presence of variable contamination with LH. High LH contents stimulate a higher number of follicles to grow, but such follicles regress during the treatment or are unable to ovulate (Rubianes et al. 1995; González-Bulnes et al. 2000a), which are possibly related to saturation of the LH receptors in theca and/or granulosa cells as described in cattle (Boland et al. 1991). These observations have favored the use of highly purified gonadotrophins, but one has to keep in mind that very low amounts of LH at the end of treatment may also induce lower ovulation rates and a higher incidence of fertilization failures (Picton et al. 1990; Cognié 1999). These aberrations may be reduced by inducing ovulation via appropriate drugs (i.e., GnRH; Menchaca et al. 2010).

The protocol of administration of gonadotrophins is also critical for the ovulatory response. Some protocols use constant dosages instead of decreasing dosage regimens (step-down). However, mean ovulation rate and mean numbers of recovered and viable embryos are usually higher in the step-down approach, which is closer to the physiological situation in which FSH secretion decreases during non-stimulated follicular phases (Gonzalez-Bulnes et al. 2004b). However, administration of high doses of nonphysiological FSH may be associated with the above limitations; the use of lower doses of FSH, although yielding lower ovulatory rates, favors embryo viability and is compatible with the application of repetitive treatments (Bruno-Galarraga et al. 2014).

The complexity of treatments with several dose of FSH has favored research on the use of combined treatments such as single eCG/FSH shots for in vitro embryo production (Gibbons et al. 2007; Forcada et al. 2011). However, this protocol yielded a low number of transferable embryos when applied in vivo (Cueto et al. 2011).

6.4 Individual and Ovarian Factors Affecting Superovulatory Response

The number of transferable embryos obtained after a superovulatory treatment of donor females, in small ruminants like in other species, is characterized by high individual variability, which is actually a limiting factor in MOET programs. The number of transferable embryos is dependent on the follicular growth, the ovulation, and the viability of the embryos collected in response to the hormonal treatment for inducing a superovulatory response (Gonzalez-Bulnes et al. 2004b; Menchaca et al. 2010).

Intensive research activities developed during the past decade have identified several features that affect the ovarian status at the onset of the superovulatory treatment.

Briefly, the ovulation rate is positively related to the number of small gonadotrophin-responsive follicles (2–3 mm in size) at the first gonadotrophin dose, in both sheep (Brebion et al. 1990; González-Bulnes et al. 2000a) and goats (Gonzalez-Bulnes et al. 2003a). However, the total number of embryos and their viability are closer related to the category of follicles 3 mm in size in sheep and 4 mm in goats; a higher number of smaller follicles usually correlates with more degenerated embryos. This finding may indicate that these follicles can grow and ovulate in response to the gonadotrophin treatment, but are not sufficiently matured to develop into a viable embryo because the recruitment might have required more time to complete maturation prior to exposure to a preovulatory LH surge and ovulation. On the other hand, follicles larger than 3 mm in diameter might be in an adequate stage of development to support growth and release of a healthy oocyte. This hypothesis is supported when considering that such follicles are the main source of estradiol and inhibin A (Gonzalez-Bulnes et al. 2003b, 2004a), which are two well-known markers of the follicular status (Ireland and Roche 1983; Campbell et al. 1995).

The final number of transferable embryos is affected by the presence or absence of a large follicle (Gonzalez-Bulnes et al. 2002a, 2003a). This effect is thought to be related to the dominant effects, since the presence of a large follicle at the first gonadotrophin injection (or two in case of codominance effects; Veiga-Lopez et al. 2006a) determines both the number of corpora lutea and the total number of recovered embryo derived from small follicles, 2–3 mm in diameter (Veiga-Lopez et al. 2005). In the absence of a large follicle, ovulation rate and the number of total embryos are related to the number of follicles 3–5 mm in diameter, suggesting that dominant follicles impair the development of gonadotrophin-dependent follicles (4–5 mm in size). Such dominance effects are primarily systemic, but there are also local effects, exerted by direct action, which are independent from systemic pathways through FSH modulation, on neighboring follicles (Gonzalez-Bulnes and Veiga-Lopez 2008).

Moreover, there is evidence that the presence of large follicles modulates the timing of the preovulatory LH surge and ovulation ultimately inducing a shorter period for final maturation and ovulation of smaller follicles (Veiga-Lopez et al. 2006a, 2008a). Some of the subordinate follicles may even grow to preovulatory size, but ovulation is disturbed or impeded (Veiga-Lopez et al. 2006b). The persistency of these follicles beyond the ovulation period contributes to decreased embryo yields by affecting rates of fertilization and viability in the oocytes from other follicles.

The presence or absence of a functional corpus luteum at the time of superovulation induction has a significant effect on oocyte and embryo yields. In the breeding season, the presence of a corpus luteum (CL) at the beginning of the progestagen treatment and its persistency at the start of a subsequent gonadotrophin treatment affect the final number of transferable embryos, likely by interaction with the dominant follicles. Ewes bearing a CL at the first gonadotrophin injection have a lower rate of degenerated embryos and show fewer deleterious effects resulting from the presence of a dominant follicle (Gonzalez-Bulnes et al. 2002b, 2005).

6.5   Strategies for Selection and Preparation of Donor Females

The effects of the ovarian status on the superovulatory response after the gonadotrophin treatment suggest (for ethical, technical, and economic reasons) the possibility of selecting females in adequate conditions prior to treatment and/or attempting to

defining adequate ovarian status. Moreover, the high individual variability in the response to superovulatory treatments is also associated with a high intraindividual repeatability in response to successive superovulatory treatments (Bari et al. 2001; Ptak et al. 2003; Bruno-Galarraga et al. 2014), which suggest the possibility of applying predictive measures for preselection of ewes with high ovulatory responses.

A predictive evaluation of the superovulatory response may be attempted by evaluating the ovarian status directly by ovarian imaging (ultrasonography) or indirectly by hormonal analyses. The use of high-resolution ultrasonography (probes with a frequency of 7.5 or higher) is useful for determining the presence or absence of large follicles and corpora lutea and the number of gonadotrophin-responsive follicles and their growth during gonadotrophin treatment (Gonzalez-Bulnes et al. 2002c, 2004b). The use of Doppler ultrasonography for examination of follicle blood flow on the final day of the superovulatory treatment appears to be predictive with regard to number and percentage of unfertilized oocytes (Oliveira et al. 2014).

Hormonal assays are a major tool in the evaluation of follicular hormones, including estradiol, inhibin, and, more recently, anti-mullerian hormone (AMH). The high correlation observed between the growth pattern of follicles yielding viable oocytes and the plasma profile inhibin A, rather than with E2, favors inhibin A measurement for surveillance of ovarian functionality in stimulated cycles of sheep (Gonzalez-Bulnes et al. 2002a; Veiga-Lopez et al. 2008b) and goats (Gonzalez-Bulnes et al. 2004c). Measurement of AMH levels is predictive of the follicular pool (Lahoz et al. 2014; Torres-Rovira et al. 2014) and therefore of the ovarian response to FSH stimulation. High numbers of oocytes are collected from lambs with high level of AMH, and after in vitro fertilization and culture development to blastocysts, the number of oocytes is higher from these animals than those derived from lambs with low levels of AMH (McGrice et al. 2016). Moreover, the measurement of AMH in lambs is promising for discriminating high and low responders (Torres-Rovira et al. 2014).

Preselection of ewes with high ovulatory responses may also be performed via exogenous FSH ovarian reserve tests (EFORTs). EFORTs are based on the administration of a single-shot treatment and the evaluation of subsequent follicular development (Torres-Rovira et al. 2014). A single eCG dose has been reported to be a useful tool to discriminate populations of prolific carriers from populations of non-prolific carriers in adult ewes (Kelly et al. 1983) and in prepubertal ewe lambs (Davis and Johnstone 1985; Gootwine et al. 1989, 1993). The use of a single-shot FSH/eCG is practical and cost-efficient for choosing donors with putatively high ovarian responses for a cost-efficient eCG treatment (Bruno-Galarraga et al. 2015).

Another approach for optimizing superovulatory yields in a group of females is the preparation of adequate ovarian conditions as determined by the presence of corpora lutea, the absence of large follicles, and/or a high number of gonadotrophin-responsive follicles. The presence of corpora lutea may be induced by pre-synchronization of the cycle with two prostaglandin doses and starting the progestagen treatment in the early luteal phase. The follicular status may be modified via direct ablation of the follicle (Gonzalez-Bulnes et al. 2004b), or the use of

Day-0 protocol, in which the superovulatory treatments initiated soon after the previous ovulation (Rubianes and Menchaca 2003) and/or by using GnRH analogues.

Administration of the GnRH antagonist or agonist analogues has a double effect and eliminates a dominant follicle and in parallel increases the recruitment of gonadotrophin-responsive follicles. Treatment with a GnRH agonist suppresses secretion of LH pulses during treatment, after an initial short stimulatory "flare effect," and thereby blocks follicle development beyond 3 mm (McNeilly and Fraser 1987). On the other hand, GnRH antagonists produce an immediate effect, without a desensitization period, by competitive blockade of the GnRH receptors, causing a rapid decline of FSH and LH levels in serum, and the loss of follicles larger than

Table 6.1 Hormonal treatments management of reproductive activity and control of the ovarian cycle in donor and recipient females

Compound	Donor/ recipient	Aim	Effect	Administration
Progesterone or analogues (progestagens)	Both	Induction and synchronization of ovulations and estrous cycles	To simulate endogenous corpus luteum	CIDR or intravaginal sponges
Prostaglandins or analogues (prostanoids)	Both	Induction and synchronization of ovulations and estrous cycles	To cause lysis of endogenous corpus luteum	Intramuscular injection (single/ double dose)
Melatonin	Both	Induction of breeding season activity	To raise melatonin levels for mimicking reproductive season	Subcutaneous implants
Equine chorionic gonadotrophin (eCG)	Both	Induction of follicular growth and ovulation	To raise endogenous levels of FSH and LH	Intramuscular injection (single dose)
Follicle-stimulating hormone (FSH)	Donor	Induction of follicular growth	To raise endogenous levels of FSH	Intramuscular injections (multiple dose)
FSH/eCG	Donor	Induction of follicular growth	To raise endogenous levels of FSH	Intramuscular injection (single dose)
GnRH analogues (agonists or antagonists)	Donor	Suppression of dominant follicles and stimulation of follicular growth	To decrease endogenous levels of FSH and mainly LH	Intramuscular injections (multiple dose)
GnRH or GnRH agonist	Donor	Induction of ovulation	To increase endogenous levels of LH	Intramuscular injection (single dose)
LH	Donor	Induction of ovulation	To increase endogenous levels of LH	Intramuscular injection (single dose)

3 mm in ewes (Campbell et al. 1998). A GnRH antagonist pretreatment is a good option to increase efficiency of superovulatory protocols in sheep (Brebion et al. 1990; Cognié 1999; Cognié et al. 2003) and goats (Cognié et al. 2003; Gonzalez-Bulnes et al. 2004d), albeit daily doses of GnRH represent a time-consuming procedure. In contrast, injection of a single dose of 1.5 mg of GnRH antagonist in sheep suppresses the effects of follicular dominance, thus allowing a significant increase (usually more than twofold) in the mean number of gonadotrophin-responsive follicles 2–3 mm in size, which grow to preovulatory size in response to the administration of exogenous FSH (Lopez-Alonso et al. 2005a, b). However, negative consequences of a high number of smaller follicles (2–3 mm) on oocyte maturation and fertilization and degeneration rates must be taken into account (Cognié et al. 2003; Gonzalez-Bulnes et al. 2004c, Gonzalez-Añover et al. 2004; Berlinguer et al. 2006) (Table 6.1).

6.6 Embryo Recovery and Transfer: Surgical and Nonsurgical Methods

Currently, embryo collection and transfer in small ruminants can be performed by surgical (Loi et al. 1998; Lehloenya and Greyling 2009; Torres and Sevellec 1987; Bruno-Galarraga et al. 2014), laparoscopic (McKelvey et al. 1986; Flores-Foxworth et al. 1992), or transcervical methods (Nagashima et al. 1987; Pereira et al. 1998; Fonseca et al. 2013, 2014).

Laparotomy—Surgical techniques for embryo collection and transfer in small ruminants, with exposure of the reproductive tract, are now used on a global scale. Laparotomy allows exact counting of the number of corpora lutea and visual inspection of the reproductive organs. The percentage of embryos recovered with this technique ranges between 40% and 80% according to the flushing methodology used, volume and number of washing attempts of the uterine horns, and operator skills and experience. Since the 1930s, when the first experiments were performed (Warwick et al. 1934), it remained in use even though it implies more risk for the treated animals as it requires general anesthesia, which in turn involves the need for animal fasting, drug administration, and surgical intervention. A consequence of this approach is that the donor can develop adhesions that can involve the ovaries, oviducts, and uterus and usually is associated with reduction in embryo recovery rates (Lehloenya and Greyling 2009). Hence, the number of collections per female is usually limited to two or three (Torres and Sevellec 1987). Other limitations are the relatively high costs of equipment and the stress of the animals due the manipulation of the exteriorized reproductive tract. Transfer of embryos by surgical laparotomy follows the same procedure used for collection: the embryos (generally one or two per recipient if in vivo-produced embryos are used, or higher numbers if in vitro-produced embryos are employed) are transferred in the exposed uterine tract with the aid of a small catheter (Tomcat catheter or similar) into the uterine horn, 2–3 cm from the uterine-oviductal junction.

Laparoscopy—Laparoscopy is an alternative technique to recover and transfer embryos from goats and sheep. It leads to fewer adhesions, and, therefore, a donor could be repeatedly used for collections for up to seven times (Flores-Foxworth et al. 1992). However, this method still requires special equipment and skilled personnel. Regardless of the good efficiency (Schiewe et al. 1984), this technique has not been extensively adopted worldwide (Schiewe et al. 1984). Limiting factors are the required refined ability of operators associated with the relatively expensive equipment to perform embryo recoveries. Both laparoscopic and laparotomy techniques are associated with prolonged fasting of donors that are usually maintained under general anesthesia. Transfer of embryos can be also performed by endoscopical examination of the uterine horn (to the horn ipsilateral to the corpus luteum). Embryos are placed in an insemination straw (0.25 ml) and are inserted in a modified insemination gun equipped in the terminal part with an 18 needle that, after penetrating the uterine wall, can facilitate the release of the embryos inside the uterine lumen.

Transcervical procedures—Recovery of embryos from sheep and goats by nonsurgical procedures (NSER) has been developed in the 1980s (Lin et al. 1979) and recently has received renewed interest. The technique is less invasive and needs a simpler anesthetic protocol (epidural block and local cervical anesthesia) than laparotomy and laparoscopy. Moreover, animals may remain in a standing position under sedation. Nonsurgical collection and transfer have been reported first in goats, in which, due to the anatomical configuration of the reproductive tract, it is easier to pass the cervical plicas compared to sheep. When using the nonsurgical technique, the cervix is clipped with nontraumatic forceps that allow traction, and a catheter is inserted through the cervix to reach the desired uterine horn (Fonseca et al. 2014). A different type of catheter can be used (usually with one or two ways) to flush the reproductive tract with different volumes of collection medium. Recovery rates range from 60% to 100%, depending on the animal and the operator skills, and are not very different from the rates obtained by laparotomy or laparoscopy. Future studies have to reveal if the technique can be used in different breeds of small ruminants and, in particular, in animals with reduced size and weight.

The absence of adhesions when using NSER is a major advantage and allows successive collections in contrast to laparotomy or laparoscopy. On the other hand, the difficulty of introducing a catheter through the cervix, mainly in sheep, and the missing option of rectal manipulation of the tract are main obstacles of NSER procedures. Transfer of embryos is performed in a similar way and a device containing the embryo is coupled to a catheter to release the embryo into the uterus. A comparison revealed that nonsurgical embryo transfer usually results in a recovery rate similar to that of surgical techniques (Fonseca et al. 2014; Zambrini et al. 2014, 2015) with pregnancy and birth rates of around 50% (Fonseca et al. 2014). Studies with higher numbers of animals could show if the technique can be used successfully in different breeds, in particular, in animals with reduced size and weight (Table 6.2).

Table 6.2 Methods of embryo recovery in sheep and goats and their efficiency

Method	Required anesthesia	Repeatability	Amount of flushed medium (ml)	Range of recovered embryos	Embryos collected/N. ovulations	Average embryos recovered
Laparotomy	Yes	1–4 times	40–60	0–30	40–100	5–12
Laparoscopy	Yes	1–8 times	40–60	0–15	0–85	3–8
Transcervical	No	1–>15 times	40–1200	0–18	0–90	4–10

6.7   Factors That Can Be Relevant for Pregnancy Success

The full realization of the potential of embryo transfer procedures in small ruminants depends on optimizing the number of progeny born from females with high genetic merits. Pregnancy rates after ET in sheep and goats vary from 29% to 75% and are affected by synchronization protocol and superovulatory response but are also directly related to the viability of the transferred embryos.

Several factors play a critical role in determining embryonic and fetal losses in the ewe. Bolet et al. (1986) suggested that these losses could be caused by at least three components: (a) paternal influences, related to the quality of semen; (b) the female, due to the quality of the ova and uterine environment; or (c) the embryo itself. There are conflicting reports on embryo survival in sheep ranging from unaffected survival rates (Armstrong and Evans 1983) to increased (Cseh and Seregi 1993) or decreased survival rates (Mutiga 1991).

Several other factors need to be included: the maternal effect of the recipient, the number and the developmental stage of the embryos that were transferred, and the method of embryo production (e.g., in vivo versus in vitro). Eventually, embryo storage and its manipulation could affect the success of as well the age of donors and culture conditions for in vitro-produced embryos (Thompson et al. 1995; Holm et al. 1996; Ptak et al. 1999; Dattena et al. 2000; Naitana et al. 1996).

6.7.1   Maternal Effect

Factors related to both embryos and recipients have been suggested to affect survival of the transferred embryos in sheep and goats, including the stage of embryo development, embryo quality, the number of corpora lutea, and age and parity of the recipients (Donaldson 1985; Alabart et al., 1995; Thompson et al. 1995; Armstrong and Evans 1983).

The term "maternal effect" indicates an influence of the dam on its offspring, either from genetic or environmental causes. Embryo transfer technology enables experimental investigation of embryo-maternal communication providing unique opportunities to study the genetic control of embryonic survival and growth.

Among the possible factors, progesterone levels have been found to play a vital role in early embryo development, implantation, and establishment of pregnancy. The plasma progesterone concentration in recipient animals is related to the number of ovulations or corpora lutea in sheep (Ashworth et al. 1989). While in goats embryo survival usually is higher with increased numbers of corpora lutea (Armstrong and Evans 1983) and plasma progesterone concentrations, little information is available for sheep. We have observed that pregnancy rates after embryo transfer were higher in ewes with more than one CL compared to animals with only one CL (data not published—personal observation). On the other hand, it is known that at least one embryo must be present in the uterine lumen by day 12.5 post-estrus to prevent luteolysis (Moore 1985). The bidirectional communication between endometrium and embryos is critical to determine the role of the uterine environment. In a large-scale study with records from 11,369 animals, the effects of age, weight, and sire on embryo and fetal survival in sheep were investigated (Shorten et al. 2013). The author concluded that, from a genetic point of view, the dam's ability to maintain a pregnancy is significantly higher than the effects of embryo competence. Therefore, a selection of dams based on their maternal performance could provide effective means to improve embryonic survival.

Moreover, Cumming et al. (1975) reported that embryonic survival from breeding to days 26–30 was greater in crossbred than in Merino twin-ovulating ewes, but did not differ between breeds of single-ovulating ewes. Naqvi et al. (2006, 2007) investigated developmental competence, birth, and survival of Garole (small-sized) lambs after transfer of two or three embryos into large-sized non-prolific recipient ewes. They found that embryos derived from prolific sheep developed to term at a higher proportion when transferred into the uterine environment of higher-body-sized non-prolific sheep, which provided more space for embryo development than small-sized Garole ewes. They also observed that the monotocous character of recipients was not a limiting factor for pregnancy success when two or three embryos had been transferred.

6.7.2   Number of Embryos Transferred

There is an economic incentive on transferring multiple embryos to reduce the number of recipient ewes or does. However, Anderson et al. (1979) reported that uterine crowding can cause an increased frequency of pregnancy losses in nulliparous recipients receiving more than one embryo.

In sheep, embryonic and fetal mortality leads to large economic losses. Embryonic and fetal losses are estimated to 30% (Bolet et al. 1986). Most embryonic losses have been reported to occur before day 18 (Hulet et al. 1956; Moore et al. 1960; Quinlivan 1966). Complete losses from day 18 to lambing were estimated to 9.4% (Hulet et al. 1956), and fetal losses from day 30 to term were only 1–5% (Quinlivan 1966). More recently O'Connell et al. (2016) reported that embryo loss mainly occurred prior to day 14 of gestation with 6% losses before day 4 and 12% loss between days 4 and 14 of gestation. It has been reported that embryonic losses increased with an increasing ovulation rate (Kleemann and Walker 2005).

Naqvi et al. (2007) found that the incidence of embryonic mortality up to day 40 of gestation was reduced when the number of transferred embryos had been increased. Embryo survival up to 40 days of gestation and up to term was 38.1% when three embryos had been transferred per ewe which was higher than after transfer of two embryos per ewe (28.6%). In the same study, all embryos were transferred to the ipsilateral uterine horn. It has also been reported that transfer of two embryos to the ipsilateral or both uterine horns does not influence survival of the embryos (Torres and Sevellec 1987), due to migration of embryos during early stages of development. A higher pregnancy rate of 55.2% has been reported in Hungarian Merino ewes following transfer of two embryos per recipient, compared to 45.6% in case of single-embryo transfers (Cseh and Seregi 1993). Mutiga (1991) reported that transfers of multiple embryos in tropical sheep increased the number of lambs born per pregnant ewes. Pregnancy rate was significantly higher after transfer of embryos pairs (64%) than single (39%) embryos in in vitro-produced embryos (Brown and Radziewic 1998). Contrary to this result, data from embryo transfer studies (Land and Wilmut 1977) have shown that doubling the number of embryos transferred resulted in a decrease of the number of lambs born.

Armstrong and Evans (1983) indicated that embryo survival can be observed in twins when embryos had been placed into the same oviduct, which suggests that synergism between embryos influences each other's survival upon transfer in the goat. A possible explanation for such cooperation includes enhanced luteotrophic or anti-luteolytic gender actions resulting in improved luteal maintenance in recipients or enhanced signaling to the endometrium involved in the process of implantation (placental attachment). Whatever the explanation, the finding has important implications by enabling the embryo-carrying capacity of the recipient pool of goats to be doubled.

In addition, transfer of two embryos into the ipsilateral uterine horn is likely to increase the amounts of interferon-t and other embryonic signaling molecules in the uterus needed to maintain pregnancy and prevent luteolysis.

6.7.3   Development Stage and Grade of Embryos

Morphological evaluation of the developmental stage takes into account age and quality of the embryo. Embryonic stages and quality are usually based on the descriptions published by the International Embryo Transfer Society (IETS) (Stringfellow and Seidel 1998).

In embryo transfer, programs in small ruminants and cattle, higher fertility rates were obtained when the transferred embryos were in a more advanced development stage (Alabart et al. 2003). These findings agree with previous work conducted by Moore and Shelton (1962) in which an increased embryonic survival was observed with an increased age of the transferred embryos.

Embryo age largely corresponds to the stage of development. Based on a large number of fresh in vivo-derived embryo transfers, it was shown that embryonic stages ranging from late morulae to expanded blastocysts result in comparable pregnancy rates, whereas after hatching lower pregnancy rates can be expected (Hasler 1998). It has also been reported that when the embryos were recovered at early stages, in vitro embryo culture until the blastocyst stage might provide advantages over traditional protocols by allowing transfer of embryos into a synchronized uterine environment. Moreover, during culture, there is the possibility to select only those embryos that have demonstrated the potential for continued development under embryonic genomic control (Johnson et al. 2007).

Transfer or recovery of embryos has been also performed by transferring at early stages (2–3.5 days post fertilization) with acceptable results (Alabart et al. 2003). Technically the method implies transfers into the oviduct when the embryos are prior to the 8-cell stage and transfers to the uterus with embryos beyond the 8-/16-cell stage. Practical advantages are usually not associated with embryo transfers at early stages, and viability is not changed compared to later embryonic stages (Ishwar and Menon 1996). In contrast, as IVP embryos are more stage sensitive than are in vivo-derived embryos, higher conception rates were achieved following transfer of expanded blastocysts compared to morulae or earlier stages (Lamb 2005; Naitana et al. 1996).

A correlation between embryo morphology and pregnancy rates has been discussed for many years (Steer et al. 1992). Within each embryonic stage, morphological quality is also closely associated with pregnancy rate, as reported in a number of studies (Donaldson 1985; Hasler 2001; Lindner and Wright 1983). Farin et al. (1995) showed that agreement among six experienced embryo evaluators was higher for in vivo-derived embryos compared to their in vitro-derived counterparts. In addition, there was a relatively high degree of agreement when evaluating excellent and degenerated (poor) embryos, but a lower agreement relative to good and fair embryo viability categories (Lindner and Wright 1983).

Although many comprehensive morphological descriptions have been published (Shea 1981; Lindner and Wright 1983), individual variation in embryo grades and quality ratings is still prevalent. It is not surprising that the individual embryologist is found to account for significant variation in the embryo's quality grading. Conversely, the embryologist has less influence on the developmental scores, suggesting that this trait is easier to describe. Quality evaluation is further hampered by loose or degenerate cells in the embryo that are often more difficult to see in the blastocyst than in the morula. No practical method to replace the visual morphological scoring method has been found so far (Betteridge and Rieger 1993). Evaluation of embryo quality is even more challenging when embryos have been produced in vitro. Timing of development is considered as predictive marker of embryo quality in these embryos. Reduced viability due to poorer quality grade resulted in slower rates of development (Walker et al. 1996; Leoni et al. 2007) and reduced pregnancy rates after transfer (Shea 1981; Lindner and Wright 1983; Hasler 1998) (Fig. 6.1).

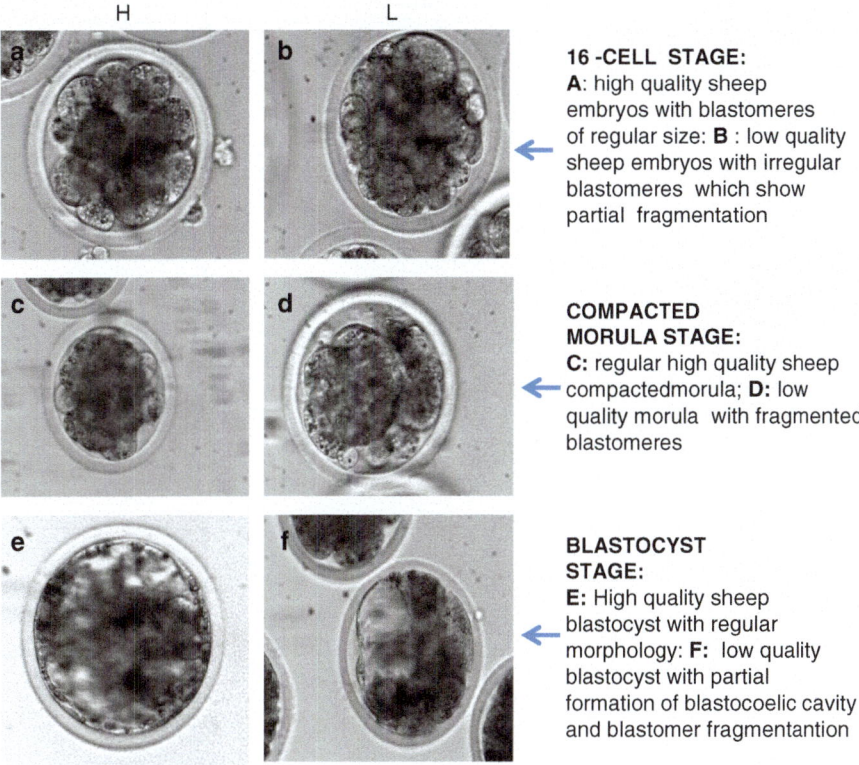

H   L

16 -CELL STAGE:
A: high quality sheep embryos with blastomeres of regular size: **B** : low quality sheep embryos with irregular blastomeres which show partial fragmentation

COMPACTED MORULA STAGE:
C: regular high quality sheep compactedmorula; **D:** low quality morula with fragmented blastomeres

BLASTOCYST STAGE:
E: High quality sheep blastocyst with regular morphology: **F:** low quality blastocyst with partial formation of blastocoelic cavity and blastomer fragmentantion

Fig. 6.1 Morphological evaluation of high-quality (H) and low-quality (L) embryos recovered from superovulated ewes. *Cell stage*: (**a**) High quality sheep embryos with blastomeres of regular size: (**b**) low quality sheep embryos with irregular blastomeres which show partial fragmentation. Compacted morula stage: (**c**) regular high quality sheep compacted morula; (**d**) low quality morula with fragmented blastomeres. Blastocyst stage: (**e**) High quality sheep blastocyst with regular morphology: (**f**) low quality blastocyst with partial formation of blastocoelic cavity and blastomer fragmentantion

6.7.4 Donor Effect on Embryo Quality

A large proportion of the variability in embryo development and quality has been attributed to the donor animal. The background for this variation cannot be fully explained, as indicated by the relatively low repeatability for both embryo stage and quality grade in the bovine (Callesen et al. 1995). Probably, donor hormone levels during the preovulatory period may affect fertilization and early embryonic development. In cattle the causes of the variation between donors (i.e., donor breed and parity, insemination bull, year and season) were insufficient and could only partially explain the variability (Callesen et al. 1995). With regard to gonadotrophin regimes, embryo quality seems closely related to the type and dosage of the stimulating

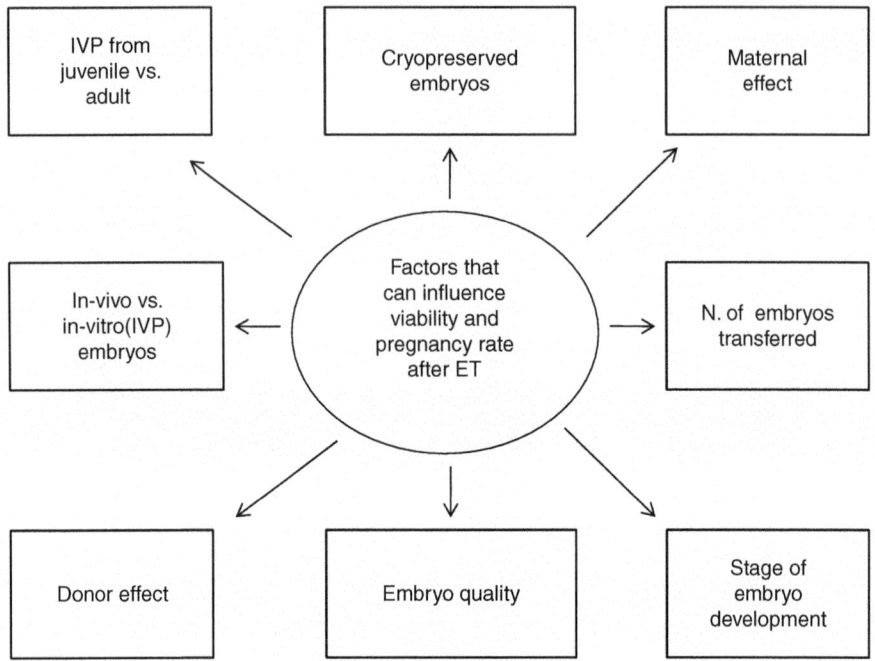

Fig. 6.2 Factors that can be relevant for pregnancy success

hormones during the superovulatory treatment. It has been shown that different FSH regimes can affect the developmental capacity and cryotolerance of ovine embryos derived from oocytes collected by ovum pick-up of donor sheep (Berlinguer et al. 2004). In particular the dosage regimes of FSH influence the developmental capacity of recovered oocytes to develop to blastocysts in vitro and their cryotolerance after vitrification procedures. Embryo quality is also significantly affected by the stimulating hormones during the superovulatory treatment. Often a high superovulatory response is followed by reduced fertilization rates and reduced embryo quality. A superovulatory treatment of ewes with eCG/FSH increased the ovarian responses compared with FSH alone, but the embryos showed reduced viability rates after vitrification (Leoni et al. 2001) (Fig. 6.2).

6.7.5 Source of Embryos: In Vivo vs. In Vitro Embryos

One of the solutions to overcome the relatively low efficiency of MOET programs is to produce and transfer in vitro-produced embryos (IVEP). The IVEP procedure does not require superovulation because oocytes are recovered directly from the follicle in hormonally or unstimulated females. IVEP can also be used in non-fertile females and pregnant, lactating, and even slaughtered females. Moreover, embryos can be produced in vitro from oocytes of prepubertal females with a technology called "juvenile in vitro embryo transfer" (JIVET) that is compatible with reduced generation intervals and concomitantly increased genetic gain. Thus, in a JIVET scheme using oocytes obtained from 3- to 4-week-old females, it is possible to increase the rate of genetic gain by approximately 5% (reviewed by Morton 2008). In practical terms, animals of high genetic merit are selected, and their oocytes are collected from live individuals (adult or juvenile) by surgical procedures or from slaughtered animals. These techniques are compatible with the production of a high number of cheap embryos, but several limitations currently prevent a more widespread application.

The collection of cumulus-oocyte complex (COCs) from living small ruminants implies a laparotomy or laparoscopy. COCs collection via laparotomy would prevent the reuse of the same donors in repeated collections, while oocyte collection can be performed repeatedly by laparoscopy ovum pick-up (LOPU) in the same animals. Furthermore, several studies indicate that the viability of in vitro-produced embryos in small ruminants is lower compared to their in vivo-produced counterparts (Cognié et al. 2003, 2004). This low viability is observed irrespective of the embryonic stage, age of donors, and technique used to obtain the oocytes (from slaughterhouse or LOPU).

6.7.6   IVP from Adult Animals

IVP embryos are usually produced from oocytes collected from slaughtered animals or from in vivo by laparoscopy and oocyte ovum pick-up (LOPU). These oocytes can be in vitro matured, fertilized, and cultured up to blastocysts. Success rates are high in all these steps with rates of IVM and cleavage being around 90% and 75%, respectively, and blastocyst rates at 30–50%, to some extent dependent on age, genetic background, nutritional management, and culture conditions. Comparative studies in sheep and goats have shown that the two species can generate embryos with similar rates of development and viability (Cox and Alfaro 2007). Hormonally stimulated ewes and goats have been subjected nine to ten times to oocyte collection by laparoscopic-guided follicular puncture. The success rates after IVM, IVF, and IVC were similar to those obtained with oocytes derived from abattoir ovaries.

Similar results have been found by Cocero et al. (2011) who showed that the development of IVP embryos up to the blastocyst stage was not different between slaughterhouse and laparoscopic ovum pick-up-derived oocytes in sheep. In goats, oocytes derived from abattoir ovaries had different oocyte maturation kinetics and a higher percentage of development up to the blastocyst stage compared to oocytes isolated laparoscopic ovum pick-up. No differences were observed in the number of blastocysts per cleaved IVP embryos (Souza-Fabjan et al. 2014). Differences have been observed when oocytes were recovered by LOPU after different stimulation regimens that can influence the quality of the derived oocytes and subsequent development in vitro (Berlinguer et al. 2004). At present, the percentage of blastocysts that can be produced from adult sheep ranges between 30% and 60% with pregnancy rates of 30–60% which is lower than after transfer of in vivo-produced embryos (60–80%). The reduced viability of in vitro-produced embryos becomes also more evident if the number of embryos transferred to recipients that can develop to term is considered. In fact, ewes that receive two or more IVP embryos often develop a single pregnancy and yield one offspring and only rarely carry twins (Papadopoulos et al. 2002). Similar data have been reported by Dattena et al. (2000) which showed a lambing rate from in vitro-produced and freshly transferred embryos of 40% (20 lambs/50 blastocysts transferred), which was significantly lower when compared to the 81.2% of in vivo-derived blastocysts (32 transferred fresh, 26 lambs born).

This reduced viability leads to embryonic losses mainly at 20–25 days of gestation, while prior to this there is no a significant reduction in the development of transferred embryos (14- and 25-day-old embryos, personal observation). The elevated embryonic and fetal losses of IVP embryos in this period could be related to alterations in angiogenesis in IVP embryos compared to in vivo embryos (Reynolds et al. 2015). An aberrant placental angiogenesis is thought to interfere with embryonic and fetal development. An increase in fetal weight is observed in IVP embryos, and consequently the reduced trophic supply of the altered placenta can interfere with regular conceptus development.

6.7.7 IVP Embryos from Juvenile Donors

The use of juvenile donors in embryo transfer (ET) programs offers considerable potential for accelerated genetic gain in domestic livestock through reduced generation intervals. This possibility has been investigated in the last years and factors such as donor selection, oocyte collection methods, and hormone stimulation methods designed to produce maximum yields of viable oocytes for young age donors have been studied. Overall the rates of juvenile ovine IVP embryos are significantly lower compared to embryos derived from adult dams and far away from that of in vivo-recovered embryos. Due to the presence of a large population of antral follicles, high numbers of oocytes can be collected from each donor, with the possibility to generate an average of eight to ten pregnancies from 6–8-week-old lambs (Armstrong et al. 1997).

The production of viable IVP embryos has become possible in younger lambs (4 weeks old) subjected to different hormonal stimulations (Ledda et al. 1997; Ptak et al. 1999). Results indicate that the number of oocytes recovered increased in lambs stimulated by hormones, while blastocyst quality seemed to be equivalent in hormonally treated and non-stimulated animals. Overall, results confirmed the reduced viability of embryos derived from juvenile animals with increased fetal losses after embryo transfer. The reduced viability seems to be related to morphological and metabolic changes observed in oocytes derived from prepubertal animals compared to their adult counterparts (Ledda et al. 2001; Leoni et al. 2015). Similar results have been observed in goats. Comparing the IVP embryos from prepubertal and adult animals, the developmental competence was primarily related to the size of follicles and oocyte diameter. Thus, oocytes derived from the largest follicle had a diameter comparable to that of adult oocytes performed in a similar way when subjected to IVM, IVF, and IVC (Paramio and Izquierdo 2014; Romaguera et al. 2011). To increase the efficiency of IVP, embryos derived from prepubertal animals, research has been undertaken to optimize donor selection and hormonal stimulation methods to reduce the variability and increase the proportion of donors responding to hormonal stimulation and to increase oocyte developmental competence. Recent improvements to JIVET, resulting from a modified hormonal stimulation regime, have eliminated the failure of donors to respond to hormonal stimulation and increased both number and developmental competence of oocytes harvested from very young prepubertal lambs (Kelly et al. 2005). This increased efficiency has facilitated incorporation of other reproductive technologies such as sperm sexing with JIVET, resulting in the birth of lambs of predetermined sex from prepubertal lambs (Morton 2008).

To respond to the increasing interest in the generation of embryos from juvenile donors, several other strategies have been explored to improve the efficiency of the JIVET system. As the technique is based on the large number of developmentally competent oocytes collected per single animal, predictive markers of the potential follicular population have been investigated, to select the best responding animals for hormonal stimulation. The concentration of AMH in lambs during the first weeks after parturition has been found to be a good predictive marker of the antral follicle population (Torres-Rovira et al. 2014; Kelly et al. 2016) that correlates well with success to hormonal stimulation. Lambs that were 3 weeks of age with high level of AMH yielded the largest number of oocytes with highest development to blastocyst when cultured in vitro. The number of antral follicles in prepubertal ewes is affected by specific gestational environmental conditions. The proportion of blastocysts calculated, as a percentage of cleaved embryos from total cumulus-oocyte complexes collected, was higher ($p < 0.05$) in females born with a female co-twin compared with those born with a male co-twin. These results indicate an enhancing effect of the female co-twin on oocyte development. Taking this into consideration could allow to selecting lambs for a JIVET program based on litter size and sex of the co-twin. In prepubertal goats and sheep, the possibility to select oocytes with high developmental competence prior to maturation has been investigated by noninvasive systems. Immature oocytes from adult and prepubertal donors can be differentially stained by Brilliant Cresyl Blue (BCB) which indicates differences in glucose-6-phosphate dehydrogenase (G6PDH) activity. The G6PDH amount is higher in growing oocytes, while it is low in fully grown oocytes, which are the ones that most frequently yield viable offspring. Oocytes selected by the BCB stain produced more blastocysts in vitro. The blastocysts were also of better quality compared to the pool of unselected oocytes. However, due to different staining and technical protocols, the BCB approach needs to be validated before it can be used for evaluating developmental capacity of ovine and capacity oocytes (Opiela and Kątska-Książkiewicz 2013) (Fig. 6.3).

Fig. 6.3 Workflow of
in vitro embryo production
(IVP) form juvenile donors
and embryo transfer
(JIVET)

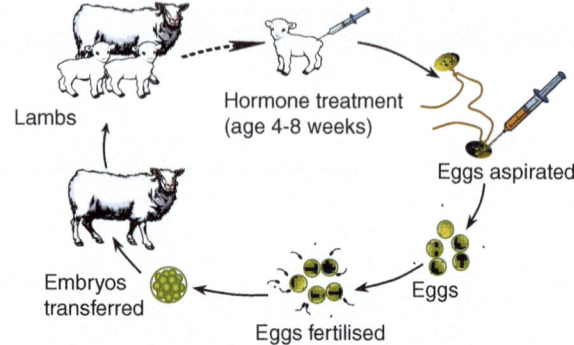

The figure was kindly provided by J. Kelly

6.8   Embryo Cryopreservation and Transfer

A successful MOET program usually includes the possibility for freezing the
embryos prior to transfer to synchronized recipients. Cryopreservation has become
an integral part of the commercial embryo transfer industry, but application in small
ruminants is based on relatively few studies (Boundy et al. 1985; Ishwar and Menon
1996), and the freezing process needs to be constantly improved and simplified
(McGinnis et al. 1993; Vajta 2000). From the practical viewpoint, embryo freezing
has many advantages: (1) freezing of embryos obtained from females with high
genetic value facilitates distribution of superior genetics from dams, which accelerates the rate of genetic improvement; (2) embryo cryopreservation facilitates international trade of valuable genetic stock which is a financially feasible and safe
alternative to live animal transport. Data on the success rates of embryo freezing
protocols in small ruminants are relatively scarce compared to cattle. The first lambs
from frozen/thawed embryos were born in 1976 (Willadsen et al. 1976), and lambs
from vitrified embryos were born in 1990 (Széll et al. 1990).

Slow-freezing protocols require a biological freezer and need more time to be
completed. The ultra-rapid technique, such as vitrification, is time and cost effective, since it does not require any special equipment, and is, therefore, well adapted
to routine field use (Baril et al. 2001). Sheep and goat embryos are able to survive
both slow-freezing and vitrification procedures (Martinez et al. 1998). Comparisons
between the different techniques are mainly based on lambing rates after embryo
transfer. However, selection of embryos for transfer is based on the stereomicroscopic evaluation of embryo morphology after thawing (Abe et al. 2002) in accordance to the guidelines of the International Embryo Transfer Society (Stringfellow
and Seidel 1998).

This selection step can be somewhat subjective as has been demonstrated by
ultrastructural studies of vitrified in vitro- and in vivo-produced bovine blastocysts
(Vajta 2000) and in controlled slow frozen in vivo-produced ovine morulae and

blastocysts (Cocero et al. 2002), which have shown that certain abnormalities remain undetected in the stereomicroscopic analysis.

Vitrification of embryos is most likely the technique that will be used in the future (Fahy and Rall 2007) and different devices and systems have been proposed, varying with regard to type and concentrations of the cryoprotectant. The 0.25 ml straw (Naitana et al. 1997) or the open-pulled straw (OPS) has been used for successful freezing of ovine morulae and blastocysts produced in vivo (Baril et al. 2001; Dattena et al. 2004; Martinez et al. 2006) or in vitro (Dattena et al. 2004).

Baril et al. (2001) reported a 50% embryo survival rate and a high pregnancy rate to term (72%) after direct transfer of vitrified ovine embryos which was similar to the results to the stepwise dilution method (72% and 60%, respectively). No differences were found between vitrified embryos transferred after in vitro removal of the cryoprotectant or directly after thawing in terms of lambing (67% vs. 75%, respectively) and embryo survival rates (lambs born/embryos transferred; 49% vs. 53%, respectively). However, the viability depends on the origin of embryos, and differences were found in the survival rates between embryos produced in vivo and embryos derived from IVP techniques. In fact, the viability is not significantly reduced after freezing of in vivo-produced embryos (70–90%), which it is significantly lower for cryopreserved IVP embryos (30–40%). Post-thaw viability is also reduced in IVP embryos generated from prepubertal oocytes.

6.9 Recipient Females

Transfer of the embryos to suitable recipients is the final step in a MOET program. The conditions of the recipient females (breed, age, nutrition and health, and reproductive status) were described as main limiting factors for the success of embryo transfer programs (Moore et al. 1959). Other variables include the aptitude of the recipient to maintain the pregnancy and the degree of synchronization between donors and recipients (Rowson and Moor 1966). Nowadays the selection of the most suitable recipients remains critical, and finding reliable criteria for ultimate recipient is a major focus of research. Selection of recipients is primarily based on direct observations either by laparoscopy or laparotomy of the corpora lutea (size, number, and vascularization of the structure) and the evaluation of uterine tone and morphology (Torres and Sevellec 1987). These approaches have the limitation of an invasive handling, which may interfere with pregnancy rates.

Alternatively the corpora lutea can be evaluated by transrectal ultrasonography, which allows to visualizing if the recipient has ovulated and evaluation of the quality of the luteal tissue, since morphological and echogenic characteristics of the corpus luteum are related to concentrations of progesterone in plasma and are reliable factors for determining luteal function in small ruminants (González-Bulnes et al. 2000b).

Conclusion

The current state of the art in multiple ovulation and embryo transfer (MOET) technology in small ruminants is steadily improving and could become one of the most applicable tools for the development of a future-oriented genetic program. New findings on the follicular wave patterns in small ruminants, the elucidation of follicular dominance, and the integration of this information into superovulation treatments are instrumental in predicting good responders and reducing variability. Protocols that control follicular dominance have been designed to allow the initiation of superstimulation precisely at the beginning of a follicular wave. These new approaches are based on the pretreatment with a gonadotrophin-releasing hormone (GnRH) antagonist prior to the FSH treatments to avoid follicular dominance and the administration of somatotropin or melatonin to improve oocyte quality and competence. These protocols will provide a rather homogeneous pool of small follicles that are gonadotrophin responsive, thereby enhancing the superovulatory response and embryo yields with a reduction of the incidence of unovulated follicles and early regression of corpora lutea. Significant improvements in the development of nonsurgical techniques are paving the way to reducing stress and costs of donors and recipient management, indicating the possible repeated use of individual donors. In addition, the progress with IVP embryos generated from adult and juvenile animals, combined with the genomic analysis of economically productive tracts, is opening new perspectives and could be instrumental for improving MOET programs in small ruminants.

References

Abe H, Matsuzaki S, Hoshi H (2002) Ultrastructural differences in bovine morulae classified as high and low qualities by morphological evaluation. Theriogenology 57:1273–1283

Abecia JA, Forcada F, González-Bulnes A (2012) Hormonal control of reproduction in small ruminants. Anim Reprod Sci 130(3–4):173–179

Alabart JL, Folch J, Fernández-Arias A, Ramón JP, Garbayo A, Cocero MJ (1995) Screening of some variables influencing the results of embryo transfer in the ewe. I. Five-day-old embryos. Theriogenology 44:1011–1026

Alabart JL, Folch J, Fernández-Arias A, Ramón JP, Garbayo A, Cocero MJ (2003) Screening of some variables influencing the results of embryo transfer in the ewe. Part II: Two-day-old embryos. Theriogenology 59(5–6):1345–1356

Anderson GB, Cupps PT, Drost M (1979) Induction of twins in cattle with bilateral and unilateral embryo transfer. J Anim Sci 49:1037–1042

Armstrong DT, Evans G (1983) Factors influencing success of embryo transfer in sheep and goats. Theriogenology 19:31–42

Armstrong DT, Kotaras PJ, Earl CR (1997) Advances in production of embryos in vitro from juvenile and prepubertal oocytes from the calf and lamb. Reprod Fertil Dev 9(3):333–339

Ashworth CJ, Sales DI, Wilmut I (1989) Evidence of an association between the survival of embryos and the periovulatory plasma progesterone concentration in the ewe. J Reprod Fertil 87:23–32

Bari F, Khalid M, Wolf B, Haresign W, Murray A, Merrell B (2001) The repeatability of superovulatory response and embryo recovery in sheep. Theriogenology 56:147–155

Baril G, Traldi AL, Cognié Y, Leboeuf B, Beckers JF, Mermillod P (2001) Successful direct transfer of vitrified sheep embryos. Theriogenology 56:299–305

Bartlewski PM, Seaton P, Szpila P, Oliveira ME, Murawski M, Schwarz T, Kridli RT, Zieba DA (2015) Comparison of the effects of pretreatment with Veramix sponge (medroxy-progesterone acetate) or CIDR (natural progesterone) in combination with an injection of estradiol-17β on ovarian activity, endocrine profiles, and embryo yields in cyclic ewes superovulated in the multiple-dose Folltropin-V (porcine FSH) regimen. Theriogenology 84(7):1225–1237

Berlinguer F, Leoni G, Bogliolo L, Pintus PP, Rosati I, Ledda S, Naitana S (2004) FSH different regimes affect the developmental capacity and cryotolerance of embryos derived from oocytes collected by ovum pick-up in donor sheep. Theriogenology 61(7–8):1477–1486

Berlinguer F, Gonzalez-Bulnes A, Succu S, Leoni GG, Veiga-Lopez A, Mossa F, Garcia-Garcia RM, Bebbere D, Galioto M, Cocero MJ, Naitana S (2006) GnRH antagonist enhance follicular growth in FSH-treated sheep but affect developmental competence of oocytes collected by ovum pick-up. Theriogenology 65(6):1099–1109

Betteridge KJ, Rieger D (1993) Embryo transfer and related techniques in domestic animals, and their implications for human medicine. Hum Reprod Update 8:147

Boland MP, Goulding D, Roche JF (1991) Alternative gonadotrophins for superovulation in cattle. Theriogenology 35:5–17

Bolet G (1986) Timing and Extent of Embryonic Mortality in Pigs Sheep and Goats: Genetic Variability. In: Sreenan JM, Diskin MG (eds) Embryonic Mortality in Farm Animals. Current Topics in Veterinary Medicine and Animal Science, vol 34. Springer, Dordrecht

Boundy T, Clarkson MJ, Winter AC (1985) Embryo transfer in sheep under practice conditions. Vet Rec 12:379–381

Brebion P, Belloc JP, Briois M. (1990) Elite Lacaune ewes pretreated with a GnRH antagonist yield more usable embryos following pFSH. In: Proceedings of the 6th meeting European Association for embryo transfer, p 12

Brown BW, Radziewic T (1998) Production of sheep embryos in-vitro and development of progeny following single and twin embryo transfers. Theriogenology 49(8):15–25

Bruno-Galarraga M, Cueto M, Gibbons A, Pereyra-Bonnet F, Catalano R, González-Bulnes A (2014) Repeatability of superovulatory response to successive FSH treatments in Merino sheep. Small Rumin Res 120:84–89

Bruno-Galarraga M, Cueto M, Gibbons A, Pereyra-Bonnet F, Subiabre M, González-Bulnes A (2015) Preselection of high and low ovulatory responders in sheep multiple ovulation and embryo transfer programs. Theriogenology 84:784–790

Callesen H, Lovendahl P, Bak A, Greve T (1995) Factors affecting the developmental stage of embryos recovered on day 7 from superovulated dairy cattle. J Anim Sci 73(6):1539–1543

Campbell BK, Scaramuzzi RJ, Webb R (1995) Control of follicle development and selection in sheep and cattle. J Reprod Fertil Suppl 49:335–350

Campbell BK, Dobson H, Scaramuzzi RJ (1998) Ovarian function in ewes made hypogonadal with GnRH antagonist and stimulated with FSH in the presence or absence of low amplitude LH pulses. J Endocrinol 156:213–222

Cocero MJ, Diaz de la Espina SM, Aguilar B (2002) Ultrastructural characteristics of fresh and frozen-thawed ovine embryos using two cryoprotectants. Biol Reprod 66:1244–1258

Cocero MJ, Alabart JL, Hammami S, Martí JI, Lahoz B, Sánchez P, Echegoyen E, Beckers JF, Folch J (2011) The efficiency of in vitro ovine embryo production using an undefined or a defined maturation medium is determined by the source of the oocyte. Reprod Domest Anim 46(3):463–470

Cognié Y (1999) State of the art in sheep-goat embryo transfer. Theriogenology 51(1):105–116

Cognié Y, Baril G, Poulin N, Mermillod P (2003) Current status of embryo technologies in sheep and goat. Theriogenology 59(1):171–188

Cognié Y, Poulin N, Locatelli Y, Mermillod P (2004) State-of-the-art production, conservation and transfer of in-vitro-produced embryos in small ruminants. Reprod Fertil Dev 16(4):437–445

Contreras-Solis I, Vasquez B, Diaz T, Letelier C, Lopez-Sebastian A, Gonzalez-Bulnes A (2009a) Efficiency of estrous synchronization in tropical sheep by combining short-interval cloprostenol-based protocols and "male effect". Theriogenology 71(6):1018–1025

Contreras-Solis I, Vasquez B, Diaz T, Letelier C, Lopez-Sebastian A, Gonzalez-Bulnes A (2009b) Ovarian and endocrine responses in tropical sheep treated with reduced doses of cloprostenol. Anim Reprod Sci 114(4):384–392

Cox JF, Alfaro V (2007) In vitro fertilization and development of OPU derived goat and sheep oocytes. Reprod Domest Anim 42(1):83–87

Cseh S, Seregi J (1993) Practical experiences with sheep embryo transfer. Theriogenology 39:207

Cueto MI, Gibbons AE, Pereyra-Bonnet F, Silvestre P, Gonzalez-Bulnes (2011) Effects of season and superovulatory treatment on embryo yields in fine-wool merinos maintained under field conditions. Reprod Domest Anim 46:770–775

Cumming IA, De Blockey MA, Winfield CG, Parr RA, Williams AH (1975) A study of the relationships of breed, time of mating, level of nutrition, live weight, body condition, and face color, to embryo survival in ewes. J Agric Sci 84:559–565

Dattena M, Ptak G, Loi P, Cappai P (2000) Survival and viability of vitrified in vitro and in vivo produced ovine blastocysts. Theriogenology 53(8):1511–1519

Dattena M, Accardo C, Pilichi S, Isachenko V, Mara L, Chessa B et al (2004) Comparison of different vitrification protocols on viability after transfer of ovine blastocysts in vitro produced and in vivo derived. Theriogenology 62:481–493

Davis GH, Johnstone PD (1985) Ovulation response to pregnant mares' serum gonadotrophin in prepubertal ewe lambs of different Booroola genotypes. Anim Reprod Sci 9:145–151

Dingwall WS, McKelvey WAC, Mylne MJA, Simm G (1994) A protocol for MOET in Suffolk ewes. In: 45 annual meeting of the European Association for animal production, pp S2–S4 (abstr)

Donaldson LE (1985) Matching of embryo stages and grades with recipient oestrous synchrony in bovine embryo transfer. Vet Rec 117:489–491

Fahy GM, Rall WF (2007) Vitrification: an overview. In: Liebermann J, Tucker MJ (eds) Vitrification in assisted reproduction: a user's manual and troubleshooting guide. Informa Healthcare, London

Farin PW, Britt JH, Shaw DW, Slenning BD (1995) Agreement among evaluators of bovine embryos produced in vivo or in vitro. Theriogenology 44:339–349

Flores-Foxworth G, McBride BM, Kraemer DC, Nuti LC (1992) A Comparison between laparoscopic and transcervical embryo collection and transfer in goats. Theriogenology 37:213 (abstr)

Fonseca JF, Zambrini FN, Alvim GP, Peixoto MGCD, Verneque RS, Viana JHM (2013) Embryo production and recovery in goats by non-surgical transcervical technique. Small Rumin Res 111:96–99

Fonseca JF, Esteves LV, Zambrini FN, Brandão FZ, Peixoto MGCD, Verneque S et al (2014) Viable offspring after successful non-surgical embryo transfer in goats. Arq Bras Med Vet Zootec 66:613–616

Forcada F, Abecia JA, Cebrián-Pérez JA, Muiño-Blanco T, Valares JA, Palacín I, Casao A (2006) The effect of melatonin implants during the seasonal anestrus on embryo production after superovulation in aged high-prolificacy Rasa Aragonesa ewes. Theriogenology 65(2):356–365

Forcada F, Ait Amer-Meziane M, Abecia JA, Maurel MC, Cebrián-Pérez JA, Muiño-Blanco T, Asenjo B, Vázquez MI, Casao A (2011) Repeated superovulation using a simplified FSH/eCG treatment for in vivo embryo production in sheep. Theriogenology 75(4):769–776

Gibbons A, Bonnet FP, Cueto MI, Catala M, Salamone DF, Gonzalez-Bulnes A (2007) Procedure for maximizing oocyte harvest for in vitro embryo production in small ruminants. Reprod Domest Anim 42:423–426

Gonzalez-Añover P, Encinas E, Garcia-Garcia RM, Veiga-Lopez A, Cocero MJ, McNeilly AS, Gonzalez-Bulnes A (2004) Ovarian response in sheep superovulated after pretreatment with growth hormone and GnRH antagonists is weakened by failures in oocyte maturation. Zygote 12:301–304

Gonzalez-Bulnes A, Veiga-Lopez A (2008) Evidence of intraovarian follicular dominance effects during controlled ovarian stimulation in a sheep model. Fertil Steril 89:1507–1513

González-Bulnes A, Santiago-Moreno J, Cocero MJ, López-Sebastián A (2000a) Effects of FSH commercial preparation and follicular status on follicular growth and superovulatory response in Spanish Merino ewes. Theriogenology 54:1055–1064

González-Bulnes A, Santiago-Moreno J, Gómez-Brunet A, López-Sebastián A (2000b) Relationship between ultrasonographic assessment of the corpus luteum and plasma progesterone concentration during the oestrous cycle in monovular ewes. Reprod Domest Anim 35:65–68

Gonzalez-Bulnes A, Santiago-Moreno J, Cocero MJ, Souza CJH, Groome NP, Garcia-Garcia RM et al (2002a) Measurement of inhibin A predicts the superovulatory response to exogenous FSH in sheep. Theriogenology 57:1263–1272

Gonzalez-Bulnes A, Garcia-Garcia RM, Santiago-Moreno J, Lopez-Sebastian A, Cocero MJ (2002b) Effects of follicular status on superovulatory response in ewes is influenced by presence of CL at first FSH dosage. Theriogenology 58:1607–1614

Gonzalez-Bulnes A, Garcia-Garcia RM, Souza CJH, Santiago-Moreno J, Lopez-Sebastian A, Cocero MJ, Baird DT (2002c) Patterns of follicular growth in superovulated sheep and influence on endocrine and ovarian response. Reprod Domest Anim 37:357–361

Gonzalez-Bulnes A, Carrizosa JA, Diaz-Delfa C, Garcia-Garcia RM, Urrutia B, Santiago-Moreno J, Cocero MJ, Lopez-Sebastian A (2003a) Effects of ovarian follicular status on superovulatory response of dairy goats to FSH treatment. Small Rumin Res 48:9–14

Gonzalez-Bulnes A, Garcia-Garcia RM, Castellanos V, Santiago-Moreno J, Ariznavarreta C, Dominguez V et al (2003b) Influence of maternal environment on the number of transferable embryos obtained in response to superovulatory FSH treatments in ewes. Reprod Nutr Dev 43:17–28

Gonzalez-Bulnes A, Souza CJH, Campbell BK, Baird DT (2004a) Systemic and intraovarian effects of dominant follicles on ovine follicular growth. Anim Reprod Sci 84:107–119

Gonzalez-Bulnes A, Baird DT, Campbell BK, Cocero MJ, García-García RM, Inskeep EK, López-Sebastián A, McNeilly AS, Santiago-Moreno J, Souza CJ, Veiga-López A (2004b) Multiple factors affecting the efficiency of multiple ovulation and embryo transfer in sheep and goats. Reprod Fertil Dev 16(4):421–435

Gonzalez-Bulnes A, Garcia-Garcia RM, Carrizosa JA, Urrutia B, Souza CJH, Cocero MJ, Lopez-Sebastian A, McNeilly AS (2004c) Plasma inhibin A determination at start superovulatory FSH treatments is predictive for embryo outcome in goats. Domest Anim Endocrinol 26:259–266

Gonzalez Bulnes A, Santiago Moreno J, Garcia-Garcia RM, Souza CJH, Lopez-Sebastian A, McNeilly AS (2004d) Effect of GnRH antagonists treatment on gonadotrophin secretion, follicular development and inhibin A secretion in goats. Theriogenology 61:977–985

Gonzalez-Bulnes A, Berlinguer F, Cocero MJ, Garcia-Garcia RM, Leoni G, Naitana S, Rosati I, Succu S, Veiga-Lopez A (2005) Induction of the presence of corpus luteum during superovulatory treatments enhances in vivo and in vitro blastocysts output in sheep. Theriogenology 64:1392–1403

Gootwine E, Bor A, Braw-Tal R (1989) Plasma FSH levels and ovarian response to PMSG in ewe lambs of related genotypes that differ in their prolificacy. Anim Reprod Sci 19:109–116

Gootwine E, Braw-Tal R, Shalhevet D, Bor A, Zenou A (1993) Reproductive performance of Assaf and Booroola-Assaf crossbred ewes and its association with plasma FSH levels and induced ovulation rate measured at prepuberty. Anim Reprod Sci 31:69–81

Gordon I (1975) Hormonal control of reproduction in sheep. Proc Br Soc Anim Prod 4:79–93

Hasler JF (1998) The current status of oocyte recovery, in vitro embryo production, and embryo transfer in domestic animals, with an emphasis on the bovine. J Anim Sci 76(Suppl 3):52–74

Hasler JF (2001) Factors affecting frozen and fresh embryo transfer pregnancy rates in cattle. Theriogenology 56:1401–1415

Hawk HW, Conley HH (1971) Sperm transport in ewes administered synthetic progestagen. J Anim Sci 33:255–256

Holm P, Walker SK, Seamark RF (1996) Embryo viability, duration of gestation and birth weight in sheep after transfer of in vitro matured and in vitro fertilized zygotes cultured in vitro or in vivo. J Reprod Fertil 107(2):175–181

Hulet CV, Voightlander HP, Pope AL, Casida LE (1956) The nature of early-season infertility in sheep. J Anim Sci 15:607–616

HYPERLINK (2013) http://www.faostat.fao.org

Ireland JJ, Roche JF (1983) Development of monovulatory antral follicles in heifers: changes in steroids in follicular fluid and receptors for gonadotrophins. Endocrinology 112:150–156

Ishwar AK, Menon MA (1996) Embryo transfer in sheep and goats: a review. Small Rumin Res 19(1):35–43

Johnson SK, Dailey RA, Inskeep EK, Lewis PE (1996) Effect of peripheral concentrations of progesterone on follicular growth and fertility in ewes. Domest Anim Endocrinol 13:69–79

Johnson N, Blake D, Farquhar C (2007) Blastocyst or cleavage-stage embryo transfer? Best Pract Res Clin Obstet Gynaecol 21(1):21–40

Kelly RW, Owens JL, Crosbie SF, McNatty KP, Hudson N (1983) Influence of Booroola Merino genotype on the responsiveness of ewes to pregnant mares serum gonadotropin, luteal tissue weights and peripheral progesterone concentrations. Anim Reprod Sci 6:199–207

Kelly JM, Kleemann DO, Walker SK (2005) Enhanced efficiency in the production of offspring from 4- to 8-week-old lambs. Theriogenology 63(7):1876–1890

Kelly JM, Kleemann DO, McGrice H, Len JA, Kind KL, van Wettere WH, Walker SK (2016) Sex of co-twin affects the in vitro developmental competence of oocytes derived from 6- to 8-week-old lambs. Reprod Fertil Dev. https://doi.org/10.1071/RD16098

Killian DB, Kiesling DO, Warren JR (1985) Lifespan of corpora lutea induced in estrous-synchronized cycling and anoestrous ewes. J Anim Sci 61:210

Kleemann DO, Walker SK (2005) Fertility in South Australian commercial Merino flocks: sources of reproductive wastage. Theriogenology 63:2075–2088

Kojima FN, Stumpf TT, Cupp AS, Werth LA, Robertson MS, Wolfe MW et al (1992) Exogenous progesterone and progestins as used in estrous synchrony do not mimic the corpus luteum in regulation in luteinizing hormone and 17b-estradiol in circulation of cows. Biol Reprod 47:1009–1017

Lahoz B, Alabart JL, Cocero MJ, Monniaux D, Echegoyen E, Sánchez P, Folch J (2014) Anti-Müllerian hormone concentration in sheep and its dependence of age and independence of BMP15 genotype: an endocrine predictor to select the best donors for embryo biotechnologies. Theriogenology 81(2):347–357

Lamb C (2005) Factors affecting pregnancy rates in an IVF embryo transfer program. In: Joint proceedings of the AETA and the CETA, pp 31–36

Land RB, Wilmut I (1977) The survival of embryos transferred in large groups to sheep of breeds with different ovulation rates. Anim Prod 24:183–187

Ledda S, Bogliolo L, Calvia P, Leoni G, Naitana S (1997) Meiotic progression and developmental competence of oocytes collected from juvenile and adult ewes. J Reprod Fertil 109(1):73–78

Ledda S, Bogliolo L, Leoni G, Naitana S (2001) Cell coupling and maturation-promoting factor activity in in vitro-matured prepubertal and adult sheep oocytes. Biol Reprod 65(1):247–252

Lehloenya KC, Greyling JPC (2009) Effect of route of superovulatory gonadotrophin administration on the embryo recovery rate of Boer goat does. Small Rumin Res 87:39–44

Leoni G, Bogliolo L, Pintus P, Ledda S, Naitana S (2001) Sheep embryos derived from FSH/eCG treatment have a lower in vitro viability after vitrification than those derived from FSH treatment. Reprod Nutr Dev 41(3):239–246

Leoni GG, Rosati I, Succu S, Bogliolo L, Bebbere D, Berlinguer F, Ledda S, Naitana S (2007) A low oxygen atmosphere during IVF accelerates the kinetic of formation of in vitro produced ovine blastocysts. Reprod Domest Anim 42(3):299–304

Leoni GG, Palmerini MG, Satta V, Succu S, Pasciu V, Zinellu A, Carru C, Macchiarelli G, Nottola SA, Naitana S, Berlinguer F (2015) Differences in the kinetic of the first meiotic division and in active mitochondrial distribution between prepubertal and adult oocytes mirror differences in their developmental competence in a sheep model. PLoS One 10(4):e0124911. https://doi.org/10.1371/journal.pone.0124911

Letelier CA, Contreras-Solis I, García-Fernández RA, Ariznavarreta C, Tresguerres JA, Flores JM, Gonzalez-Bulnes A (2009) Ovarian follicular dynamics and plasma steroid concentrations are not significantly different in ewes given intravaginal sponges containing either 20 or 40 mg of fluorogestone acetate. Theriogenology 71(4):676–682

Leyva V, Buckrell BC, Walton JS (1998) Regulation of follicular activity and ovulation in ewes by exogenous progestagen. Theriogenology 50:395–416

Lin A, Lee K, Chang S, Lee P (1979) Non-surgical embryo transfer in goats. Memoir Coll Agr 19:25–33

Lindner GM, Wright RW (1983) Bovine embryo morphology and evaluation. Theriogenology 20:407

Lindsell CE, Rajkumar K, Manning AW, Emery SK, Mapletoft RJ, Murphy BD (1986) Variability in FSH: LH ratios among batches of commercially available gonadotrophins. Theriogenology 25:167 (abstr)

Loi P, Ptak G, Dattena M, Ledda S, Naitana S, Cappai P (1998) Embryo transfer and related technologies in sheep reproduction. Reprod Nutr Dev 38(6):615–628

Lopez-Alonso C, Encinas T, Garcia-Garcia RM, Veiga-Lopez A, Ros JM, McNeilly AS, Gonzalez-Bulnes A (2005a) Administration of single short-acting doses of GnRH antagonist modifies pituitary and follicular function in sheep. Domest Anim Endocrinol 29(3):476–487

Lopez-Alonso C, Encinas T, Veiga-Lopez A, Garcia-Garcia RM, Cocero MJ, Ros JM, McNeilly AS, Gonzalez-Bulnes A (2005b) Follicular growth, endocrine response and embryo yields in sheep superovulated with FSH after pretreatment with a single short-acting dose of GnRH antagonist. Theriogenology 64:1833–1843

Martinez AG, Matkovic M (1998) Cryopreservation of ovine embryos: slow freezing and vitrification. Theriogenology 49:1039–1049

Martinez AG, Valcarcel A, Furnus CC, de Matos DG, Iorio G, de las Heras MA (2006) Cryopreservation of in vitro-produced ovine embryos. Small Rumin Res 63:288–296

Mayorga I, Mara L, Sanna D, Stelletta C, Morgante M, Casu S, Dattena M (2011) Good quality sheep embryos produced by superovulation treatment without the use of progesterone devices. Theriogenology 75(9):1661–1668

McEvoy TG, Robinson JJ, Aitken RP, Robertson IS (1998) Melatonin treatment of embryo donor and recipient ewes during anestrus affects their endocrine status, but not ovulation rate, embryo survival or pregnancy. Theriogenology 49(5):943–955

McGinnis LK, Duplantis SC, Youngs CR (1993) Cryopreservation of sheep embryos using ethylene glycol. Anim Reprod Sci 30:273 280

McGrice H, Kelly JM, Kind KL, Kleemann DO, Hampton AJ, Hannemann P, Walker SK, van Wettere WHEJ (2016) Plasma anti-Mullerian hormone as a predictive marker of juvenile in vitro embryo production outcomes in Merino ewe lambs. In: 18th international congress on animal reproduction (ICAR), June 26– 30th, 2016, W 125

McKelvey WAC, Robinson JJ, Aitken RP, Robertson LS (1986) Repeated recoveries of embryos from ewes by laparoscopy. Theriogenology 25:855–865

McNeilly AS, Fraser HM (1987) Effect of gonadotrophin-releasing hormone agonist-induced suppression of LH and FSH on follicle growth and corpus luteum function in the ewe. J Endocrinol 115:273–282

Menchaca A, Rubianes E (2004) New treatments associated with timed artificial insemination in small ruminants. Reprod Fertil Dev 16(4):403–413

Menchaca A, Vilariño M, Crispo M, de Castro T, Rubianes E (2010) New approaches to superovulation and embryo transfer in small ruminants. Reprod Fertil Dev 22(1):113–118

Moore NW (1985) The use of embryo transfer and steroid hormone replacement therapy in the study of prenatal mortality. Theriogenology 23:121

Moore NW, Shelton JN (1962) The application of the technique of egg transfer to sheep breeding. Aust J Agric Res 13:718–724

Moore NW, Rowson LEA, Short RV (1959) Egg transfer in sheep. Factors affecting the survival and development of transferred eggs. J Reprod Fertil 1:332–339

Moore NW, Rowson LE, Short RV (1960) Egg transfer in sheep. Factors affecting the survival and development of transferred eggs. J Reprod Fertil 1:332–349

Morton KM (2008) Developmental capabilities of embryos produced in vitro from prepubertal lamb oocytes. Reprod Domest Anim 43(Suppl 2):137–143

Mutiga ER (1991) Increasing reproductive rates in tropical sheep by means of embryo transfer. Theriogenology 36:681–687

Nagashima H, Matsui K, Sawasaki T, Kano Y (1987) Nonsurgical collection of embryos in Shiba goats. Jikken Dobutsu 36:51–56

Naitana S, Loi P, Ledda S, Cappai P, Dattena M, Bogliolol L, Leoni G (1996) Effect of biopsy, vitrification on in vitro survival of ovine embryos at different stages of development. Theriogenology 46:813–824

Naitana S, Ledda S, Loi P, Leoni G, Bogliolo L, Dattena M, Cappai P (1997) Polyvinyl alcohol as a defined substitute for serum in vitrification and warming solutions to cryopreserve ovine embryos at different stages of development. Anim Reprod Sci 48(2–4):247–256

Naqvi SMK, Joshi A, Kumar D, Gulyani R, Maurya VP, Saha S, Mittal JP, Singh VK (2007) Developmental competence, birth and survival of lambs following transfer of twin or triple embryos of dwarf size prolific donor into large size non-prolific recipient sheep. J Cell Anim Biol 1(5):82–86

O'Connell AR, Demmers KJ, Smaill B, Reader KL, Juengel JJ (2016) Early embryo loss, morphology, and effect of previous immunization against androstenedione in the ewe. Theriogenology. https://doi.org/10.1016/j.theriogenology.2016.04.069

Oliveira ME, Feliciano MA, D'Amato CC, Oliveira LG, Bicudo SD, Fonseca JF, Vicente WR, Visco E, Bartlewski PM (2014) Correlations between ovarian follicular blood flow and superovulatory responses in ewes. Anim Reprod Sci 144(1–2):30–37

Opiela J, Kątska-Książkiewicz L (2013) The utility of Brilliant Cresyl Blue (BCB) staining of mammalian oocytes used for in vitro embryo production (IVP). Reprod Biol 13(3):177–183

Papadopoulos S, Rizos D, Duffy P, Wade M, Quinn K, Boland MP, Lonergan P (2002) Embryo survival and recipient pregnancy rates after transfer of fresh or vitrified, in vivo or in vitro produced ovine blastocysts. Anim Reprod Sci 74(1–2):35–44

Paramio MT, Izquierdo D (2014) Current status of in vitro embryo production in sheep and goats. Reprod Domest Anim 49(Suppl 4):37–48

Pereira RJTA, Sohnrey B, Holtz W (1998) Nonsurgical embryo collection in goats treated with prostaglandin F2-alpha and oxitocin. J Anim Sci 76:360–363

Picton HM, Tsonis CG, McNeilly AS (1990) The antagonistic effect of exogenous LH pulses on FSH-stimulated preovulatory follicle growth in ewes chronically treated with a gonadotrophin-releasing hormone agonist. J Endocrinol 127:273–283

Ptak G, Loi P, Dattena M, Tischner M, Cappai P (1999) Offspring from one-month-old lambs: studies on the developmental capability of prepubertal oocytes. Biol Reprod 61(6):1568–1574

Ptak G, Tischner M, Bernabé N, Loi P (2003) Donor-dependent developmental competence of oocytes from lambs subjected to repeated hormonal stimulation. Biol Reprod 69:278–285

Quinlivan TD (1966) Estimates of pre- and perinatal mortality in the New Zealand Romney Marsh ewe. J Reprod Fertil 11:379–390

Reynolds LP, Haring JS, Johnson ML, Ashley RL, Redmer DA, Borowicz PP, Grazul-Bilska AT (2015) Placental development during early pregnancy in sheep: estrogen and progesterone receptor messenger RNA expression in pregnancies derived from in vivo-produced and in vitro-produced embryos. Domest Anim Endocrinol 53:60–69

Ritar AJ, Maxwell WM, Salamon S (1984) Ovulation and LH secretion in the goat after intravaginal progestagen sponge-PMSG treatment. J Reprod Fertil 72(2):559–563

Robinson TJ, Moore NW, Holst PJ, Smith JF (1967) The evaluation of several progestogens administered in intravaginal sponges for the synchronization of estrus in the entire cyclic merino ewe. In: Robinson TJ (ed) Control of the ovarian cycle in the sheep. White and Bull PTY Ltd., Australia, pp 76–91

Romaguera R, Moll X, Morató R, Roura M, Palomo MJ, Catalá MG, Jiménez-Macedo AR, Hammami S, Izquierdo D, Mogas T, Paramio MT (2011) Prepubertal goat oocytes from large follicles result in similar blastocyst production and embryo ploidy than those from adult goats. Theriogenology 76(1):1–11

Rowson LE, Moor RM (1966) Embryo transfer in the sheep: the significance of synchronizing oestrus in the donor and recipient animal. J Reprod Fertil 11:207–212

Rubianes E, Menchaca A (2003) The pattern and manipulation of ovarian follicular growth in goats. Anim Reprod Sci 78:271–287

Rubianes E, Ibarra D, Ungerfeld R, de Castro T, Carbajal B (1995) Superovulatory response in anestrous ewes is affected by the presence of a large follicle. Theriogenology 43:465–472

Rubianes E, Ungerfeld R, Viñoles C, Rivero A, Adams GP (1997) Ovarian response to gonadotropin treatment initiated relative to wave emergence in ultrasonographically monitored ewes. Theriogenology 47:1479–1488

Rutigliano HM, Adams BM, Jablonka-Shariff A, Boime I, Adams TE (2014) Effect of time and dose of recombinant follicle stimulating hormone agonist on the superovulatory response of sheep. Theriogenology 82(3):455–460

Scaramuzzi RJ, Downing JA, Campbell BK, Cognié Y (1988) Control of fertility and fecundity of sheep by means of hormonal manipulation. Austr. J Biol Sci 41:37–45

Schiewe MC, Bush M, Stuart LS, Wildt DE (1984) Laparoscopic embryo transfer in domestic sheep: a preliminary study. Theriogenology 22:675–682

Shea BF (1981) Evaluating the bovine embryo. Theriogenology 15:31

Shorten PR, O'Connell AR, Demmers KJ, Edwards SJ, Cullen NG, Juengel JL (2013) Effect of age, weight, and sire on embryo and fetal survival in sheep. J Anim Sci 91(10):4641–4653

Souza-Fabjan JM, Locatelli Y, Duffard N, Corbin E, Touzé JL, Perreau C, Beckers JF, Freitas VJ, Mermillod P (2014) In vitro embryo production in goats: slaughterhouse and laparoscopic ovum pick up-derived oocytes have different kinetics and requirements regarding maturation media. Theriogenology 81(8):1021–1031

Steer CV, Mills CL, Tan SL et al (1992) The cumulative embryo score: a predictive embryo scoring technique to select the optimal number of embryos to transfer in an in-vitro fertilization and embryo transfer programme. Hum Reprod 7:117–119

Stringfellow DA, Seidel SM (eds) (1998) Manual of the international embryo transfer society. Savoy, IL, USA, p 106

Széll A, Zhang J, Hudson R (1990) Rapid cryopreservation of sheep embryos by direct transfer into liquid nitrogen vapour at −180 °C. Reprod Fertil Dev 2:613–618

Thompson JGE, Simpson AC, James RW, Tervit HR (1990) The application of progesterone-containing CIDR devices to superovulated ewes. Theriogenology 33:1297–1304

Thompson JGE, Bell ACS, McMillan WH, Peterson AJ, Tervit HR (1995) Donor and recipient ewe factors affecting in vitro development and post-transfer survival of cultured sheep embryos. Anim Reprod Sci 40:269–227

Torres S, Sevellec C (1987) Repeated superovulation and surgical recovery of embryos in the ewe. Reprod Nutr Dev 27:859–863

Torres-Rovira L, González-Bulnes A, Succu S, Spezzigu A, Manca M, Leoni G et al (2014) Predictive value of antral follicle count and anti-Müllerian hormone for follicle and oocyte developmental competence during the early prepubertal period in a sheep model. Reprod Fertil Dev 26:1094–1106

Ungerfeld R, Rubianes E (1999) Effectiveness of short-term progestogen primings for the induction of fertile oestrus with eCG in ewes during late seasonal anoestrus. Anim Sci 68:349–353

Vajta G (2000) Vitrification of the oocytes and embryos of domestic animals. Anim Reprod Sci 60/61:357–364

Veiga-Lopez A, Gonzalez-Bulnes A, Garcia-Garcia RM, Dominguez V, Cocero MJ (2005) The effects of previous ovarian status on ovulation rate and early embryo development in response to superovulatory FSH treatments in sheep. Theriogenology 63:1973–1983

Veiga-Lopez A, Cocero MJ, Dominguez V, McNeilly AS, Gonzalez-Bulnes A (2006a) Follicular wave status at the beginning of the FSH treatment modifies reproductive features in superovulated sheep. Reprod Biol 6:243–264

Veiga-Lopez A, Gonzalez-Bulnes A, Tresguerres JAF, Dominguez V, Ariznavarreta C, Cocero MJ (2006b) Causes, characteristics and consequences of anovulatory follicles in superovulated sheep. Domes Anim Endocrinol 30:76–87

Veiga-Lopez A, Encinas T, McNeilly AS, Gonzalez-Bulnes A (2008a) Timing of preovulatory LH surge and ovulation in superovulated sheep are affected by follicular status at start of the FSH treatment. Reprod Domest Anim 43:92–98

Veiga-Lopez A, Dominguez V, Souza CJH, Garcia-Garcia RM, Ariznavarreta C, Tresguerres JAF, McNeilly AS, Gonzalez-Bulnes A (2008b) Features of follicle-stimulating hormone-stimulated follicles in a sheep model: keys to elucidate embryo failure in assisted reproductive technique cycles. Fertil Steril 89:1328–1337

Viñoles C, Meikle A, Forsberg M, Rubianes E (1999) The effect of subluteal levels of exogenous progesterone on follicular dynamics and endocrine patterns during the early luteal phase of the ewe. Theriogenology 51:1351–1361

Viñoles C, Forsberg M, Banchero G, Rubianes E (2001) Effect of long-term and short-term progestagen treatment on follicular development and pregnancy rate in cyclic ewes. Theriogenology 55:993–1004

Walker SK, Hill JL, Kleemann DO, Nancarrow CD (1996) Development of ovine embryos in synthetic oviductal fluid containing amino acids at oviductal fluid concentrations. Biol Reprod 55(3):703–708

Warwick BL, Berry RO, Horlacher WR (1934) Results of mating rams to angora female goats. In: Proceedings of the American Society of animal production, pp 225–227

Welch RAS, Andrewes WD, Barnes DR, Bremer K, Harvey TG (1984). CIDR dispensers for oestrus and ovulation control in sheep. In: Proceedings of the 10th international congress on animal reproduction & artificial insemination, Urbana, IL, USA, vol 3, pp 354–355

Willadsen SM, Polge C, Rowson LEA (1976) Deep freezing of sheep embryos. J Reprod Fertil 46:151

Zambrini FN, Guimaraes JD, Esteves LV, Castro ACR, Fonseca JF (2014) Cervix dilation and transcervical embryo recovery in cervical Santa Inês sheep. Anim Reprod 11:419

Zambrini FN, Guimaraes JD, Prates JF, Esteves LV, Souza-Fabjan JMG, Brandao FZ et al (2015) Superovulation and non-surgical embryo recovery in Santa Inês ewes. Anim Reprod 12:720

Zhang L, Chai M, Tian X, Wang F, Fu Y, He C, Deng S, Lian Z, Feng J, Tan DX, Liu G (2013) Effects of melatonin on superovulation and transgenic embryo transplantation in small-tailed han sheep (Ovis aries). Neuro Endocrinol Lett 34(4):294–301

Embryo Transfer Technologies in Pigs

7

Curtis R. Youngs

Abstract

Embryo transfer (ET) became a reality in the swine industry with the birth of the first live ET piglets in 1950. Since that pioneering achievement more than 68 years ago, significant developments in porcine ET and its related technologies have occurred. Although the volume of commercial ET activity with pigs is low compared to that reported for cattle, substantial porcine ET activity is taking place in private companies and institutes engaged in biomedical research. In vitro production of pig embryos has greatly surpassed that of in vivo-derived embryos, and development of nonsurgical methods for transfer of swine embryos has opened the door to potential widespread commercial application of porcine ET. The historical inability to cryopreserve pig embryos has been overcome to a great extent with development of protocols for vitrification of porcine embryos. The creation of genetically modified pigs via somatic cell nuclear transfer or genome-editing technologies depends upon successful ET, and the needs of the biomedical research community likely will be the impetus for further refinements in pig ET technologies.

7.1 History of Embryo Transfer in Pigs

Nearly six decades passed between the birth of the first mammalian embryo transfer (ET) offspring at the University of Cambridge, England, on May 29, 1890 (rabbits; Heape 1890), and the birth of the first pig ET offspring born on March 27, 1950, at the Pig Breeding Research Institute in Poltava, Ukraine (Kvasnitski 1950; for an English translation please see Kvasnitski (2001)). Kvasnitski's groundbreaking

C. R. Youngs
Department of Animal Science, Iowa State University, Ames, IA, USA
e-mail: cryoungs@iastate.edu

© Springer International Publishing AG, part of Springer Nature 2018
H. Niemann, C. Wrenzycki (eds.), *Animal Biotechnology 1*,
https://doi.org/10.1007/978-3-319-92327-7_7

Fig. 7.1 One of the four purebred Mirgorod piglets from the first successful pig embryo transfer performed by Kvasnitski (1950). (Photo courtesy of the wife of Dr. Kvasnitski)

experiment involved transfer of embryos into four recipients. One of four recipients farrowed (25% farrowing rate), and embryo survival in the pregnant recipient was 44% (four piglets born from nine embryos transferred). Figure 7.1 shows one of the piglets from the very first ET litter.

Ten years passed before the next known report on pig ET (Pomeroy 1960). In that interesting experiment, embryos were flushed from only one oviduct of each donor female, and embryos from a different donor were transferred into the flushed oviduct (allowing a donor to also serve as a recipient). Only one of six recipients farrowed (17% farrowing rate), and three piglets were born (only two of which were from the eight transferred embryos [25% embryo survival rate]). Two years later, the true potential of ET as a genetic improvement tool for pigs was convincingly demonstrated (Hancock and Hovell 1962). Three of six ET recipients farrowed (50% farrowing rate), litter size averaged 10.3 piglets, and embryo survival in pregnant recipients was 74%.

7.2 The Commercial Swine ET Industry

The primary reason that ET was originally considered for pig production was to enhance genetic improvement, and that reason is still valid today. The repeated collection of preimplantation embryos from genetically superior sows/gilts, and subsequent transfer of harvested embryos to lower genetic merit (but reproductively sound) recipients, enables genetically superior females to produce more progeny in 1 year than would be possible through natural mating. Using only a small number of highly selected females as parents of the next generation of piglets equates to increased genetic selection intensity, and intense selection will lead to an accelerated rate of genetic improvement. Given the corporate structure of swine breeding companies in many parts of the world today, the use of ET in central nucleus herds seems wise and inevitable given the large number of sows over which the costs of ET may be spread.

Embryo transfer can also facilitate swine genetic improvement by introducing germplasm resources from other locales. The swine industry recognized the value of ET for that purpose many decades ago, as evidenced by two early reports on transport of pig embryos from one country to another (Baker and Dziuk 1970; Wrathall et al. 1970). Moving valuable genetic resources as embryos instead of live animals can be done at a lower cost and with less risk of disease transmission (Youngs 2007), and it also eliminates potential stress associated with transport of live animals (an important animal welfare consideration).

Despite strong incentives to utilize ET in the swine industry, commercial pig ET activity remains low. Each year the IETS data retrieval committee gathers data on commercial ET activity in various livestock species. Unlike the data for bovine ET which tend to be complete, ET data on swine tends to be incomplete/underreported. Swine ET data compiled over a 20-year period (1997–2016) for in vivo-derived (IVD) preimplantation embryos are presented in Table 7.1. The largest number of embryo recoveries recorded during that time period was 701 (in 1998), the largest

Table 7.1 Commercial swine embryo transfer data reported to the data retrieval committee of the International Embryo Transfer Society[a,b]

Calendar year	Number of embryo collections reported	Number of embryos recovered	Number of embryos recovered per donor	Number of fresh embryos transferred	Number of frozen embryos transferred
1997	5	105	21	0	0
1998	701	11,264	16	2111	214
1999	241	8071	33	2529	0
2000	260	9043	35	7091	0
2001	469	13,607	29	13,589	0
2002	264	5008	19	0	0
2003	345	9609	28	8349	0
2004	488	11,833	24	4302	0
2005	501	11,806	17	8406	2280
2006	150	2840	19	4419	0
2007	173	4009	23	3675	0
2008	149	3800	26	3092	0
2009	149	3800	26	3092	0
2010	0	0	0	0	0
2011	53	997	19	1381	0
2012	114	2478	22	2478	0
2013	27	397	15	397	0
2014	0	0	0	0	0
2015	39	395	10	441	0
2016	49	413	28	397	0

[a]Data were compiled from the annual reports of the International Embryo Technology Society (IETS) data retrieval committee published each year in the December issue of the IETS Embryo Transfer Newsletter
[b]Data from commercial swine embryo transfer activity in some years is incomplete, leading to apparent discrepancies/mistakes in the data (e.g., the number of embryos transferred some years appears larger than the number of embryos recovered because embryo recovery data were not provided by all ET service providers)

number of embryos recovered was 13,607 (in 2001), the highest average number of embryos recovered per donor was 35 (in 2000), and the largest number of fresh embryos transferred was 13,589 (in 2001). Interestingly, no commercial swine ET activity was reported for calendar year 2014.

There are a number of reasons why commercial swine ET is not as prevalent as ET in other livestock species such as cattle. Firstly, the pig is a litter-bearing species with a relatively young age at puberty and relatively short gestation length. Those combined factors serve as a disincentive for pork producers to spend time and money performing ET. Secondly, embryo collection and transfer methods historically have been surgical, and the necessity of performing surgery limits the practicality of ET under field conditions. Thirdly, it had been extremely difficult to successfully freeze pig embryos for approximately three decades, and that inability to cryopreserve embryos essentially mandated that all ETs be conducted only with freshly collected embryos. Fortunately, for the latter two points, there have been significant technological advancements during the past decade.

7.3 Collection of In Vivo-Derived Embryos

To increase the efficiency of ET in pigs, it is important to regulate the time of estrus in donor and recipient pigs. In contrast to some species where exogenous prostaglandin F2α (PGF) is often used to regulate cyclicity, the use of PGF is limited in pigs because receptors for PGF do not appear on the corpora lutea until approximately day 12 of the estrous cycle (Guthrie and Polge 1976). Thus, the primary means to regulate the time of estrus in pigs is via use of the progesterone analogue altrenogest (Kraeling and Webel 2015). The timing of estrus in donors and recipients influences the success of ET in pigs, and asynchronous ET (placement of embryos into a uterus that is not as far along in the estrous cycle as the donor) often yields higher pregnancy rates than synchronous ET (reviewed in Youngs 2001).

Another approach to potentially increase the efficiency of ET in pigs is to superovulate donor females with exogenous gonadotropins. Although follicle stimulating hormone (FSH) could theoretically be used to induce superovulation, the hormone of choice seems to be the lower-cost equine chorionic gonadotropin (eCG, also known as pregnant mare serum gonadotropin [PMSG]) because it exhibits FSH-like activity and therefore stimulates ovarian follicular growth and development. Because the response to superovulation in the pig is highly variable and excessive response may lead to reduced embryo viability, many choose not to perform superovulation (Youngs 2001).

Embryo development in the pig is somewhat different than that of other mammalian livestock species. Embryos typically enter the uterus at the 4-cell stage of embryonic development, and embryos continue their development in the tip of the uterine horn until approximately day 6 at which time embryo migration begins. Transuterine embryo migration occurs beginning on day 9. Embryos are typically recovered from donors 5 to 6 days after the onset of estrus. Retrograde flushing of the reproductive tract is not advisable because of the strong nature of the uterine-tubal junction.

Surgical recovery of porcine embryos, patterned after the method of Hancock and Hovell (1962), is illustrated in Fig. 7.2. Depending on the day of embryo recovery, sterile flushing medium is introduced into either the oviduct (for recovery of early-stage embryos) or the tip of the uterine horn. Medium is "flushed" through the reproductive tract and collected into an embryo recovery dish with the aid of a glass cannula placed into the uterine horn beyond the suspected location of the embryos. Repeated embryo collections from the same donor are possible if the uterus is kept moist during the procedure.

In non-superovulated donors, fertilization rate is expected to be near 95% and the number of embryos recovered should closely follow the number of corpora lutea counted at the time of flushing. Porcine embryos appear different than those from cattle. In addition to the darker colored cytoplasm and presence of more lipid droplets within the cytoplasm, the zona pellucida appears "dirty" due to the presence of multiple spermatozoa trapped in the zona pellucida (see Fig. 7.3).

Fig. 7.2 Surgical embryo recovery in the pig

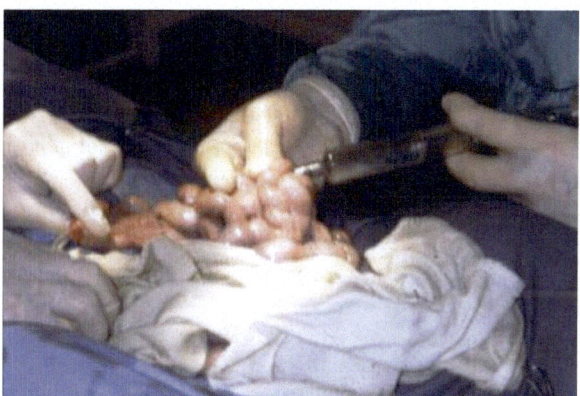

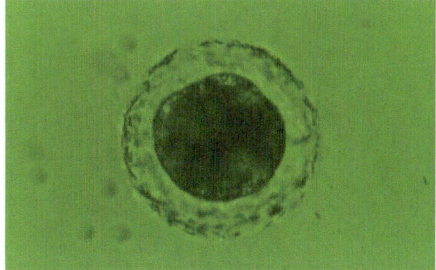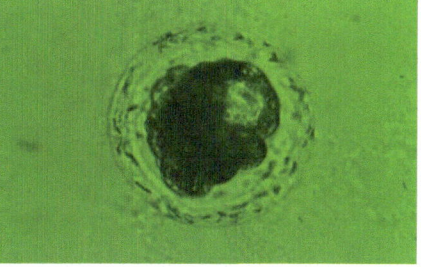

Fig. 7.3 In vivo-derived porcine embryos showing spermatozoa trapped in the zona pellucida due to the relatively slow zona reaction in this species. A compact morula (left) and early blastocyst (right) are shown

7.4   Recent Developments in Swine ET

7.4.1   Nonsurgical Transfer of Porcine Embryos

There has long been interest in performing nonsurgical ET in pigs. One early study of nonsurgical transfer of porcine embryos led to establishment of a day 17 "pregnancy" (Polge and Day 1968), while another study nearly 20 years later (Sims and First 1987) reported pregnancies that presumably did not persist to term. Birth of the first live piglets resulting from nonsurgical ET (Reichenbach et al. 1993), however, provided strong encouragement for future studies. Numerous other reports of porcine nonsurgical ET quickly followed (Hazeleger and Kemp 1994; Galvin et al. 1994; Yonemura et al. 1996, 2003; Li et al. 1996; Hazeleger et al. 2000; Martinez et al. 2004; Ducro-Steverink et al. 2004; Suzuki et al. 2004).

Successful reports of nonsurgical ET in pigs led to the commercial development and marketing of a variety of different porcine nonsurgical ET apparatuses (see, e.g., Fig. 7.4). In general, the approach used to perform nonsurgical ET in pigs is to place an artificial insemination (AI) catheter into the cervix where it is "locked" in place. A smaller diameter flexible catheter is then introduced into the lumen of the AI catheter, and the flexible catheter is guided through the interdigitating prominences of the cervix, through the uterine body, and into a uterine horn. Typically, all embryos are deposited into one uterine horn because transuterine migration of embryos is a normal reproductive phenomenon in pigs (Anderson and Parker 1976). Deposition of embryos into the uterine body is avoided, however, because it leads to lower pregnancy and embryo survival rates than deposition of embryos in the uterine horn (Wallenhorst and Holtz 1999).

The nonsurgical ET technique that perhaps has been most widely investigated is a deep intrauterine ET method (Martinez et al. 2013). Results of various studies conducted by this research group have been summarized recently (Martinez et al. 2016a). Compared with the early pioneering studies, significant progress has been made. However, some issues with this nonsurgical ET method remain to be resolved, such as failure to pass the ET catheter through the cervix (Martinez et al. 2016b) and failure to properly position the catheter during transfer (15.9% of transfers—Martinez et al. 2015). Results obtained under fairly large-scale field studies show

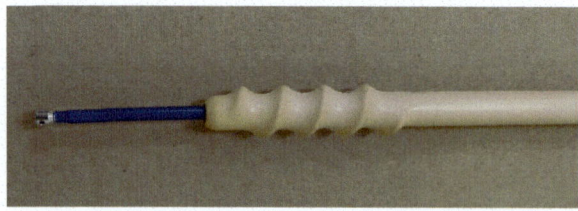

Fig. 7.4 Example of commercially manufactured apparatus for nonsurgical embryo transfer in pigs. Please note the small diameter flexible catheter protruding from the lumen of the artificial insemination catheter. The flexible catheter is long enough to be inserted deeply into a uterine horn

great promise for adoption of this method; farrowing rates above 60% with average litter sizes exceeding nine piglets were reported.

7.4.2 Cryopreservation of Porcine Embryos

The birth of the first mammalian offspring derived from transfer of frozen-thawed embryos occurred in mice (Whittingham et al. 1972). Nearly two decades passed, however, before the first piglet was born after transfer of frozen-thawed embryos (Hayashi et al. 1989). It was well documented that porcine embryos are extremely sensitive to temperatures below 15 °C (Wilmut 1972; Polge et al. 1974), and it was hypothesized that the chilling sensitivity of pig embryos was due to its high lipid content (Niemann 1985). Research on the biochemical composition of pig embryos (Youngs et al. 1994b) discovered that the predominant fatty acid present in porcine blastocysts was oleic acid (which has a melting point of 13.4 °C). Further evidence in support of lipid being responsible for the difficulty associated with cryopreservation emerged when the removal of lipid from porcine embryos via centrifugation led to the birth of live piglets (Nagashima et al. 1995). In addition, expanded blastocysts (containing less lipid) were found to be more cryotolerant than earlier developmental stages (Berthelot et al. 2003).

Porcine embryos have been cryopreserved using slow cooling equilibrium methods as well as ultrarapid cooling methods more commonly known as vitrification (reviewed in Youngs et al. 2010). Studies on slow cooling approaches for cryopreservation of IVD porcine embryos are extremely sparse, but pregnancy rates ranging from 14% to 100% were reported. Ultrarapid methods for cryopreservation yielded pregnancy rates ranging from 60% to 80%, and vitrification is now recognized as the preferred method for cryopreservation of porcine embryos (including those produced in vitro—Maehara et al. 2012) because of greater success and consistency of results. Live piglets have also been produced from vitrified-warmed oocytes subsequently used for in vitro embryo production (Somfai et al. 2014). At present, however, there is no single universally accepted protocol for vitrification of porcine embryos.

7.5 Swine ET as a Research Tool

In addition to its use for improving pork production, ET is used in pigs to gain greater insights into fundamental biological mechanisms. For example, the Chinese Meishan pig is known to have larger litters than other breeds such as the Yorkshire. At the time of importation of the Meishan breed into the United States for study, the mechanism for increased prolificacy was unknown. A reciprocal ET study with Meishan and Yorkshire embryos (Youngs et al. 1994a) revealed that the prolificacy of the Meishan results not only from an inherently slower development rate of the Meishan preimplantation embryo but also from a suppressive (growth-slowing) effect of the Meishan uterus. Other researchers have utilized reciprocal ET studies to study porcine fetal development (Miles et al. 2012) and piglet growth, lactation performance, and milk composition (Miles et al. 2015).

7.6 Swine ET as a Tool for Production of Healthy Pigs

In many parts of the world, corporate swine breeding companies operate large production facilities where thousands of sows are housed on a single site. The high density of animals in these production facilities can represent a potential risk for disease transmission if the animals are not properly managed. Diseases such as PRRS (porcine reproductive and respiratory syndrome), PED (porcine epidemic diarrhea), and PCV2 (porcine circovirus type 2; Bielanski et al. 2013) are of great significance to these breeding companies.

During a disease outbreak, ET may represent a way to "rescue" the genetics of valuable females. Preimplantation embryos that are properly washed and handled in accordance with guidelines promulgated by the International Embryo Transfer Society (Stringfellow 2011) represent negligible risk of concomitant disease transmission at the time of ET into healthy recipients at a disease-free site.

Briefly, guidelines for sanitary handling of embryos include examination of embryos at a magnification of at least 50X to verify the presence of an intact zona pellucida free of adherent material, five-step washing of no more than ten embryos from a single donor (a limitation for pigs compared with other species) using a 1 to 100 dilution of washing solution, two-step washing procedure in 0.25% trypsin to remove any viruses that may be attached to the zona pellucida, five-step washing procedure (similar to that previously described), and finally inspection of washed embryos at ≥50X magnification to ensure embryos have an intact zona pellucida free of any mucus or cellular debris.

7.7 Swine ET as a Tool for Development of Assisted Reproductive Technologies

With each passing year, researchers continue to refine assisted reproductive technologies (ART). The development of methods for in vitro production (IVP) of embryos was important not only in and of itself but also because it served as a springboard for development of affiliated reproductive technologies such as somatic cell nuclear transfer (SCNT) and genome editing of zygotes.

7.7.1 In Vitro Fertilization

Since the birth of the first in vitro-produced (IVP) piglet (Cheng et al. 1986), there has been continual scientific effort to refine and enhance methodology for in vitro maturation (IVM) of oocytes, in vitro fertilization (IVF) of IVM oocytes, and in vitro culture (IVC) of presumptive zygotes produced via IVF. The chronology of major developments related to porcine IVP embryos has been reviewed recently (Grupen 2014). Despite valiant research efforts, two problems still remain (Kikuchi et al. 2016) with the in vitro production of porcine embryos: (1) polyspermy and (2) imbalance of nuclear and cytoplasmic maturation.

A recent meta-analysis of ET data from porcine IVF studies (Liu et al. 2015) revealed that factors influencing the success of ET with IVP embryos are not the same as those reported for IVD embryos (Youngs 2001). Although lacking some data for some traits across all 246 published articles, the meta-analysis revealed an average of 106 and 31 embryos transferred per recipient prior to day 4 and on days 4–7, respectively. Average pregnancy rate was 44%, and an average pregnancy loss of 14% was observed. Recipients exhibited a 43% farrowing rate with an average litter size of 5.1 piglets. Piglet production efficiency, calculated as the total number of piglets produced divided by the total number of embryos transferred during the experiment, was 5.7%. This compared to a piglet production efficiency of 20.3% for IVD embryos.

7.7.2 Somatic Cell Nuclear Transfer

Pig nuclear transfer was originally performed using embryonic blastomeres as donor cells (Prather et al. 1989), but this technique suffered from the fact that the genetic merit of cells of an embryo is unknown (i.e., mating genetically superior boars and sows together is no guarantee that the resultant piglets will be genetically outstanding). The first successful somatic cell nuclear transfer (SCNT) in pigs was reported more than 15 years ago (Onishi et al. 2000).

Two recent studies examined the overall efficiency of SCNT in pigs. Data from 274 studies that generated 18,649 SCNT embryos during a 3-year period were analyzed in one report (Kurome et al. 2013). An average of 97 embryos were transferred to each of 193 recipients: 109 recipients (56%) became pregnant, and 85 (78%) gave birth. Of 318 piglets produced (1.7% piglet production efficiency), 75 (24%) were stillborn, and 243 (76%) were born alive. Of the 243 born alive, 100 (31% of total) died soon after birth, 39 (12%) were killed by their mother or died from infections unrelated to SCNT, 7 (2%) were utilized immediately after birth for investigations, and 97 (31%) were clinically healthy and exhibited normal development.

The meta-analysis previously mentioned (Liu et al. 2015) also examined various parameters associated with SCNT experiments. An average of 203 and 62 embryos were transferred per recipient prior to day 4 and on days 4–7, respectively. Average pregnancy rate was 45%, and an average pregnancy loss of 25% was observed. Recipients exhibited a 36% farrowing rate with an average litter size of 6.3 piglets. Piglet production efficiency was 1.0%.

7.7.3 Genome Editing of Zygotes

Genome editing is a term used to describe the specific and selective editing of the genetic makeup of an organism. When utilized in zygotes (recently fertilized eggs) created through IVM/IVF technology, genome editing allows researchers to knock out a gene, repair a gene, or introduce a novel genetic change at a specific location in the genome. These genetic alterations can be incorporated into the animal's germ line and thus be transmitted to progeny. A recent review of genome-editing technology (Tan et al. 2016) revealed that more than 300 pigs, cattle, sheep, and goats have been generated using genome editing. Genome editing has the potential to greatly enhance the rate of swine genetic improvement, and readers are referred to a separate chapter in this volume for a more detailed discussion of this topic.

References

Anderson LL, Parker RO (1976) Distribution and development of embryos in the pig. J Reprod Fertil 46:363–368. https://doi.org/10.1530/jrf.0.0460363

Baker RD, Dziuk PJ (1970) Aerial transport of fertilized pig ova. Can J Anim Sci 50(1):215–216

Berthelot F, Martinat-Botté F, Vajta G, Terqui M (2003) Cryopreservation of porcine embryos: state of the art. Livest Prod Sci 83:73–83. https://doi.org/10.1016/S0301-6226(03)00038-1

Bielanski A, Algire J, Lalonde A, Garceac A, Pollard JW, Plante C (2013) Nontransmission of porcine circovirus 2 (PCV2) by embryo transfer. Theriogenology 80(2):77–83. https://doi.org/10.1016/j.theriogenology.2013.03.022

Cheng WTK, Polge C, Moor RM (1986) In vitro fertilization of pig and sheep oocytes. Theriogenology 25(1):146 (abstr)

Ducro-Steverink DWB, Peters CGW, Maters CC, Hazeleger W, Merks JWM (2004) Reproduction results and offspring performance after non-surgical embryo transfer in pigs. Theriogenology 62(3/4):522–531. https://doi.org/10.1016/j.theriogenology.2003.11.010

Galvin JM, Killian DB, Stewart ANV (1994) A procedure for successful nonsurgical embryo transfer in swine. Theriogenology 41(6):1279–1289. https://doi.org/10.1016/0093-691X(94)90486-3

Grupen CG (2014) The evolution of porcine embryo in vitro production. Theriogenology 81(1):24–37. https://doi.org/10.1016/j.theriogenology.2013.09.022

Guthrie HD, Polge C (1976) Luteal function and oestrus in gilts treated with a synthetic analogue of prostaglandin F-2α (ICI 79,939) at various times during the oestrus cycle. J Reprod Fertil 48(2):423–425. https://doi.org/10.1530/jrf.0.0480423

Hancock JL, Hovell GJR (1962) Egg transfer in the sow. J Reprod Fertil 4:195–201

Hayashi S, Kobayashi K, Mizuno J, Saitoh K, Hirano S (1989) Birth of piglets from frozen embryos. Vet Rec 125:43–44

Hazeleger W, Kemp B (1994) Farrowing rate and litter size after transcervical embryo transfer in sows. Reprod Dom Anim 29(6):481–487. https://doi.org/10.1111/j.1439-0531.1994.tb00597.x

Hazeleger W, Noordhuizen JPTM, Kemp B (2000) Effect of asynchronous non-surgical transfer of porcine embryos on pregnancy rate and embryonic survival. Livest Prod Sci 64(2/3):281–284. https://doi.org/10.1016/S0301-6226(99)00147-5

Heape W (1890) Preliminary note on the transplantation and growth of mammalian ova within a uterine foster-mother. Proc R Soc Lond 48:457–458

Kikuchi K, Kaneko H, Nakai M, Somfai T, Kashiwazaki N, Nagai T (2016) Contribution of in vitro systems to preservation and utilization of porcine genetic resources. Theriogenology 86(1):170–175. https://doi.org/10.1016/j.theriogenology.2016.04.029

Kraeling RR, Webel SK (2015) Current strategies for reproductive management of gilts and sows in North America. J Anim Sci Biotechnol 6:3 http://www.jasbsci.com/content/6/1/3

Kurome M, Geistlinger L, Kessler B, Zakhartchenko V, Klymiuk N, Wuensch A, Richter A, Baehr A, Kraehe K, Burkhardt K, Flisikowski K, Flisikowska T, Merkl C, Landmann M, Durkovic M, Tschukes A, Kraner S, Schindelhauer D, Petri T, Kind A, Nagashima H, Schnieke A, Zimmer

R, Wolf E (2013) Factors influencing the efficiency of generating genetically engineered pigs by nuclear transfer: multi-factorial analysis of a large data set. BMC Biotechnol 13:43 http://www.biomedcentral.com/1472-6750/13/43

Kvasnitski AV (1950) [The research on interbreed ova transfer in pigs.] Socialist Livestock Breeding Journal, Semi-annual report of the Ukrainian Ministry of Agriculture 1950; November issue, pp 12–15 (in Ukrainian)

Kvasnitski AV (2001) Research on interbreed ova transfer in pigs. Theriogenology 56(8):1285–1289

Li J, Rieke A, Day BN, Prather RS (1996) Technical note: porcine non-surgical embryo transfer. J Anim Sci 74(9):2263–2268. https://doi.org/10.2527/1996.7492263x

Liu Y, Li J, Løvendahl P, Schmidt M, Larsen K, Callesen H (2015) In vitro manipulation techniques of porcine embryos: a meta-analysis related to transfers, pregnancies and piglets. Reprod Fertil Dev 27:429–439. https://doi.org/10.1071/RD13329

Maehara M, Matsunari H, Honda K, Nakano K, Takeuchi Y, Kanai T, Matsuda T, Matsumura Y, Hagiwara Y, Sasayama N, Shirasu A, Takahashi M, Watanabe M, Umeyama K, Hanazono Y, Nagashima H (2012) Hollow fiber vitrification provides a novel method for cryopreserving in vitro maturation/fertilization-derived porcine embryos. Biol Reprod 87(6):133., 8 pages. https://doi.org/10.1095/biolreprod.112.100339

Martinez EA, Caamaño JN, Gil MA, Rieke A, McCauley TC, Cantley TC, Vazquez JM, Roca J, Vazquez JL, Didion BA, Murphy CN, Prather RS, Day BN (2004) Successful nonsurgical deep uterine embryo transfer in pigs. Theriogenology 61(1):137–146. https://doi.org/10.1016/S0093-691X(03)00190-0

Martinez EA, Cuello C, Parrilla I, Rodriguez-Martinez H, Roca J, Vazquez JL, Vazquez JM, Gil MA (2013) Design, development, and application of a non-surgical deep uterine embryo transfer technique in pigs. Anim Front 3(4):40–47. https://doi.org/10.2527/af.2013-0032

Martinez EA, Martinez CA, Nohalez A, Sanchez-Osorio J, Vazquez JM, Roca J, Parrilla I, Gil MA, Cuello C (2015) Nonsurgical deep uterine transfer of vitrified, in vivo-derived, porcine embryos is as effective as the default surgical approach. Sci Rep 5:10587. https://doi.org/10.1038/srep10587

Martinez EA, Cuello C, Parrilla I, Martinez CA, Nohalez A, Vazquez JL, Vazquez JM, Roca J, Gil MA (2016a) Recent advances toward the practical application of embryo transfer in pigs. Theriogenology 85(1):152–161. https://doi.org/10.1016/j.theriogenology.2015.06.002

Martinez EA, Nohalez A, Martinez CA, Parrilla I, Vila J, Colina I, Diaz M, Reixach J, Vazquez J, Roca J, Cuello C, Gil M (2016b) The recipients' parity does not influence their reproductive performance following non-surgical deep uterine porcine embryo transfer. Reprod Dom Anim 51(1):123–129. https://doi.org/10.1111/rda.12654

Miles JR, Vallet JL, Ford JJ, Freking BA, Cushman RA, Oliver WT, Rempel LA (2012) Contributions of the maternal uterine environment and piglet genotype on weaning survivability potential: I. Development of neonatal piglets after reciprocal embryo transfers between Meishan and White crossbred gilts. J Anim Sci 90(7):2181–2192. https://doi.org/10.2527/jas.2011-4724

Miles JR, Vallet JL, Ford JJ, Freking BA, Oliver WT, Rempel LA (2015) Contributions of the maternal uterine environment and piglet genotype on weaning survivability potential: II. Piglet growth, lactation performance, milk composition, and piglet blood profiles during lactation following reciprocal embryo transfers between Meishan and White crossbred gilts. J Anim Sci 93(4):1555–1564. https://doi.org/10.2527/jas.2014-8426

Nagashima H, Kashiwazaki N, Ashman RJ, Grupen CG, Nottle MB (1995) Cryopreservation of porcine embryos. Nature 374(6521):416. https://doi.org/10.1038/374416a0

Niemann H (1985) Sensitivity of pig morulae to DMSO/PVP or glycerol treatment and cooling to 10°C. Theriogenology 23(1):213 (abstr)

Onishi A, Iwamoto M, Akita T, Mikawa S, Takeda K, Awata T, handed H, Perry AC (2000) Pig cloning by microinjection of fetal fibroblast nuclei. Science 289(5482):1188–1190. https://doi.org/10.1126/science.289.5482.1188

Polge C, Day BN (1968) Pregnancy following non-surgical egg transfer in pigs. Vet Rec 82:712

Polge C, Wilmut I, Rowson LEA (1974) The low temperature preservation of cow, sheep and pig embryos. Cryobiology 560:82 (abstr)

Pomeroy RW (1960) Infertility and neonatal mortality in the sow. III. Neonatal mortality and foetal development. J Agric Sci (Camb) 54:31–56

Prather RS, Sims MM, First NL (1989) Nuclear transplantation in early pig embryos. Biol Reprod 41(3):414–418. https://doi.org/10.1095/biolreprod41.3.414

Reichenbach HD, Modl J, Brem G (1993) Piglets born after transcervical transfer of embryos into recipient gilts. Vet Rec 133(2):36–39

Sims MM, First NL (1987) Nonsurgical embryo transfer in swine. J Anim Sci 65(Suppl 1):386 (abstr)

Somfai T, Yoshioka K, Tanihara F, Kaneko H, Noguchi J, Kashiwazaki N, Nagai T, Kikuchi K (2014) Generation of live piglets from cryopreserved oocytes for the first time using a defined system for in vitro embryo production. PLoS One 9(5):e97731. https://doi.org/10.1371/journal.pone.0097731

Stringfellow DA (2011) Recommendations for the sanitary handling of in-vivo derived embryos. In: Stringfellow DA, Givens MD (eds) Manual of the International Embryo Transfer Society, 4th Ed. IETS, Champaign, pp 65–68

Suzuki C, Iwamura S, Yoshioka K (2004) Birth of piglets through the non-surgical transfer of blastocysts produced in vitro. J Reprod Dev 50(4):487–491. https://doi.org/10.1262/jrd.50.487

Tan W, Proudfoot C, Lillico SG, Whitelaw CBA (2016) Gene targeting, genome editing: from Dolly to editors. Transgenic Res. https://doi.org/10.1007/s11248-016-9932-x

Wallenhorst S, Holtz W (1999) Transfer of pig embryos to different uterine sites. J Anim Sci 77(9):2327–2329. https://doi.org/10.2527/1999.7792327x

Whittingham DG, Leibo SP, Mazur P (1972) Survival of mouse embryos frozen to −196°C and −269°C. Science 178(4059):411–414. https://doi.org/10.1126/science.178.4059.411

Wilmut I (1972) The low temperature preservation of mammalian embryos. J Reprod Fertil 31:513–514

Wrathall AE, Done JT, Stuart P, Mitchell D, Betteridge KJ, Randall GCB (1970) Successful intercontinental pig conceptus transfer. Vet Rec 87:226–228

Yonemura I, Fujino Y, Irie S, Miura Y (1996) Transcervical transfer of porcine embryos under practical conditions. J Reprod Dev 42:89–94. https://doi.org/10.1262/jrd.42.89

Yonemura I, Miyamoto K, Nishida M (2003) Non-surgical transfer of porcine embryos. Theriogenology 59:378 (abstr)

Youngs CR (2001) Factors influencing the success of embryo transfer in the pig. Theriogenology 56(8):1311–1320. https://doi.org/10.1016/S0093-691X(01)00632-X

Youngs CR (2007) Methods for disease-free introduction of swine germplasm resources. In: Proc. seminar on reproduction: optimizing genetics, health, and production, American Association of Swine Veterinarians, March 4, Orlando, FL, pp 7–12

Youngs CR, Christenson LK, Ford SP (1994a) Investigations into the control of litter size in swine: III. A reciprocal embryo transfer study of early conceptus development. J Anim Sci 72(3): 725–731. https://doi.org/10.2527/1994.723725x

Youngs CR, Knight TJ, Batt SM, Beitz DC (1994b) Phospholipid, cholesterol, triacylglycerol, and fatty acid composition of porcine blastocysts. Theriogenology 41(1):343 (abstr)

Youngs CR, Leibo SP, Godke RA (2010) Embryo cryopreservation in domestic mammalian livestock species. CAB Reviews: Perspectives in Agriculture, Veterinary Science, Nutrition and Natural Resources 5(60):1–11. https://doi.org/10.1079/PAVSNNR20105060

Equine Embryo Transfer

8

H. Sieme, J. Rau, D. Tiedemann, H. Oldenhof, L. Barros,
R. Sanchez, M. Blanco, G. Martinsson, C. Herrera,
and D. Burger

Abstract

Embryo transfer has become a commonly used procedure in equine breeding worldwide. It allows for efficient use of valuable mares and mares in athletic competition. In addition, mares with reproductive problems can donate embryos to healthy recipients. This review describes techniques for embryo collection and transfer to the recipient, methods of transportation and cryopreservation, and superovulation procedures. Effects of specific procedures on success rates for embryo collection and pregnancies are discussed, as well as factors affecting the resulting offspring. Furthermore, an outlook is given on recent biotechnological technologies like preimplantation diagnostics and in vitro embryo production.

H. Sieme (✉) · J. Rau · D. Tiedemann · H. Oldenhof · L. Barros
Unit for Reproductive Medicine—Clinic for Horses, University of Veterinary Medicine Hannover, Hannover, Germany
e-mail: Harald.Sieme@tiho-hannover.de; Janina.Rau@tiho-hannover.de; daniela.tiedemann@ewe.net; Harriette.Oldenhof@tiho-hannover.de; Lawrence.Barros@bol.com.br

R. Sanchez · M. Blanco
Unit for Reproductive Medicine—Clinic for Horses, University of Veterinary Medicine Hannover, Hannover, Germany

PS Pferdehaltung, Neustadt-Glewe, Germany
e-mail: Roberto.Sanchez.Arbouin@tiho-hannover.de; miguelblanco__vet@hotmail.com

G. Martinsson
National Stud Lower Saxony, Celle, Germany
e-mail: Gunilla.Martinsson@lgst-celle.niedersachsen.de

C. Herrera
Clinic for Reproductive Medicine, Vetsuisse Faculty, University of Zürich, Zürich, Switzerland

D. Burger
Swiss Institute of Equine Medicine ISME, University of Berne and Agroscope, Avenches, Switzerland
e-mail: dominik.burger@vetsuisse.unibe.ch

© Springer International Publishing AG, part of Springer Nature 2018
H. Niemann, C. Wrenzycki (eds.), *Animal Biotechnology 1*,
https://doi.org/10.1007/978-3-319-92327-7_8

179

8.1 Introduction

Equine embryo transfer is a biotechnological procedure, which has become common practice in horse breeding, and the number of foals from ET in commercial operations is steadily increasing. According to data collected by the International Embryo Technology Society (IETS), a total of 21.321 equine embryos were transferred in 2015 in 11 countries (http://www.iets.org/pdf/comm_data/IETS_Data_Retrieval_2015_V2.pdf). Probably actual numbers of transfers are higher, because a few countries, with high embryo transfer activity, did not disclose their ET records. Using embryo transfer, it is possible to obtain more than one foal per year from a good mare. Furthermore, mares in training and competition can be used for breeding (Campbell 2014). Embryo collection can be performed using 1- or 2-year-old mares (Panzani et al. 2007). This shortens the generation interval of mares with a promising pedigree. Another advantage of embryo transfer is that it can be used in mares that cannot carry a foal themselves due to general health problems or reproductive weaknesses (e.g., repeated history of pregnancy loss). Embryo collection and subsequent cryopreservation of embryos can be performed throughout the whole year as long as the donor mare is cycling.

It has been a long way from the first successful nonsurgical recovery of blastocysts from the uterine lumen of native pony mares (Oguri and Tsutsumi 1972), soon followed by the birth of the first equid offspring after reciprocal, interspecies transfer of horse, donkey, mule, or hinny zygotes (Allen and Rowson 1972), followed by birth of the first foal produced by embryo transfer (Oguri and Tsutsumi 1974) to present practices where embryo transfer programs are offered commercially. Milestones of these developments included effective collection methods (Oguri and Tsutsumi 1972), nonsurgical transfer (Oguri and Tsutsumi 1974), superovulatory treatment (McCue 1996), transport of cooled embryos (Carney et al. 1991), and successful cryopreservation of embryos (Yanamoto et al. 1982). Further success of embryo transfer practices resulted from the establishment of transfer centers with large recipient herds, as well as breeding associations which accepted that mares could have more than one foal per year.

8.2 Techniques of Equine Embryo Transfer

8.2.1 Collection and Quality Evaluation of Equine Embryos

Collection of equine embryos by uterine flushing is generally performed between days 6.5 and 9 after ovulation (=day 0). Equine embryos have a rather extended oviductal passage and only enter the uterus 144–156 h after ovulation (Battut et al. 1997). A uterine collection attempt earlier performed than this is mostly unsuccessful. From day 9 after ovulation on, the embryo becomes relatively large, and the possibility to damage it during collection, storage, or transfer increases significantly (McKinnon and Squires 1988a, b; Carnevale et al. 2000). Thus embryo collection is commonly performed on day 7 or 8 because recovery rates are high and the (expanded) blastocysts can be easily recovered and handled. For embryo recovery, the tube of an embryo collection tube system (commercially available) is carefully introduced through the cervix into the uterine body. Inflating the balloon at the end of the tube keeps the catheter in place and closes the uterine lumen at the uterine side of the cervix. Depending on the size of the mare and uterus, repeatedly 1–2 L of medium is flushed into the uterus. After a short transrectal massage, the medium is collected and—depending on the method—passed through a filter, which retains the embryo. Different collection media can be used, e.g., phosphate-buffered saline with fetal bovine serum or bovine serum albumin, ready-to-use Ringer lactate, or special embryo collection media, which are commercially available. An important feature of the collection media is that they prevent the embryo from sticking to the plastic of the tube system. The media should be pre-warmed to 32–35 °C. The flushing steps are repeated 3–4 times. If the embryo is not recovered, the same procedure can be repeated after treatment with oxytocin. This is supposed to increase collection rates by about 10% due to induced uterine contractions (McCue et al. 2003; Squires et al. 2003). If successful, the embryo can be found under a stereo microscope depending on the method—directly in the in-line collection filter system or after sedimentation and filtration of the fluids. Size, developmental stage, and quality of the embryo should be assessed. The size usually varies between 150 μm (day 6) and 1200 μm (day 8). Discrepancies are possible due to asynchronous double ovulations or delayed embryonic development. The embryos are usually at the morula, blastocyst, or expanded blastocyst stage. The quality is judged on a scale from 1 to 4 (excellent to degenerated, McKinnon and Squires 1988a, b) (Fig. 8.1).

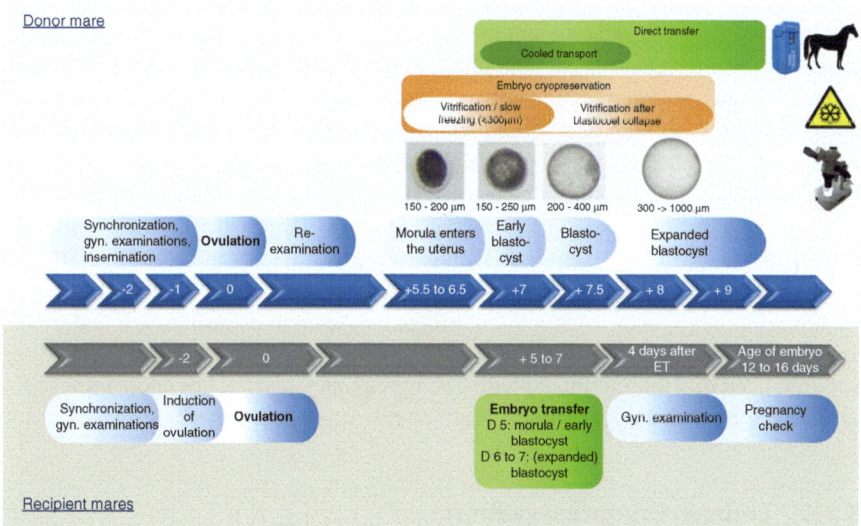

Fig. 8.1 Chronology of equine embryo transfer procedures. Typically donor and recipient mares follow a timed embryo transfer process which is orientated to the day of ovulation (day 0). The day of embryo collection from the donor mare is mainly defined by the desired embryo treatment (direct transfer, cooled transport, cryopreservation). Factors causing a delay in fertilization or embryonic development of 0.5–1 day should be considered for planning embryo collection. Recipient mares should ideally ovulate 24–48 h after the donor

8.2.2   Embryo Treatment and Direct Transfer

In the early years, transfer of embryos was conducted under general anesthesia and laparotomy (Allen 1982). Subsequently the uterus was visualized and grasped manually after flank incision of the sedated, standing horse (Iuliano et al. 1985; Squires et al. 1985). Finally, noninvasive transcervical approaches, as performed nowadays, have been advanced to highly efficient methods with regard to time and costs. Surgical transfer methods have become obsolete, also for ethical reasons.

Prior to transfer of the embryo, it is washed several times in embryo medium to eliminate adhesive cells or microorganisms and to provide a stable environment for the embryo. If the recipient mare is nearby available and has been synchronized to the donor mare, the embryo can be transferred directly. The embryo is placed in a 0.25 or 0.5 ml straw with the embryo medium (Jasko 2002). The transfer pipette, which is similar to an insemination pipette, is aseptically and atraumatically passed by manual palpation through the cervix into the uterine body, and the embryo is expelled. Alternatively the use of a speculum and a specially adapted forceps has been reported (Wilsher and Allen 2004).

8.2.3   Cooled Transport

One of the most important developments in equine embryo transfer was the finding that embryos can be stored or transported at 5 °C for 24 h without impairing viability (Carnevale et al. 1987; Moussa et al. 2004; McCue et al. 2011). Especially in North and South America, the number of transported equine embryos has greatly increased (Squires et al. 1999), but also in Europe this technique has been instrumental for making embryo transfer programs more widely accepted. The advantages of embryo cooling and subsequent transfer reduce factors encountered with availability of recipient mares, i.e., owners from mares enrolled in an embryo collection program are not required to provide the recipient mare. Instead, embryos can be shipped to institutions with large recipient herds. Different media are available for transportation. Using Ham's F-10 medium for 24-h storage results in similar pregnancy rates as observed in direct (fresh) transfer (Carney et al. 1991). Since this medium requires special ambient conditions, its practical use is limited, and therefore other commercial holding media are more generally used (Embryo Holding Solution®, ViGro Holding Plus®) with good pregnancy results. For cooled transportation several passive cooling devices are commercially available.

8.2.4   Embryo Cryopreservation

Freezing equine embryos after collection and prior to transfer has even further increased the options of embryo transfer technology. Embryos can be transported worldwide, and the time of embryo collection is not limited to the desired breeding time during the year. As long as the mare is cycling, embryos can be readily collected and stored frozen. Transfer then takes place whenever it is convenient. This has a positive impact on the planning of the breeding season. If more than one embryo is obtained from a single collection (multiple ovulation), but only one mare is available as recipient, the other embryo(s) can be stored frozen. Especially for endangered breeds or breeds with small numbers, cryopreservation of embryos allows the establishment of gene banks and maintenance of biodiversity. By using embryo cryopreservation, female genetics of a breed are also partially preserved, while so far usually only male genetics can be preserved as frozen semen.

The first foal from a frozen embryo was born in 1982 (Yanamoto et al. 1982). Since then, great efforts have been made to modify freezing methods to obtain improved survival of frozen/thawed quality embryos and to yield higher pregnancy rates. Pregnancy rates of 50–60% with the traditional slow freezing method can only be achieved if the embryo is smaller than 300 µm in diameter (Czlonkowska et al. 1985; Slade et al. 1985; Skidmore et al. 1991; Squires et al. 2003). In order to collect embryos of this small size, the uteri had to be flushed on day 6–7 after ovulation. However, this rather early approach was usually associated with lower recovery rates of embryos. Freezing larger embryos generally leads to a high rate of embryonic apoptosis (Tharasanit et al. 2005). On days 6 or 7, a glycoprotein capsule forms around the equine embryo (Betteridge et al. 1982; Flood et al. 1982), which is critical for embryo survival (Stout et al. 2005). This capsule is considered to be responsible for the higher apoptosis rate after freezing. Capsule thickness is correlated with bad freezing results probably due to the fact that the capsule hinders penetration of cryoprotectants (Legrand et al. 1999; Bruyas et al. 2000). Glycerol is used as standard cryoprotectant. Alternatives like combinations of glycerol with 1,2-propanediol or sucrose (Ferreira et al. 1997) and ethylene glycol (Bruyas et al. 2000) have been tested, but did not improve cryosurvival.

For conventional slow freezing, the temperature reduction must be tightly controlled, which is only possible by using expensive controlled freezing equipment. Due to these limitations of the slow freezing method, vitrification has become more commonly and generally applied for equine embryos. During vitrification, the embryo and surrounding media are transformed into a glassy state, thus avoiding the formation of ice crystals. Vitrification solutions contain extraordinary high concentrations of cryoprotectants, which allow for a fast removal of the water in the cells. After exposure to vitrification solutions, the embryo is directly plunged into liquid nitrogen, which is associated with an ultrarapid cooling rate. For small embryos (<300 µm Ø), similar pregnancy results can be achieved when compared to the slow freezing method (Oberstein et al. 2001; Moussa et al. 2005; Eldridge-Panuska et al. 2005). Vitrification kits are commercially available and protocols can be performed without special equipment. This renders vitrification more practical and it is increasingly established as the standard method.

Larger embryos (>300 µm) at the expanded blastocyst stage have been considered to be unsuitable for cryopreservation (Eldridge-Panuska et al. 2005; Barfield et al. 2009). However, recently progress has been made allowing successful cryopreservation of such embryos. It has been demonstrated that the collapse of the blastocoel prior to cryopreservation is essential for the survival of large embryos (>300 µm Ø) after warming. To achieve collapse, embryos are punctured using a micromanipulation unit and a fine micropipette. The blastocoel fluid is aspirated, and the embryos are immediately vitrified using a similar technique than the one used for small embryos. First studies reported pregnancy rates between 70% and 83% after transfer of these embryos (Choi et al. 2011; Diaz et al. 2016; Sanchez et al. 2017). If a micromanipulator is not available, it is also possible to manually collapse large equine embryos before vitrification using a 25-gauge needle to puncture the capsule and remove fluid (Ferris et al. 2016). In the near future, cryopreservation of large embryos will probably render equine embryo transfers less expensive (no recipient synchronization necessary) and more efficient (later collection possible, transfer at perfect time) which in turn will lead to establishing an international equine embryo market.

8.3   Influences on Embryo Collection Success

It is sometimes difficult to assess whether the collection procedure was unsuccessful or if there was no embryo to begin with. Assuming that the collection has been performed correctly by skilled practitioners, an embryo is generally recovered if present in the uterus (Hartman 2011), and recovery rates are indicative for insemination success (Campbell 2014). Very important factors influencing the success rates are insemination management and general fertility of the mare and stallion (Squires et al. 1999; Stout 2003). A satisfactory collection rate can only be expected from healthy mares without fertility problems in their breeding history, and rates drop dramatically if the mare is older than 13–15 years (Squires et al. 1985; Vogelsang and Vogelsang 1989; McCue et al. 2010; Marinone et al. 2015). Overall, embryo recovery rates are considered to be about 50% per collection attempt (Squires and McCue 2007). From reproductively healthy young mares bred with semen of good quality, embryo collection rates reach 70–75%, whereas only 20–30% of collection attempts are successful in mares with a history of reproductive restrictions (McCue and Squires 2015). For competing mares it needs to be considered that extensive exercise, heat, and stress potentially impair fertility and consequently embryo collection rates and quality (Mortensen et al. 2009; Kelley et al. 2011; Smith et al. 2012). Semen quality significantly influences embryo recovery rates (Love et al. 2015). Mares bred with fresh or cooled-transported semen provide a 20% higher rate than mares inseminated with frozen/thawed semen, respectively. Embryos of smaller sizes can be recovered if donor mares are bred with frozen/thawed semen. This might be due to a delay in fertilization or embryonic development of 0.5–1 day, which should be considered for planning embryo collection (McCue and Squires 2015). Embryo recovery rates are similar for collections from day 7 to day 10 after ovulation. Collection on day 6 has significantly lower success rates (Jacob et al. 2012).

8.3.1   Multiple Ovulations and Superovulation

The mare usually ovulates one follicle per cycle. Nevertheless, double or even triple ovulations are possible. These occur with a probability of ~22% in thoroughbreds and ~15% in warmblood breeds (Newcombe 1995). Some mares or mare families and also some breeds are significantly more likely to have multiple ovulations (Panzani et al. 2014). Increasing age of the mare also increases the probability of more than one oocyte being ovulated (Marinone et al. 2015; Panzani et al. 2014). Twin pregnancies after multiple ovulations are undesired and require adequate management. Multiple ovulations may increase recovery rates (Squires et al. 1985; Squires et al. 1987; Nagao et al. 2012). Induction of multiple ovulations would therefore greatly increase the overall success rates of embryo transfer programs. Unfortunately, superovulatory treatments in horses yield only very low success rates compared to other species. The anatomical structure of the equine ovary with the follicles on the inside and the relative low sensitivity of the equine FSH receptors (Combarnous et al. 1998) are likely involved in the failure of superovulating mares. Double ovulations can be induced using different protocols (McCue 1996), including equine chorion gonadotropin (Day 1940), gonadotropin-releasing hormone (Ginther and Bergfelt 1990), porcine follicle-stimulating hormone (Fortune and Kimmich 1993), equine pituitary gonadotropins (Hofferer et al. 1991), deslorelin acetate (Nagao et al. 2012), active (McKinnon et al. 1992) and passive (McCue et al. 1993) immunization against inhibin, or equine follicle-stimulating hormone (Niswender et al. 2003). Currently, the best results with regard to double ovulation induction and increased embryo recovery rates are achieved by treatment with recombinant equine follicle-stimulating hormone (reFSH, DeLuca et al. 2008) alone or together with recombinant equine luteinizing hormone (reLH, Meyers-Brown et al. 2011). Administration of the correct doses and proportions is important to avoid ovulation failures or premature luteinization (Briant et al. 2004). Induction of more than two or three follicles does not lead to an increase in embryo recovery rates, as no more follicles on the same ovary can ovulate due to space limitation (Riera et al. 2006; Allen 2005). Therefore, the goal of stimulatory treatment should be to improve the induction of double ovulation.

8.4   Influences on Pregnancy Rates

8.4.1   Embryo Quality

The quality of the transferred embryo has a significant influence on pregnancy rates (McKinnon and Squires 1988a, b; McCue and Squires 2015) and on rates of early pregnancy loss (Carnevale et al. 2000). Most of the embryos collected are evaluated to be excellent or good (Grade 1–2). This is likely due to the fact that poor-quality embryos do not reach the uterus or degenerate prior to collection (Carnevale et al. 2000). Recovery of morulae or very small blastocysts on days 7 or 8 after ovulation is an indicator for delayed embryo development. This also results in lower pregnancy rates after transfer (Squires et al. 1999). Embryo quality and pregnancy rates decrease with increasing age of the donor mare (Cuervo-Arango et al. 2017). However, cooled transport of embryos does not affect pregnancy rates (McCue and Squires 2015).

8.4.2   Transfer Method

The influence of the transfer method on pregnancy rates is dependent on the technical skill of the person performing the transfer (Squires et al. 1999; Allen 2005; Cuervo-Arango et al. 2017), while the available equipment and techniques themselves do not seem to make a difference with regard to the outcome (Jasko 2002). Transcervical embryo placement entails the risk to contaminate the uterus (Allen 2005); high thus strict hygienic measures are needed to avoid the risk of a uterine reaction. The manual dilation of the cervix during transfer can induce release of oxytocin or PGF2α (Handler et al. 2003; Kask et al. 1997) which in turn may lead to an impaired luteal function or even luteolysis. Consequently, cervix manipulation and dilation should be as little as possible, and flunixin meglumine can be administered as a precaution, although effects of this drug on pregnancy rates after embryo transfer are conversely discussed (Koblischke et al. 2008; Okada et al. 2018). There is no solid evidence so far that the nonsurgical transfer actually influences luteal function. Nevertheless, in practice nonsteroidal anti-inflammatory drugs, corticosteroids, antibiotics, and progesterone are often supplemented to recipients to avoid early pregnancy loss caused by inflammation or infection of the genital tract and luteolysis.

8.4.3 Cycle Synchronicity of Donor and Recipient

One of the major influences on transfer success, i.e., pregnancy rate, is the synchronicity of donor and recipient mare. As mentioned earlier, freezing the embryo or using mares from a recipient herd can avoid the cycle synchronization. In these cases it is still crucial to transfer the embryo at the correct time after ovulation. The basic requirement to achieve this is a proper mare management and ovulation control. Similar pregnancy rates have been achieved, if the recipient mare ovulates 1 day before or up to 5 days after the donor mare. Ovulation out of this time frame leads to a dramatic drop in pregnancy rates (Jacob et al. 2012). Ideally, the recipient mare ovulates 24–48 h after the donor.

Cycle length variation and especially the differences in reaction to hormonal treatments sometimes make it difficult to achieve synchronicity with just one recipient mare. Using two or three recipient mares treated with different hormonal regimes (Allen 2005) ensures to have at least one matching recipient when performing a direct transfer. Many synchronization protocols have been created for donor and recipient mares (Rocha Filho et al. 2004; Greco et al. 2012; McCue and Squires 2015; Pinto et al. 2017; Oliveira Neto et al. 2018). Generally, estrus is induced with PGF2α while the mares are in the active luteal phase. Daily follicle controls should be performed as soon as both mares show estrus symptoms. Ovulation is induced in the donor mare when she presents a follicle of at least 35 mm using human chorion gonadotropin or GnRH implants. Ovulation should occur 30–48 h after application and needs to be monitored. In the recipient mare, ovulation should be induced 24 h after the donor mare or as soon as ovulation has been diagnosed in the donor. Assuming the recipient mare reacts normally and also ovulates after 30–48 h, the desired synchronicity is achieved. Besides the recipient's day of ovulation relative to the donor mare, the size and expected developmental stage of the collected embryo should be considered for recipient selection. Especially embryos of an early developmental stage (morulae, early blastocysts) should be transferred to recipients at day 5 after ovulation.

8.4.4   Age, Fertility, and Size of the Recipient

Donor mares are usually selected very carefully by their breeding value. The selection of the recipient mare should be done with the same care. Age, general health, fertility, and size of the recipient have an impact on transfer success and the developing foal. The recipients should be in the range of 3 and 12 years old. In older mares the risk of pregnancy loss increases due to endometrial degeneration (Ricketts and Alonso 1991). The recipient should be free of any abnormalities (malpositioned vulva, abnormal cervix), signs of degeneration of the endometrium (endometrial cysts), or signs of inflammation (fluid in the uterus, contamination). The tone of the cervix and uterus is critically important. Reduced tone usually leads to a higher risk of pregnancy losses (Carnevale et al. 2000). The likelihood of pregnancy increased with increasing length of the recipient's preceding estrus which is correlated with the duration of endometrial edema (Cuervo-Arango et al. 2017). Complete breeding soundness examination (uterine culture/cytology or endometrial biopsy) is recommended prior to the start of synchronization treatment. The mammary gland should also be examined to avoid problems during nursing of the foal.

The size of the recipient mare relatively to the donor mare and stallion can affect intrauterine development of the fetus and postnatal development of the foal. Transfer of embryos into smaller mares potentially leads to delayed growth, which might not be completely compensated after birth (Allen et al. 2004). The transfer of thoroughbred embryos into ponies can lead to physical and ethological immaturity after birth (Ousey et al. 2004). Smaller size differences between donor and recipient are compensated after birth. Other influences on the foals have been described (Peugnet et al. 2014), but all possible effects are not fully understood yet. Preliminary data show that contrary to general breeders' opinion, the character of embryo transfer offspring is not significantly affected by the recipient mare, and behavioral traits might be associated mostly with genetics of the parents (Burger et al. 2008). Further investigations on possible impact of recipient mares' behavior and metabolism are needed as well as on epigenetic effects (Chavatte-Palmer et al. 2016).

8.5   Benefits of New Technologies for Embryo Transfer

Equine embryo transfer allows the direct evaluation of the embryo. Besides the previously described quality assessment, it is also possible to perform genetic analysis of the embryo prior to transfer. This technique is known as preimplantation genetic diagnosis (PGD) and generally involves the collection of a few cells from the embryo using an inverted microscope equipped with a micromanipulation system. After collection, the cells are genetically analyzed; this procedure does not compromise the viability of the embryo. Genetic diagnosis can also be used for sex determination before transfer (Herrera et al. 2014), which is greatly desired in some breeds. The embryos can also be analyzed to determine the breeding value or the incidence of genetic disorders (Guignot et al. 2015).

The alternative to embryo collection is producing embryos in vitro. Oocytes are collected from mares using ovum pickup (OPU, Galli et al. 2001). In vitro fertilization by coincubation of oocyte and sperm has been unsuccessful in the horse. Instead, fertilization is achieved using intracytoplasmic sperm injection (ICSI, Hinrichs 2010). The combination of these two techniques enables breeding of mares and stallions that would otherwise not be suitable. Embryos from mares with reproductive organ pathologies can be produced. Additionally, immotile stallion sperm or sperm of low availability can be used very effectively. A recent increase in the efficiency of these two techniques has allowed its application even in commercial programs (Choi et al. 2002; Dell'Aquila et al. 1997; Galli et al. 2007; Galli et al. 2014).

References

Allen WR (ed) (1982) Embryo transfer in the horse. CRC Press, Boca Raton, FL

Allen WR (2005) The development and application of the modern reproductive technologies to horse breeding. Reprod Domest Anim 40:310–329

Allen WR, Rowson LEA (1972). Transfer of ova between horses and donkeys. In: Proc., 7th int. congress on animal reproduction and artificial insemination, 6–9 June 1972, Munich, pp 484–487

Allen WR, Wilsher S, Tiplady C, Butterfield RM (2004) The influence of maternal size on pre- and postnatal growth in the horse: III. Postnatal growth. Reproduction 127:67–77

Barfield JP, McCue PM, Squires EL, Seidel GE Jr (2009) Effect of dehydration prior to cryopreservation of large equine embryos. Cryobiology 59:36–41

Battut I, Colchen S, Fieni F, Tainturier D, Bruyas JF (1997) Success rates when attempting to nonsurgically collect equine embryos at 144, 156 or 168 hours after ovulation. Equine Vet J Suppl (25):60–62

Betteridge KJ, Eaglesome MD, Mitchell D, Flood PF, Beriault R (1982) Development of horse embryos up to twenty two days after ovulation: observations on fresh specimens. J Anat 135:191–209

Briant C, Toutain PL, Ottogalli M, Magallon T, Guillaume D (2004) Kinetic studies and production rate of equine (e) FSH in ovariectomized pony mares. Application to the determination of a dosage regimen for eFSH in a superovulation treatment. J Endocrinol 182:43–54

Bruyas JF, Sanson JP, Battut I, Fieni F, Tainturier D (2000) Comparison of the cryoprotectant properties of glycerol and ethylene glycol for early (day 6) equine embryos. J Reprod Fertil Suppl (56):549–560

Burger D, Schauer SN, Waegeli S, Aurich C, Gerber V, Thun R (2008) Influence of the recipient mare on size and character traits of adult offspring in a warmblood embryo transfer program— preliminary results, 7. Havemeyer embryo transfer symposium, Cambridge, July 2008

Campbell ML (2014) Embryo transfer in competition horses: managing mares and expectations. Equine Vet Educ 26:322–327

Carnevale EM, Squires EL, Mckinnon AO (1987) Comparison of Ham's F10 with CO_2 or Hepes buffer for storage of equine embryos at 5 C for 24 H. J Anim Sci 65:1775–1781

Carnevale EM, Ramirez RJ, Squires EL, Alvarenga MA, Vanderwall DK, McCue PM (2000) Factors affecting pregnancy rates and early embryonic death after equine embryo transfer. Theriogenology 54:965–979

Carney NJ, Squires EL, Cook VM, Seidel GE Jr, Jasko DJ (1991) Comparison of pregnancy rates from transfer of fresh versus cooled, transported equine embryos. Theriogenology 36:23–32

Chavatte-Palmer P, Robles M, Tarrade A, Duranthon V (2016) Gametes, embryos and their epigenome: considerations for equine embryo technologies. J Equine Vet Sci 41:13–21

Choi YH, Love CC, Love LB, Varner DD, Brinsko S, Hinrichs K (2002) Developmental competence in vivo and in vitro of in vitro-matured equine oocytes fertilized by intracytoplasmic sperm injection with fresh or frozen-thawed spermatozoa. Reproduction 123:455–465

Choi YH, Velez IC, Riera FL, Roldan JE, Hartman DL, Bliss SB, Blanchard TL, Hayden SS, Hinrichs K (2011) Successful cryopreservation of expanded equine blastocysts. Theriogenology 76:143–152

Combarnous Y, Richard F, Martinat N (1998) Mammalian follicle stimulating hormone receptors and their ligands. Eur J Obstet Gynecol Reprod Biol 77:125–130

Cuervo-Arango J, Claes AN, Ruijter-Villani M, Stout TA (2017) Likelihood of pregnancy after embryo transfer is reduced in recipient mares with a short preceding oestrus. Equine Vet J 50(3):386–390

Czlonkowska M, Boyle MS, Allen WR (1985) Deep freezing of horse embryos. J Reprod Fertil 75:485–490

Day F (1940) Clinical and experimental observations on reproduction in the mare. J Agric Sci Camb 30:244–261

Dell'aquila ME, Cho YS, Minoia P, Traina V, Fusco S, Lacalandra GM, Maritato F (1997) Intracytoplasmic sperm injection (ICSI) versus conventional IVF on abattoir-derived and in vitro-matured equine oocytes. Theriogenology 47:1139–1156

DeLuca CA, McCue PM, Patten ML, Squires EL (2008) Comparison of three doses of reFSH for superovulation of mares. Theriogenology 70:587–588

Diaz F, Bondiolli K, Paccamonti D, Gentry GT (2016) Cryopreservation of Day 8 equine embryos after blastocyst micromanipulation and vitrification. Theriogenology 85:894–903

Eldridge-Panuska WD, Di Brienza VC, Seidel GE Jr, Squires EL, Carnevale EM (2005) Establishment of pregnancies after serial dilution or direct transfer by vitrified equine embryos. Theriogenology 63:1308–1319

Ferreira JC, Meira C, Papa FO, Landin E Alvarenga FC, Alvarenga MA, Buratini J (1997) Cryopreservation of equine embryos with glycerol plus sucrose and glycerol plus 1,2-propanediol. Equine Vet J Suppl:88–93

Ferris RA, McCue PM, Trundell DA, Morrissey JK, Barfield JP (2016) Vitrification of large equine embryos following manual or micromanipulator-assisted blastocoele collapse. J Equine Vet Sci 41:64–65

Flood PF, Betteridge KJ, Diocee MS (1982) Transmission electron microscopy of horse embryos 3-16 days after ovulation. J Reprod Fertil Suppl 32:319–327

Fortune JA, Kimmich TL (1993) Purified pig FSH increases the rate of double ovulations in the mare. Equine Vet J Suppl 15:95–98

Galli C, Crotti G, Notari C, Turini P, Duchi R, Lazzari G (2001) Embryo production by ovum pick up from live donors. Theriogenology 55:1341–1357

Galli C, Colleoni S, Duchi R, Lagutina I, Lazzari G (2007) Developmental competence of equine oocytes and embryos obtained by in vitro procedures ranging from in vitro maturation and ICSI to embryo culture, cryopreservation and somatic cell nuclear transfer. Anim Reprod Sci 98:39–55

Galli C, Duchi R, Colleoni S, Lagutina I, Lazzari G (2014) Ovum pick up, intracytoplasmic sperm injection and somatic cell nuclear transfer in cattle, buffalo and horses: from the research laboratory to clinical practice. Theriogenology 81:138–151

Ginther OJ, Bergfelt DR (1990) Effect of GnRH treatment during the anovulatory season on multiple ovulation rate and on follicular development during the ensuing pregnancy in mares. J Reprod Fertil 88:119–126

Greco GM, Burlamaqui FG, Pinna AE, Queiroz FR, Cunha MS, Brandão FZ (2012) Use of long-acting progesterone to acyclic embryo recipient mares. Rev Bras Zootec Anim 41(3):607–611

Guignot F, Reigner F, Perreau C, Tartarin P, Babilliot JM, Bed'hom B, Vidament M, Mermillod P, Duchamp G (2015) Preimplantation genetic diagnosis in Welsh pony embryos after biopsy and cryopreservation. J Anim Sci 93:5222–5231

Handler J, Konigshofer M, Kindahl H, Schams D, Aurich C (2003) Secretion patterns of oxytocin and PGF2alpha-metabolite in response to cervical dilatation in cyclic mares. Theriogenology 59:1381–1391

Hartman DL (2011) Embryo transfer. In: McKinnon AO, Squires EL, Vaala WE, Varner DD (eds) Equine reproduction, 2nd edn. Blackwell Publishing Ltd., Ames, IA

Herrera C, Morikawa MI, Bello MB, Von Meyeren M, Centeno JE, Dufourq P, Martinez MM, Llorente J (2014) Setting up equine embryo gender determination by preimplantation genetic diagnosis in a commercial embryo transfer program. Theriogenology 81:758–763

Hinrichs K (2010) In vitro production of equine embryos: state of the art. Reprod Domest Anim 45(Suppl 2):3–8

Hofferer S, Duchamp G, Palmer E (1991) Ovarian response in mares to prolonged treatment with exogenous equine pituitary gonadotrophins. J Reprod Fertil Suppl 44:341–349

Iuliano MF, Squires EL, Cook VM (1985) Effect of age of equine embryos and method of transfer on pregnancy rate. J Anim Sci 60:258–263

Jacob JC, Haag KT, Santos GO, Oliveira JP, Gastal MO, Gastal EL (2012) Effect of embryo age and recipient asynchrony on pregnancy rates in a commercial equine embryo transfer program. Theriogenology 77:1159–1166

Jasko DJ (2002) Comparison of pregnancy rates following non-surgical transfer of day 8 embryos using various transfer devices. Theriogenology 58:713–716

Kask K, Odensvik K, Kindahl H (1997) Prostaglandin F2alpha release associated with an embryo transfer procedure in the mare. Equine Vet J 29:286–289

Kelley DE, Gibbons JR, Smith R, Vernon KL, Pratt-Phillip SE, Mortensen CJ (2011) Exercise affects both ovarian follicular dynamics and hormone concentrations in mares. Theriogenology 76:615–622

Koblischke P, Kindahl H, Budik S, Aurich J, Palm F, Walter I, Kolodziejek J, Nowotny N, Hoppen HO, Aurich C (2008) Embryo transfer induces a subclinical endometritis in recipient mares which can be prevented by treatment with non-steroid anti-inflammatory drugs. Theriogenology 70:1147–1158

Legrand E, Bencharif D, Battut I, Taintuier D, Bruyas JF (1999) Horse embryo freezing: influence of thickness of the capsule. In: Proceedings of the 15th scientific meeting of the European Embryo Transfer Association, Lyon, France, pp 184–185

Love CC, Noble JK, Standridge SA, Bearden CT, Blanchard TL, Varner DD, Cavinder CA (2015) The relationship between sperm quality in cool-shipped semen and embryo recovery rates in horses. Theriogenology 84:1587–1593

Marinone AI, Losinno L, Fumuso E, Rodriguez EM, Redolatti C, Cantatore S, Cuervo-Arango J (2015) The effect of mare's age on multiple ovulation rate, embryo recovery, post-transfer pregnancy rate, and interovulatory interval in a commercial embryo transfer program in Argentina. Anim Reprod Sci 158:53–59

McCue PM (1996) Superovulation. Vet Clin North Am Equine Pract 12:1–11

McCue PM, Squires EL (2015) Equine embryo transfer, 1st edn. CRC Press, Boca Raton, FL

McCue PM, Hughes JP, Lasley BL (1993) Effect on ovulation rate of passive immunisation of mares against inhibin. Equine Vet J 25:103–106

McCue PM, Niswender KD, Macon KA (2003) Modification of the flush procedure to enhance embryo recovery. J Equine Vet Sci 23:336–337

McCue PM, Ferris RA, Lindholm AR, DeLuca CA (2010) Embryo recovery procedures and collection success: results of 492 embryo-flush attempts. In: Proceedings of the 56th annual convention of the American Association of Equine Practitioners

McCue PM, DeLuca CA, Wall JJ (2011) Cooled transported embryo technology. In: McKinnon AO, Squires EL, Vaala WE, Varner DD (eds) Equine reproduction, 2nd edn. Wiley-Blackwell, Ames, IA, pp 2880–2886

McKinnon AO, Squires EL (1988a) Equine embryo transfer. Vet Clin North Am Equine Pract 4:305–333

McKinnon AO, Squires EL (1988b) Morphologic assessment of the equine embryo. J Am Vet Med Assoc 192:401–406

McKinnon AO, Brown RW, Pashen RL, Greenwood PE, Vasey JR (1992) Increased ovulation rates in mares after immunisation against recombinant bovine inhibin alpha-subunit. Equine Vet J 24:144–146

Meyers-Brown G, Bidstrup LA, Famula TR, Colgin M, Roser JF (2011) Treatment with recombinant equine follicle stimulating hormone (reFSH) followed by recombinant equine luteinizing hormone (reLH) increases embryo recovery in superovulated mares. Anim Reprod Sci 128:52–59

Mortensen CJ, Choi YH, Hinrichs K, Ing NH, Kraemer DC, Vogelsang SG, Vogelsang MM (2009) Embryo recovery from exercised mares. Anim Reprod Sci 110:237–244

Moussa M, Tremoleda JL, Duchamp G, Bruyas JF, Colenbrander B, Bevers MM, Daels PF (2004) Evaluation of viability and apoptosis in horse embryos stored under different conditions at 5° C. Theriogenology 61:921–932

Moussa M, Bersinger I, Doligez P, Guignot F, Duchamp G, Vidament M, Mermillod P, Bruyas JF (2005) In vitro comparisons of two cryopreservation techniques for equine embryos: slow-cooling and open pulled straw (OPS) vitrification. Theriogenology 64:1619–1632

Nagao JF, Neves Neto JR, Papa FO, Alvarenga MA, Freitas-Dell'Aqua CP, Dell'Aqua JA (2012) Induction of double ovulation in mares using deslorelin acetate. Anim Reprod Sci 136:69–73

Newcombe JR (1995) Incidence of multiple ovulation and multiple pregnancy in mares. Vet Rec 137:121–123

Niswender KD, Alvarenga MA, McCue PM, Hardy QP, Squires EL (2003) Superovulation in cycling mares using equine follicle stimulating hormone (eFSH). J Equine Vet Sci 23:497–500

Oberstein N, O'Donovan MK, Bruemmer JE, Seidel GE Jr, Carnevale EM, Squires EL (2001) Cryopreservation of equine embryos by open pulled straw, cryoloop, or conventional slow cooling methods. Theriogenology 55:607–613

Oguri N, Tsutsumi Y (1972) Non-surgical recovery of equine eggs, and an attempt at non-surgical egg transfer in horses. J Reprod Fertil 31:187–195

Oguri N, Tsutsumi Y (1974) Non-surgical egg transfer in mares. J Reprod Fertil 41:313–320

Okada CTC, Segabinazzi LG, Crespilho AM, Dell'Aqua JA Jr, Alvarenga MA (2018) Effect of the flunixin meglumine on pregnancy rates in an equine embryo transfer program. J Equine Vet 62:40–43

Oliveira Neto IV, Canisso IF, Segabinazzi LG, Dell'Aqua CPF, Alvarenga MA, Papa FO, Dell'Aqua JA Jr (2018) Synchronization of cyclic and acyclic embryo recipient mares with donor mares. Anim Reprod Sci 190:1–9

Ousey JC, Rossdale PD, Fowden AL, Palmer L, Turnbull C, Allen WR (2004) Effects of manipulating intrauterine growth on postnatal adrenocortical development and other parameters of maturity in neonatal foals. Equine Vet J 36:616–621

Panzani D, Rota A, Pacini M, Vannozzi I, Camillo F (2007) One year old fillies can be successfully used as embryo donors. Theriogenology 67:367–371

Panzani D, Rota A, Marmorini P, Vannozzi I, Camillo F (2014) Retrospective study of factors affecting multiple ovulations, embryo recovery, quality, and diameter in a commercial equine embryo transfer program. Theriogenology 82:807–814

Peugnet P, Wimel L, Duchamp G, Sandersen C, Camous S, Guillaume D, Dahirel M, Dubois C, Jouneau L, Reigner F, Berthelot V, Chaffaux S, Tarrade A, Serteyn D, Chavatte Palmer P (2014) Enhanced or reduced fetal growth induced by embryo transfer into smaller or larger breeds alters post-natal growth and metabolism in pre-weaning horses. PLoS One 9:e102044

Pinto MR, Miragaya MH, Burns P, Douglas R, Neild DM (2017) Strategies for increasing reproductive efficiency in a commercial embryo transfer program with high performance donor mares under training. J Equine Vet Sci 54:93–97

Ricketts SW, Alonso S (1991) The effect of age and parity on the development of equine chronic endometrial disease. Equine Vet J 23:189–192

Riera F, Roldan J, Hinrichs K (2006) Patterns of embryo recovery in mares with unilateral and bilateral double ovulations. Anim Reprod Sci 94:398–399

Rocha Filho AN, Pessoa MA, Gioso MM, Alvarenga MA (2004) Transfer of equine embryos into anovulatory recipients supplemented with short or long-acting progesterone. Anim Reprod Sci 1:91–95

Sanchez R, Blanco M, Weiss J, Rosati I, Herrera C, Bollwein H, Burger D, Sieme H (2017) Influence of embryonic size and manipulation on pregnancy rates of mares after transfer of cryopreserved equine embryos. J Equine Vet Sci 49:54–59

Skidmore JA, Boyle MS, Allen WR (1991) A comparison of two different methods of freezing horse embryos. J Reprod Fertil Suppl 44:714–716

Slade NP, Takeda T, Squires EL, Elsden RP, Seidel GE Jr (1985) A new procedure for the cryopreservation of equine embryos. Theriogenology 24:45–58

Smith RL, Vernon KL, Kelley DE, Gibbons JR, Mortensen CJ (2012) Impact of moderate exercise on ovarian blood flow and early embryonic outcomes in mares. J Anim Sci 90:3770–3777

Squires EL, McCue PM (2007) Superovulation in mares. Anim Reprod Sci 99:1–8

Squires EL, Garcia RH, Ginther OJ (1985) Factors affecting the success of equine embryo transfer. Equine Vet J Suppl 3:920–925

Squires EL, Mckinnon AO, Carnevale EM, Morris R, Nett TM (1987) Reproductive characteristics of spontaneous single and double ovulating mares and superovulated mares. J Reprod Fertil Suppl 35:399–403

Squires EL, McCue PM, Vanderwall D (1999) The current status of equine embryo transfer. Theriogenology 51:91–104

Squires EL, Carnevale EM, McCue PM, Bruemmer JE (2003) Embryo technologies in the horse. Theriogenology 59:151–170

Stout TAE (2003) Selection and management of the embryo transfer donor mare. Pferdeheilkunde 19:685–688

Stout TA, Meadows S, Allen WR (2005) Stage-specific formation of the equine blastocyst capsule is instrumental to hatching and to embryonic survival in vivo. Anim Reprod Sci 87:269–281

Tharasanit T, Colenbrander B, Stout TA (2005) Effect of cryopreservation on the cellular integrity of equine embryos. Reproduction 129:789–798

Vogelsang SG, Vogelsang MM (1989) Influence of donor parity and age on the success of commercial equine embryo transfer. Equine Vet J 21:71–72

Wilsher S, Allen WR (2004) An improved method for nonsurgical embryo transfer in the mare. Equine Vet Educ 16:39–44

Yanamoto Y, Oguri N, Tsutsumi Y (1982) Experiments in the freezing and storage of equine embryos. J Reprod Fertil Suppl 32:399–403

Endoscopy in Cattle Reproduction

9

Vitezslav Havlicek, Gottfried Brem, and Urban Besenfelder

Abstract

Final follicle maturation, ovulation and early embryo development are highly dynamic processes which ultimately result in establishment of pregnancy and the birth of healthy offspring. Any intrinsic or extrinsic changes of the environmental conditions, in vivo and in vitro, including deviations caused by exogenous hormonal stimulation may have negative effects on conceptus development. To date, many technologies have provided important information contributing to our knowledge of early embryo development. Among these techniques, the application of endoscopy for the study of reproductive processes, characterised by a minimal invasive transvaginal entry into the peritoneal cavity, plays a significant role. Once established, endoscopy allows the direct visualisation of the surface of ovaries, oviducts and uterine horns in accordance to pathophysiological changes and enables the collection and transfer of oocytes and embryos at various developmental stages. This technology is particularly suitable for combining in vivo and in vitro embryo culture in order to pinpoint critical checkpoints on this process. This type of translocation from laboratory to the animal and back provides a unique chance to create novel designs and to increase understanding of early reproductive events.

V. Havlicek · G. Brem · U. Besenfelder (✉)
Reproduction Centre for Cattle—Wieselburg, University of Veterinary Medicine-Vienna, Vienna, Austria
e-mail: Vitezslav.Havlicek@vetmeduni.ac.at; Gottfried.Brem@vetmeduni.ac.at; Urban.Besenfelder@boku.ac.at

© Springer International Publishing AG, part of Springer Nature 2018
H. Niemann, C. Wrenzycki (eds.), *Animal Biotechnology 1*,
https://doi.org/10.1007/978-3-319-92327-7_9

9.1 Introduction

Embryo transfer has become an integral part of science and farm breeding management. Many decades of embryo transfer activities in different species historically attest to the impact of this reproductive technology. Apart from the routine application of embryo transfer technologies, continuous improvement of these techniques opens new avenues especially when combining different disciplines. Much work has been done in many species. The focus of this chapter is on the use of endoscopy as a tool to manipulate and to better understand early reproductive events, particularly in cattle. The use of endoscopy in the bovine species will be highlighted in this chapter.

Based on the current developmental progress in science and technology, much and detailed information can be obtained by subdividing this topic into *cells/ embryos* and *animals/donor and recipient management*.

9.1.1 The Embryo

Embryo blastomeres represent targets for many diagnostic tools for genetic selection and reduction of the generation interval which significantly enhances efficacy and accelerates progress for research and commercial purposes. Besides the production of embryos via superovulation programmes, the in vitro culture of embryos allows visualisation of developmental steps from oocyte maturation and fertilisation to embryo cleavage up to the blastocyst stage and allows access to stages of early embryo development which are normally hidden and inaccessible when conventionally collecting embryos from the uterine horns. Although in vivo and in vitro methods are used to generate large numbers of embryos, the outcome of embryo culture is highly variable and ranges from early embryo death (Wiltbank et al. 2016) to normal development of the conceptus, implantation and birth of offspring. In order to understand the key mechanisms involved in early embryogenesis, many disciplines have been attracted, and major contributions have been made such as increasing visualisation by histological examinations (embryo structures, cell cleavage, degeneration, atresia; Abe and Hoshi 2003; Leidenfrost et al. 2011), facilitating and accelerating genetic evaluation using expression analyses (Gad et al. 2012), epigenetic approaches (Salilew-Wondim et al. 2015; Shojaei Saadi et al. 2014a), embryo genotyping (Thomasen et al. 2016; Shojaei Saadi et al. 2014b) and focusing on specified pathway analyses (Demant et al. 2015; Van Hoeck et al. 2011; Aardema et al. 2013).

9.1.2 Donor and Recipient Management

Many protocols have been experimentally proven to optimise embryo collection and to promote pregnancy following embryo transfer. To achieve this, detection strategies have been established for a better assessment of animal synchronisation, including precise estimation of heat and time of ovulation (Roelofs et al. 2010), development of the corpus luteum (Bollwein et al. 2013) and evaluation of the hormonal status and, in turn, the use of hormones for optimal synchronisation (Bó et al. 2010; Pereira et al. 2013). Besides slaughter, surgical collection allows the recovery of uterine as well as tubal-stage embryos. However, it is an invasive procedure which requires particular facilities and expertise (Newcomb and Rowson 1975; Wolfe et al. 1990). Embryo transfer using the surgical route provides an unbiased visual access to the ovary and its corpus luteum, the oviduct and the uterine horns. Hence, this technique allows a controlled manipulation, i.e. the successful and precise transfer of embryos very close to the uterotubal junction. Overall, surgical manipulation has been described as being superior to non-surgical transfer (reviewed by Hasler 2006, Newcomb et al. 1980); however, due to the easy applicability, including ethical aspects, non-invasive techniques are the most preferred. The application of ultrasound is appropriate for detection of ovarian structures such as number and size of follicles and corpora lutea and for pregnancy detection. Surgery provides unimpeded optical access to reproductive organs, whereas ultrasound presents us with an echo modified by the density of the tissue, thus producing a depth effect. Consequently, ultrasonography is well accepted, being more reliable for differentiation of ovarian structures than palpation (Ginther 2014; Smith et al. 2014). Overall, monitoring cyclic activity by ultrasound images has a beneficial effect on optimization of the management of bovine recipients (Guimarães et al. 2015).

9.1.3 Endoscopy

The use of endoscopy provides more options in animal reproduction. It is a minimally invasive procedure, providing direct access to the reproductive organs. However, currently the frequency of routine application in cattle is limited only to a small number of teams (Reichenbach et al. 1994; Santl et al. 1998; Besenfelder and Brem 1998; Wirtu et al. 2010). Initially, there were some successive attempts to perform endoscopic access via the lumbar and mid-ventral area (Sirard and Lambert 1985; Fayrer-Hosken et al. 1989; Laurincík et al. 1991). But this route has been replaced by the transvaginal entry into the abdominal cavity (Reichenbach et al. 1993). In this respect, some groundbreaking work has been carried out in other species such as rabbits, swine and small ruminants (Besenfelder and Brem 1993; Besenfelder et al. 1994, 1997) all of which have contributed to the refinement of this technique in the bovine species.

There are some fundamental characteristics of this technique which make it an attractive method for cattle reproduction:

- The endoscope consists of an optical axis having a small diameter causing minimal lesions.
- Fibre optic for visualisation and illumination of organs.
- Extra channels and tubes for assisting manipulation.
- Minimal anaesthetic intervention.
- Unique in situ approach, avoiding displacement of organs.
- Prevention or minimising post-traumatic damages.
- Repetitive use of the same animal possible.
- Applicable for multiple purposes (OPU, in vivo culture, embryo collection, embryo transfer).

As endoscopy can be used to study final follicular growth, ovulation and early embryo development, it combines development of early embryos derived from in vivo or in vitro for scientific as well as commercial purposes especially when looking at generating a high amount of developmentally competent embryos and, moreover, can be used as a tool to study the loss of embryos during the early stages of development (Lucy 2001; Diskin and Morris 2008).

9.2 Development of Laparoscopic Access to Ovaries and Oviducts

Attempts have been made to gain access to the oviducts of cattle in order to recover early stage embryos or ensure optimal culture conditions for in vitro matured and fertilised embryos. Initially, access to the bovine oviducts was performed surgically (Trounson et al. 1977) or through surgically prepared and cannulated fallopian tubes of a recipient cow (Jillella et al. 1977). Subsequently, the surgical procedure in cattle was replaced by the use of laparoscopy to minimise invasive access, manipulation stress and postsurgical complications and care. Laparoscopy provides a novel visual and manipulative approach to the bovine genital tract which serves as a basis for many applications such as recording physiological processes including changes during the oestrous cycle and pregnancy, recovery of oocytes from ovaries (Lambert et al. 1986; Sirard and Lambert 1985) and the recovery and transfer of early stage embryos via the oviducts. Fayrer-Hosken et al. (1989) developed a technique for transferring embryos into the oviduct using a bronchoscope. The manipulation of oviducts, ligaments and adjacent organs was done using a Semm's atraumatic forceps. The laparoscope and forceps were placed in the right paralumbar fossa. Mesovarium and fimbria of the oviduct ipsilateral to the ovulation were grasped by atraumatic forceps, and a Tom Cat catheter loaded with embryos was inserted into the oviduct. The transfer of two- to four-cell stage bovine embryos into the oviducts of four synchronised cows resulted in the birth of one healthy calf. Later, Reichenbach et al. (1993, 1994) described a simplified method for repeated laparoscopic exploration of reproductive organs and for aspirating oocytes from follicles of cows and heifers. The most important improvements they suggested were based on the transvaginal entry into the peritoneal cavity. A universal tube together with a blunt trocar was placed via the vagina in the middorsal area of the fornix. The blunt trocar was replaced by a traumatic trocar and introduced through the vaginal wall into the peritoneal cavity. When the traumatic trocar was pulled out of the universal tube, the slight peritoneal negative pressure caused suction of air into the cavity, necessary to have sufficient space for further manipulation under optical control. The bi-tubular system bearing the endoscope and the aspiration line was inserted into the universal tube. The ovaries were presented and navigated by slowly twisting in front of the endoscope via rectal manipulation. This procedure allowed the determination of the ovarian status and the correct positioning of follicles for ovum pickup. Reichenbach and co-workers concluded that this technique was suitable for repeated oocyte recovery without affecting animal fertility.

When using this transvaginal procedure for accessing uterine horns and ovaries, Besenfelder and Brem (1998) showed that it was possible to access the oviducts as well. In order to maximise the benefit from the oviductal impact on the developing embryo, special attention was paid to the preparation of animals. Animals were deprived of feed for 12 h and restrained in a crush. An epidural anaesthesia using procaine hydrochloride guaranteed rectal manipulation by suppressing rectal contractions. After introduction of the universal tube into the cavity, the amount of air which inflates the abdomen was gradually regulated to avoid the formation of a voluminous peritoneal space which has been considered to negatively affect careful manipulation of reproductive organs. The ultimate objective was to act in situ, to accurately determine the ovarian response to hormonal treatment and to avoid any kind of disproportional manipulation causing hyperaemia or bleeding (see Fig. 9.1). After manipulation, the air from the peritoneal cavity was released using a vacuum

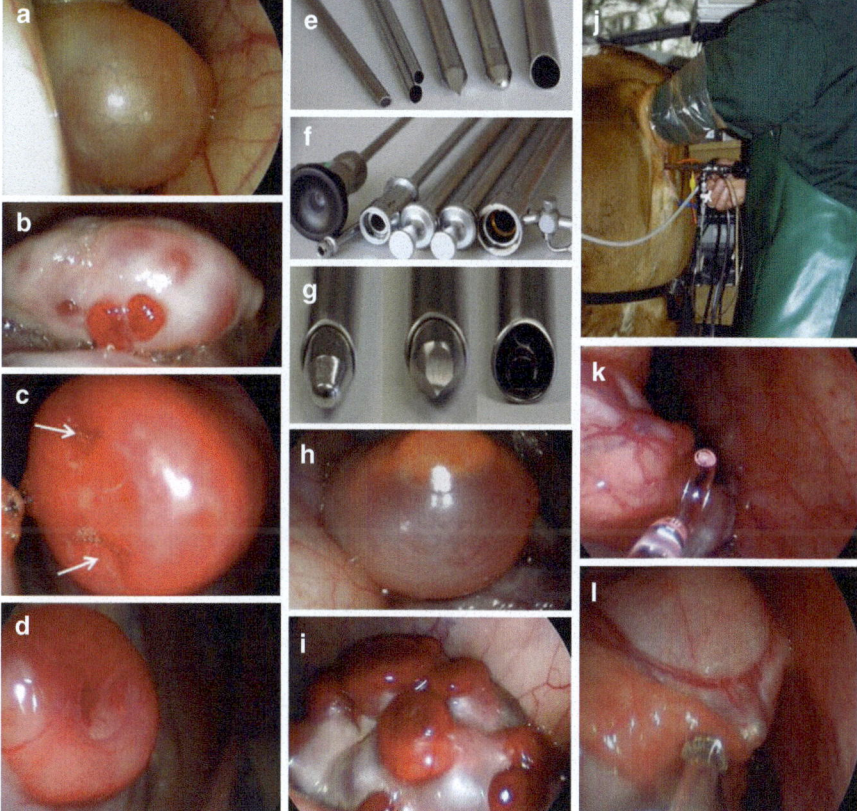

Fig. 9.1 Representative images of transvaginal endoscopy in cattle: (**a**) Graafian follicle showing luteinisation at the follicular basis; (**b**) ovary with a double ovulation at Day 1 and (**c**) at Day 3; (**d**) crown area of a Day 5 corpus luteum where the pulsation of blood vessels in the central dent could be observed; (**e**) tip and (**f**) rear of the endoscopic equipment (endoscope, bi-tubular working channel, traumatic and blunt trocar and universal tube); (**g**) universal tube plus an inserted blunt trocar, universal tube with an inserted traumatic trocar, universal tube and an inserted bi-tubular channel bearing the endoscope; (**h**) follicle 12 h before the expected time of ovulation and a regressing corpus luteum next to each other; (**i**) superstimulated ovary displaying Day 3 corpora lutea; (**j**) view of the donor animal during flushing showing the endoscope plus the uterine flushing catheter; (**k**) transfer glass capillary positioned parallel with the ampulla and (**l**) metal flushing catheter during introduction into the ampulla

pump, and the metal tube containing the endoscope was pulled out. Any further therapeutic treatment was not necessary (Besenfelder and Brem 1998; Besenfelder et al. 2001).

9.3   Transfer of Single Embryos into the Oviduct

Our first studies focused on the feasibility of transvaginal endoscopy as a method enabling the transfer of early stages of in vitro produced bovine embryos into the oviducts (Besenfelder and Brem 1998). The manipulation system consisted of a 1 ml syringe connected to a perfusor tube and a curved glass capillary. This system was entirely filled with medium and the embryos were loaded in the glass capillary. After insertion of the endoscope into the abdominal cavity, the side of ovulation was identified, and the quality of the corpus luteum (morphology and age) was estimated. The glass capillary was inserted via the infundibulum about 5 cm deep into the ipsilateral oviduct, and embryos were deposited in about 50 µl of medium in the ampulla. After overcoming initial problems, the duration of endoscopic manipulation now takes about 10 min.

First experiments describing the transfer of early stages of in vitro produced embryos in 24 animals resulted in the birth of eight calves. All of the short-term in vitro cultured and transferred embryos delivered to term with calves having a normal birth weight. In contrast, one embryo which had been cultured to the morula stage before transfer yielded an oversized calf which had to be delivered by caesarean section. These first data confirmed that the minimal invasive access to oviduct used is suitable for successful transfer of early tubal-stage embryos into a physiological environment (Besenfelder and Brem 1998).

9.4   Recovery of Early Embryo Stages from the Oviduct

Based upon the success of these first transfer experiments, the next target pursued aimed at collecting early stage bovine embryos from heifers by endoscopic flushing (Besenfelder et al. 2001).

For this purpose, the manipulation system was slightly adapted in order to allow oviduct flushing. The flushing system consisted of a 20 ml syringe, a perfusor tube and a curved metal tube with an olive in the front acting as palpation marker during fixation in the oviduct. After introduction of the metal flushing tube into the ampulla, the tube was fixed inside by slight pressure of thumb and forefinger around the olive. Later, the metal tube has been modified in order to enable flushing without digital fixation. For that purpose, numerous lateral holes were drilled and covered by a silicon tube. Consequently, increasing flushing pressure resulted in the inflation of the silicone tube similar to form a balloon which hermetically seals the oviduct and avoids reflux of flushing medium during the oviductal flushing process.

The embryos were flushed orthograde into the tip of the uterine horns where a uterine embryo flushing catheter was fixed as is normally done in MOET programmes for collection of Day 7 embryos. The oviducts were flushed with 40–60 ml of medium via the uterotubal junction in the direction of the tip of uterine horn. Flushing medium with embryos passed into the uterine embryo flushing catheter connected to an embryo filter. First signs of successful tubal flushing were obtained when a medium flow from the uterine horns into the embryo filter could be observed. Additionally, the uterine horns were flushed with 300–500 ml medium each to ensure that a maximal number of embryos located close to the tip of uterine horns were recovered.

The oviductal flushing has been developed stepwise (Besenfelder et al. 2001). First, single-ovulating animals were flushed following unilateral flushing of superovulated donors. The procedure became successful through continuous refinement and tuning of the system and practice. Ultimately, the oviducts of superovulated animals were flushed bilaterally. In total, it was shown that nearly all oocytes and embryos could be recovered from oviduct flushing. In some cases, mainly depending on the success of hormonal treatment, it was difficult to record the exact numbers of those corpora lutea which were close together on the surface of the ovary forming one big confluent luteal area.

Once developed and established, this flushing method has been used for studies examining the effect of different hormonal treatments, developmental kinetics and repeated collection on embryo recovery (Besenfelder et al. 2008). In a study in which 119 superovulated animals using either FSH or eCG were bilaterally flushed, more than 1400 oocytes/embryos at various stages were collected. The flushing of all these animals illustrated the correlation between hormones and ovarian responses, including different sizes and appearances of follicles and corpora lutea and embryo cleavage during the first days of development. There was no negative effect of repeated flushing which confirmed the usefulness and applicability of this method.

9.5 In Vivo Culture of Bovine Embryos

During the last two decades, the production of bovine embryos for commercial purposes significantly increased. According to the data collated by the International Embryo Transfer Society (www.iets.com), more than one million bovine embryos are now produced annually. Over 600,000 transferable in vivo-derived embryos were collected in 2014, and another almost 600,000 bovine embryos were produced in vitro (Perry 2015). The increase in the in vitro production of bovine embryos has been most dramatic in the last 10 years, mainly due to an exponential increase in the use of the technology in Brazil, which now accounts for most of the commercial in vitro embryo transfer activity. Moreover, in the context of genomic selection in animal breeding, there is a boom awaited in the field of IVP mainly to reduce generation interval and increase selection pressure (Ponsart et al. 2013).

Prerequisites for successful and efficient production of embryos derived in vitro are the collection of high numbers of COCs and developmentally competent embryos leading to pregnancies and the birth of healthy calves. Rizos et al. (2002b) emphasised that the intrinsic quality of the oocyte is the main factor affecting blastocyst yields, while the conditions of embryo culture play a pivotal role in determining blastocyst quality.

It has been well accepted that the in vitro environment markedly differs from the physiological milieu provided by oviducts and uterus, each of them capable of precisely responding to various changing demands such as the dynamic metabolic requirements according to each embryonic stage (Killian 2004).

The overall objective will always accomplish IVP broadly similar to oviductal performance by mimicking temporal changes in embryo requirements and oviductal fluid composition (Felmer et al. 2011; Wydooghe et al. 2014), using conditioned media (Lopera-Vásquez et al. 2016) or coculture systems with various cell types (Schmaltz-Panneau et al. 2015). Thirty years of research in this field have led to much progress, however, and from the current point of view, it seems to be an almost insurmountable challenge to copy in vitro the multitasking feature of the oviducts since regulatory key mechanisms are still not understood. Consequently, in vitro produced embryos qualitatively lag behind their in vivo counterparts which can be seen in many details such as morphology (Rizos et al. 2002a), altered gene expression (Tesfaye et al. 2004), embryo metabolism (De Souza et al. 2015), increasing cryo-sensitivity (Pollard and Leibo 1994) and embryo/foetal development after transfer and calves after birth (Young et al. 2001).

Being aware of these high demands, it is advisable to directly use tubal features for the improvement of embryo quality by transferring in vitro-derived embryos into the oviducts of temporary recipients.

In first studies attempting to benefit from the fallopian tube, an interspecies transfer was conducted. Bovine embryos were transferred into rabbit oviducts and temporarily cultured in vivo (Rowson and Adams 1972; Sirard et al. 1985). Other researchers cultured in vitro matured/fertilised bovine embryos in the oviducts of ewes (Sirard et al. 1988; Enright et al. 2000). The ewes were hormonally synchronised and prepared by an intravaginal progestogen-releasing device. Early stages of bovine embryos embedded in agar chips or without agar were surgically transferred into the oviducts prior to the ligation close to the uterotubal junction. The embryos were recovered 4–5 days later by the same surgical procedure (reviewed by Lazzari et al. 2010). The practical relevance of in vivo cultured bovine embryos in surrogate sheep oviducts was described by Galli et al. (2001, 2003). It was shown that this technology could be very efficient for the production of large numbers of embryos for commercial purposes. Moreover, the produced embryos were comparable to MOET embryos especially when they had been frozen-thawed before transfer (Galli et al. 2001, 2003). Nevertheless, the use of progestogen-supplemented, ligated heterologous sheep oviducts for in vivo culture of bovine embryos actually does not provide the basic scientific approach necessary to reveal species-specific particularities for bovine reproduction.

This in vivo approach has been accomplished in the bovine species by merging the endoscopic transfer with embryo collection procedure (Havlicek et al. 2005a). Unlike the hormonal treatment of sheep, bovine recipients are synchronised according to the developmental stage of the embryos. Embryos are transferred during the early growth of a corpus luteum which involves changes and modifications of the tubal epithelium to maximally meet the needs of the embryo. Hence, embryos are placed preferentially ipsilateral to ovulation.

In our first experiments, about 2500 embryos were transferred in groups of 10–50 embryos each after 1–3 days of in vitro culture into the oviducts of synchronised heifers. Recollection was performed 4–6 days later. The recovery rate of embryos at Day 7 revealed the different migration of embryos from oviducts in uterine horns. After solely flushing of uterine horns followed by a second flush using a combined flushing of both oviducts and uterine horns, only about half of embryos were found in the uterus, whereas the other half of the embryos remained in the oviducts. Hence, combined flushing of oviducts and uterine horns was recommended for further and effective embryo recollection after in vivo culture in the oviduct (Havlicek et al. 2005a, b).

In the following studies, Wetscher et al. (2005a, b) examined factors such as temperature, embryo structure, developmental stage, gamete co-incubation and in vitro maturation influencing in vivo culture efficiency:

1. Our first in vivo culture results revealed variable success reflecting the high demand of embryos at a very sensitive stage. Changes of medium for the transfer, a short-term decrease of temperature of the medium and long duration in which the embryos are kept outside the incubator prior to transfer usually decreased blastocyst rates.
2. Tubal migration of transferred embryos is affected by their morphology. The recollection rate increased with the size of a solid matrix around the embryo. Therefore, the best recovery rates were obtained when zygotes were embedded in sodium alginate or transferred in cumulus cells. In contrast, a lower proportion of denuded zygotes or embryos in medium containing 6 mg/ml hyaluronan were recollected, probably due to disturbed migration caused by disoriented beating activity of the ciliated cells in the oviduct. There is much evidence that embryos were expelled into the abdominal cavity.
3. Embryo transfers on Day 1 and 2 resulted in a lower recovery rates on Day 7 compared to transfers on Days 3 and 4. In the periovulation period, the oviducts appear to be hyperactive compared to Days 3 and later (Ruckebusch and Bayard 1975).
4. Blastocyst rates correlate with the stage of transferred oocytes and embryos to synchronous recipient animals. The transfer of more advanced embryonic stages into the oviduct resulted in significantly higher blastocyst rates compared to the transfer of very early stages (Havlicek et al. 2005b; Wetscher et al. 2005b).
5. Gamete intrafallopian transfer (GIFT) seems not to work very well in cattle for starting in vivo culture. The transfer of a mix of in vitro matured COCs with capacitated spermatozoa did not result in an acceptable amount of blastocysts. However, when COCs and spermatozoa were co-incubated for at least 3–4 h before transfer, there was a significant success of blastocyst development.
6. In vitro matured oocytes are not compatible with those matured in vivo, since only a few blastocysts were obtained after the transfer of in vitro matured oocytes into the oviduct of inseminated heifers. Most of the oocytes did not show any

sign of fertilisation. Compared to the high numbers and concentration of spermatozoa necessary to successfully accomplish in vitro fertilisation of in vitro matured oocytes, it seems most likely that failures in fertilisation were caused by the low number of spermatozoa in the oviduct available for fertilisation of in vitro matured oocytes (Ward et al. 2002). Moreover, there is also evidence that initiation of zona hardening occurs immediately when oocytes are in contact with the epithelial cells and are exposed to the oviduct-specific glycoprotein (Coy and Avilés 2010).

7. The zona pellucida undergoes physical changes not only during fertilisation but also during the oviductal passage. Mertens et al. (2007) showed that embryos which migrated through the oviducts into the uterine horns and developed into morulae and blastocysts had a thicker ZP compared to in vitro cultured embryos. Histological examinations revealed an increase in the reticular part, the pores in the ZP were smaller in size and the surface was covered by granules. In contrast, the ZP of in vitro produced embryos showed signs of degeneration (Mertens et al. 2007). Besides the fact that the ZP texture reflects active molecule transportation between the embryo and its environment, this structure also plays a role in the context of reducing sanitary risks when transferring embryos (Van Soom et al. 2010).

Further studies benefited from these first trials and aimed at examining the most critical developmental stages during the early embryo culture period, in vivo culture of embryos in heifers and cows (dried-off vs. milking) and embryo development under superovulation conditions (see Table 9.1).

In a large-scale study, numerous embryos at various stages were produced in vitro and then cultured in vivo for the remaining time until blastocyst stage at Day 7 and vice versa embryos at different stages were collected from the oviducts and in vitro cultured until the blastocyst stage. Using expression profile analyses, it was shown that the most critical developmental steps were found to be around fertilisation, during embryo genome activation around the 8-cell stage and during blastocyst formation. The source of oocyte collection and maturation had detrimental effect on embryo quantity (Gad et al. 2012).

In order to examine the influence of lactation on early embryo development, about 2800 in vitro-derived embryos were transferred to heifers, dairy milking cows and cows which were dried-off immediately after parturition. Embryos were recovered on Day 7. It was demonstrated that the reproductive tract of post-partum dairy cows was less capable of providing the adequate environment for an optimal embryo development compared to heifers (Rizos et al. 2010) and dry Holstein cows (Maillo et al. 2012). Moreover, even the superstimulated reproductive tract significantly affected embryo development during the first 7 days compared to in vivo culture of embryos derived from superovulation donors and transfer into non-superovulated (mono-ovulatory) animals (Gad et al. 2011). Overall, it has been shown that early developing embryos respond rapidly to even small environmental changes which in turn provide a useful indicator system for inadequate culture conditions.

Table 9.1 Recovery and blastocyst development of bovine embryos after in vivo culture in cattle

Transferred embryos						
Origin	Stage	Day of transfer/ recipients	No. of transferred embryos	No. of recovered embryos at Day 7 (%)	No. of blastocysts (%)	Authors
In vitro	2–16-cell	D1–4/ heifers	100	31 uterus/34 oviduct	17 (26.2)	Havlicek et al. (2005a)
In vitro	2–4-cell	D1–2/ heifers	162	75 (46.3)	10 (13.3)	
In vitro	4–8-cell	D3/heifers	199	68 (34.2)	25 (36.8)	
In vitro	1–8-cell	D1–3/ heifers	1358	390 (28.7): recovery from the uterus	48 (12.3)	Havlicek et al. (2005b)
In vitro	1–8-cell	D1–3/ heifers	671	390 (58.1): recovery from the oviduct and uterus	105 (26.9)	
In vitro	COCs	D1/heifers	456	348 (76.3)	3 (0.9)	Wetscher et al. (2005b)
In vitro	GIFT	D1/heifers	514	351 (68.3)	70 (19.9)	
In vitro	4–8-cell	D3/heifers	682	545 (79.9)	304 (43.3)	
In vitro	GIFT	D1/heifers	425	315 (74.1)	114 (36.2)	Havlicek et al. (2010)
In vitro	4–8-cell	D2–3/ heifers	441	264 (59.9)	108 (40.9)	
In vitro	2–4-cell	D2/heifers	1000	790 (79.0)	273 (35.5)	Rizos et al. (2010)
In vitro	2–4-cell	D2/lactating cows	800	458 (57.2)	73 (15.9)	
In vitro	2–4-cell	D2/heifers	2004	1629 (81.3)	953 (58.5)	Carter et al. (2010)
In vitro	2–4-cell	D2/heifers high P4	1673	1240 (75.7)	742 (59.8)	
In vivo	2–4-cell	D2/heifers	164	146 (89.0)	76 (52.1)	Gad et al. (2011)
In vitro	4-cell	D2/heifers	642	642 (88.5)	223 (39.3)	Gad et al. (2012)
In vitro	16-cell	D4/heifers	811	811 (76.4)	350 (56.5)	
In vitro	2–4-cell	D2/lactating cows	435	289 (65.6)	97 (32.6)	Maillo et al. (2012)
In vitro	2–4-cell	D2/ non-lactating cows	627	403 (63.9)	203 (49.3)	

Conclusion

The application of endoscopy to embryo collection and transfer provides unrivalled access to the reproductive tract as well as facilitating the collection and manipulation of various different oocyte and embryo types resulting in increased knowledge and understanding of embryo development under optimal environmental conditions. While the use of endoscopy requires particular expertise and experience, this technique can be easily combined with other routine reproductive technique in order to generate large numbers of developmentally competent embryos for both science and commercial application.

References

Aardema H, Lolicato F, van de CHA L, Brouwers JF, Vaandrager AB, van HTA T, Roelen BAJ, Vos PLAM, Helms JB, Gadella BM (2013) Bovine cumulus cells protect maturing oocytes from increased fatty acid levels by massive intracellular lipid storage. Biol Reprod 88(164):1–15. https://doi.org/10.1095/biolreprod.112

Abe H, Hoshi H (2003) Evaluation of bovine embryos produced in high performance serum-free media. J Reprod Dev 49:193–202

Besenfelder U, Brem G (1993) Laparoscopic embryo transfer in rabbits. J Reprod Fertil 99:53–56

Besenfelder U, Brem G (1998) Tubal transfer of bovine embryos: a simple endoscopic method reducing long-term exposure of in vitro produced embryos. Theriogenology 50:739–745

Besenfelder U, Zinovieva N, Dietrich E, Sohnrey B, Holtz W, Brem G (1994) Tubal transfer of goat embryos using endoscopy. Vet Rec 135:480–481

Besenfelder U, Moedl J, Mueller M, Brem G (1997) Endoscopic embryo collection and embryo transfer into the oviduct and the uterus of pigs. Theriogenology 47:1051–1060

Besenfelder U, Havlicek V, Mösslacher G, Brem G (2001) Collection of tubal stage bovine embryos by means of endoscopy. A technique report. Theriogenology 55:837–845

Besenfelder U, Havlicek V, Moesslacher G, Gilles M, Tesfaye D, Griese J, Hoelker M, Hyttel PM, Laurincik J, Brem G, Schellander K (2008) Endoscopic recovery of early preimplantation bovine embryos: effect of hormonal stimulation, embryo kinetics and repeated collection. Reprod Domest Anim 43:566–572

Bó GA, Guerrero DC, Tríbulo A, Tríbulo H, Tríbulo R, Rogan D, Mapletoft RJ (2010) New approaches to superovulation in the cow. Reprod Fertil Dev 22:106–112

Bollwein H, Lüttgenau J, Herzog K (2013) Bovine luteal blood flow: basic mechanism and clinical relevance. Reprod Fertil Dev 25:71–79

Carter F, Rings F, Mamo S, Holker M, Kuzmany A, Besenfelder U, Havlicek V, Mehta JP, Tesfaye D, Schellander K, Lonergan P (2010) Effect of elevated circulating progesterone concentration on bovine blastocyst development and global transcriptome following endoscopic transfer of in vitro produced embryos to the bovine oviduct. Biol Reprod 83:707–719

Coy P, Avilés M (2010) What controls polyspermy in mammals, the oviduct or the oocyte? Biol Rev Camb Philos Soc 85:593–605

De Souza DK, Salles LP, Rosa e Silva AA (2015) Aspects of energetic substrate metabolism of in vitro and in vivo bovine embryos. Braz J Med Biol Res 48:191–197

Demant M, Deutsch DR, Froehlich T, Wolf E, Arnold GJ (2015) Proteome analysis of early lineage specification in bovine embryos. Proteomics 15:688–701

Diskin MG, Morris DG (2008) Embryonic and early foetal losses in cattle and other ruminants. Reprod Domest Anim 43:260–267

Enright BP, Lonergan P, Dinnyes A, Fair T, Ward FA, Yang X, Boland MP (2000) Culture of in vitro produced bovine zygotes in vitro vs in vivo: implications for early embryo development and quality. Theriogenology 54:659–673

Fayrer-Hosken RA, Younis AI, Brackett BG, McBride CE, Harper KM, Keefer CL, Cabaniss DC (1989) Laparoscopic oviductal transfer of in vitro matured and in vitro fertilized bovine oocytes. Theriogenology 32:413–420

Felmer RN, Arias ME, Muñoz GA, Rio JH (2011) Effect of different sequential and two-step culture systems on the development, quality, and RNA expression profile of bovine blastocysts produced in vitro. Mol Reprod Dev 78:403–414

Gad A, Besenfelder U, Rings F, Ghanem N, Salilew-Wondim D, Hossain MM, Tesfaye D, Lonergan P, Becker A, Cinar U, Schellander K, Havlicek V, Hölker M (2011) Effect of reproductive tract environment following controlled ovarian hyperstimulation treatment on embryo development and global transcriptome profile of blastocysts: implications for animal breeding and human assisted reproduction. Hum Reprod 26:1693–1707

Gad A, Hoelker M, Besenfelder U, Havlicek V, Cinar U, Rings F, Held E, Dufort I, Sirard MA, Schellander K, Tesfaye D (2012) Molecular mechanisms and pathways involved in bovine embryonic genome activation and their regulation by alternative in vivo and in vitro culture conditions. Biol Reprod 87(100):1–13. https://doi.org/10.1095/biolreprod.112.099697

Galli C, Crotti G, Notari C, Turini P, Duchi R, Lazzari G (2001) Embryo production by ovum pick up from live donors. Theriogenology 55:1341–1357

Galli C, Duchi R, Crotti G, Turini P, Ponderato N, Colleoni S, Lagutina I, Lazzari G (2003) Bovine embryo technologies. Theriogenology 59:599–616

Ginther OJ (2014) How ultrasound technologies have expanded and revolutionized research in reproduction in large animals. Theriogenology 81:112–125 Review

Guimarães CR, Oliveira ME, Rossi JR, Fernandes CA, Viana JH, Palhao MP (2015) Corpus luteum blood flow evaluation on day 21 to improve the management of embryo recipient herds. Theriogenology 84:237–241

Hasler JF (2006) The Holstein cow in embryo transfer today as compared to 20 years ago. Theriogenology 65:4–16

Havlicek V, Wetscher F, Huber T, Brem G, Mueller M, Besenfelder U (2005a) In vivo culture of IVM/ IVF embryos in bovine oviducts by transvaginal endoscopy. J Vet Med A Physiol Pathol Clin Med 52:94–98

Havlicek V, Lopatarova M, Cech S, Dolezel R, Huber T, Pavlok A, Brem G, Besenfelder U (2005b) In vivo culture of bovine embryos and quality assessment of in vivo vs. in vitro produced embryos. Vet Med–Czech 50:149–157

Havlicek V, Kuzmany A, Cseh S, Brem G, Besenfelder U (2010) The effect of long-term in vivo culture in bovine oviduct and uterus on the development and cryo-tolerance of in vitro produced bovine embryos. Reprod Domest Anim 45:832–837

Jillella D, Eaton RJ, Baker AA (1977) Successful transfer of a bovine embryo through a cannulated fallopian tube. Vet Rec 100:385–386

Killian GJ (2004) Evidence for the role of oviduct secretions in sperm function, fertilization and embryo development. Anim Reprod Sci 82–83:141–153

Lambert RD, Sirard MA, Bernard C, Béland R, Rioux JE, Leclerc P, Ménard DP, Bedoya M (1986) In vitro fertilization of bovine oocytes matured in vivo and collected at laparoscopy. Theriogenology 25:117–133

Laurincík J, Pícha J, Píchová D, Oberfranc M (1991) Timing of laparoscopic aspiration of preovulatory oocytes in heifers. Theriogenology 35:415–423

Lawson RA, Rowson LE, Adams CE (1972) The development of cow eggs in the rabbit oviduct and their viability after re-transfer to heifers. J Reprod Fertil 28:313–315

Lazzari G, Colleoni S, Lagutina I, Crotti G, Turini P, Tessaro I, Brunetti D, Duchi R, Galli C (2010) Short-term and long-term effects of embryo culture in the surrogate sheep oviduct versus in vitro culture for different domestic species. Theriogenology 73:748–757

Leidenfrost S, Boelhauve M, Reichenbach M, Güngör T, Reichenbach H-D, Sinowatz F, Wolf E, Habermann FA (2011) Cell arrest and cell death in mammalian preimplantation development: lessons from the bovine model. PLoS One 6:e22121. https://doi.org/10.1371/journal.pone.0022121

Lopera-Vásquez R, Hamdi M, Fernandez-Fuertes B, Maillo V, Beltrán-Breña P, Calle A, Redruello A, López-Martín S, Gutierrez-Adán A, Yañez-Mó M, Ramirez MÁ, Rizos D (2016) Extracellular vesicles from BOEC in in vitro embryo development and quality. PLoS One 11:e0148083. https://doi.org/10.1371/journal.pone.0148083

Lucy MC (2001) Reproductive loss in high-producing dairy cattle: where will it end? J Dairy Sci 84:1277–1293

Maillo V, Rizos D, Besenfelder U, Havlicek V, Kelly AK, Garrett M, Lonergan P (2012) Influence of lactation on metabolic characteristics and embryo development in postpartum Holstein dairy cows. J Dairy Sci 95:3865–3876

Mertens E, Besenfelder U, Gilles M, Holker M, Rings F, Havlicek V, Schellander K, Herrler A (2007) Influence of in vitro culture of bovine embryos on the structure of the zona pellucida. Reprod Fertil Dev 19:211–212

Newcomb R, Rowson LE (1975) A technique for the simultaneous flushing of ova from the bovine oviduct and uterus. Vet Rec 96:468–469

Newcomb R, Christie WB, Rowson LE (1980) Fetal survival rate after the surgical transfer of two bovine embryos. J Reprod Fertil 59:31–36

Pereira MHC, Sanches CP, Guida TG, Rodrigues ADP, Aragon FL, Veras MB, Borges PT, Wiltbank MC, Vasconcelos JLM (2013) Timing of prostaglandin F2 treatment in an estrogen-based protocol for timed artificial insemination or timed embryo transfer in lactating dairy cows. J Dairy Sci 96:2837–2846

Perry G (2015) 2014 statistics of embryo collection and transfer in domestic farm animals. Embryo Transfer Newsletter 33:9–18

Pollard JW, Leibo SP (1994) Chilling sensitivity of mammalian embryos. Theriogenology 41:101–106

Ponsart C, Le Bourhis D, Knijn H, Fritz S, Guyader-Joly C, Otter T, Lacaze S, Charreaux F, Schibler L, Dupassieux D, Mullaart E (2013) Reproductive technologies and genomic selection in dairy cattle. Reprod Fertil Dev 26:12–21

Reichenbach HD, Wiebke NH, Besenfelder UH, Moedl J, Brem G (1993) Transvaginal laparoscopic guided aspiration of bovine follicular oocytes: preliminary results. Theriogenology 39:295 (Abstr.)

Reichenbach HD, Wiebke NH, Moedl J, Zhu J, Brem G (1994) Laparoscopy through the vaginal fornix of cows for the repeated aspiration of follicular oocytes. Vet Rec 135:353–356

Rizos D, Fair T, Papadopoulos S, Boland MP, Lonergan P (2002a) Developmental, qualitative, and ultrastructural differences between ovine and bovine embryos produced in vivo or in vitro. Mol Reprod Dev 62:320–327

Rizos D, Ward F, Duffy P, Boland MP, Lonergan P (2002b) Consequences of bovine oocyte maturation, fertilization or early embryo development in vitro versus in vivo: implications for blastocyst yield and blastocyst quality. Mol Reprod Dev 61:234–248

Rizos D, Carter F, Besenfelder U, Havlicek V, Lonergan P (2010) Contribution of the female reproductive tract to low fertility in postpartum lactating dairy cows. J Dairy Sci 93:1022–1029

Roelofs J, López-Gatius F, Hunter RH, van Eerdenburg FJ, Hanzen C (2010) When is a cow in estrus? Clinical and practical aspects. Theriogenology 74:327–344

Ruckebusch Y, Bayard F (1975) Motility of the oviduct and uterus of the cow during the oestrous cycle. J Reprod Fertil 43:23–32

Salilew-Wondim D, Fournier E, Hoelker M, Saeed-Zidane M, Tholen E, Looft C, Neuhoff C, Besenfelder U, Havlicek V, Rings F, Gagné D, Sirard MA, Robert C, Shojaei Saadi HA, Gad A, Schellander K, Tesfaye D (2015) Genome-wide DNA methylation patterns of bovine blastocysts developed in vivo from embryos completed different stages of development in vitro. PLoS One 10:e0140467. https://doi.org/10.1371/journal.pone.0140467

Santl B, Wenigerkind H, Schernthaner W, Moedl J, Stojkovic M, Prelle K, Holtz W, Brem G, Wolf E (1998) Comparison of ultrasound-guided vs laparoscopic transvaginal ovum pick-up (OPU) in Simmental heifers. Theriogenology 50:89–100

Schmaltz-Panneau B, Locatelli Y, Uzbekova S, Perreau C, Mermillod P (2015) Bovine oviduct epithelial cells dedifferentiate partly in culture, while maintaining their ability to improve early embryo development rate and quality. Reprod Domest Anim 50:719–729

Shojaei Saadi HA, O'Doherty AM, Gagné D, Fournier E, Grant JR, Sirard MA, Robert C (2014a) An integrated platform for bovine DNA methylome analysis suitable for small samples. BMC Genomics 15:451

Shojaei Saadi HA, Vigneault C, Sargolzaei M, Gagné D, Fournier E, de Montera B, Chesnais J, Blondin P, Robert C (2014b) Impact of whole-genome amplification on the reliability of pre-transfer cattle embryo breeding value estimates. BMC Genomics 15:889 http://www.biomed-central.com/1471-2164/15/889

Sirard MA, Lambert RD (1985) In vitro fertilization of bovine follicular oocytes obtained by laparoscopy. Biol Reprod 33:487–494

Sirard MA, Lambert RD, Ménard DP, Bedoya M (1985) Pregnancies after in-vitro fertilization of cow follicular oocytes, their incubation in rabbit oviduct and their transfer to the cow uterus. J Reprod Fertil 75:551–556

Sirard MA, Parrish JJ, Ware CB, Leibfried-Rutledge ML, First NL (1988) The culture of bovine oocytes to obtain developmentally competent embryos. Biol Reprod 39:546–552

Smith RF, Oultram J, Dobson H (2014) Herd monitoring to optimise fertility in the dairy cow: making the most of herd records, metabolic profiling and ultrasonography (research into practice). Animal 8(Suppl 1):185–198 Review

Tesfaye D, Ponsuksili S, Wimmers K, Gilles M, Schellander K (2004) A comparative expression analysis of gene transcripts in post-fertilization developmental stages of bovine embryos produced in vitro or in vivo. Reprod Domest Anim 39:396–404

Thomasen JR, Willam A, Egger-Danner C, Sørensen AC (2016) Reproductive technologies combine well with genomic selection in dairy breeding programs. J Dairy Sci 99:1331–1340

Trounson AO, Willadsen SM, Rowson LE (1977) Fertilization and development capability of bovine follicular oocytes matured in vitro and in vivo and transferred to the oviducts of rabbits and cows. J Reprod Fertil 51:321–327

Van Hoeck V, Sturmey RG, Bermejo-Alvarez P, Rizos D, Gutierrez-Adan A et al (2011) Elevated non-esterified fatty acid concentrations during bovine oocyte maturation compromise early embryo physiology. PLoS One 6:e23183. https://doi.org/10.1371/journal.pone.0023183

Van Soom A, Wrathall AE, Herrler A, Nauwynck HJ (2010) Is the zona pellucida an efficient barrier to viral infection? Reprod Fertil Dev 22:21–31

Ward F, Enright B, Rizos D, Boland M, Lonergan P (2002) Optimization of in vitro bovine embryo production: effect of duration of maturation, length of gamete co-incubation, sperm concentration and sire. Theriogenology 57:2105–2117

Wetscher F, Havlicek V, Huber T, Mueller M, Brem G, Besenfelder U (2005a) Effect of morphological properties of transferred embryonic stages on tubal migration implications for in vivo culture in the bovine oviduct. Theriogenology 64:41–48

Wetscher F, Havlicek V, Huber T, Gilles M, Tesfaye D, Griese J, Wimmers K, Schellander K, Müller M, Brem G, Besenfelder U (2005b) Intrafallopian transfer of gametes and early stage embryos for in vivo culture in cattle. Theriogenology 64:30–40

Wiltbank MC, Baez GM, Garcia-Guerra A, Toledo MZ, Monteiro PL, Melo LF, Ochoa JC, Santos JE, Sartori R (2016) Pivotal periods for pregnancy loss during the first trimester of gestation in lactating dairy cows. Theriogenology 86:239–253

Wirtu G, MacLean R, Galiguis J, Paccamonti D, Eilts B, Godke R, Besenfelder U, Dresser B, Gentry G (2010) Endoscope-guided transfer of sperm-injected oocytes into the oviducts of eland and bongo antelopes. Reprod Fertil Dev 22:259 (Abstr.)

Wolfe DF, Riddell MG, Mysinger PW, Stringfellow DA, Carson RL, Garrett PD (1990) A caudal flank approach for the collection of oviductal-stage bovine embryos. Theriogenology 34:167–174

Wydooghe E, Heras S, Dewulf J, Piepers S, Van den Abbeel E, De Sutter P, Vandaele L, Van Soom A (2014) Replacing serum in culture medium with albumin and insulin, transferrin and selenium is the key to successful bovine embryo development in individual culture. Reprod Fertil Dev 26:717–724

Young LE, Fernandes K, McEvoy TG, Butterwith SC, Gutierrez CG, Carolan C, Carolan C, Broadbent PJ, Robinson JJ, Wilmut I, Sinclair KD (2001) Epigenetic change in IGF2R is associated with fetal overgrowth after sheep embryo culture. Nat Genet 27:153–154

Transvaginal Ultrasound-Guided Oocyte Retrieval (OPU: Ovum Pick-Up) in Cows and Mares

10

Peter E. J. Bols and Tom A. E. Stout

Abstract

For about three decades, transvaginal ultrasound-guided oocyte retrieval (OPU, ovum pick-up) has been successfully adapted from human reproductive medicine to the use in cattle and later on in the horse. Over time, it turned out to be a reliable and minimally invasive method to collect (immature) oocytes from genetically high valuable donors on a repeated basis. While a large part of the success of this procedure relies on the availability of a reliable in vitro embryo production system, a major prerequisite remains the collection of good-quality oocytes. The current chapter will focus specifically on oocyte retrieval technology. Following a detailed description of OPU equipment, the technical and biological factors affecting oocyte retrieval in living donors are discussed extensively with particular interest on the need of donor preparation by hormonal stimulation. Attention will also be given to donor health issues related to repeated oocyte retrieval. Finally, a state of the art of OPU in the mare is given describing additional physiological aspects of the equine oocyte and embryo implying additional challenges both for oocyte retrieval and in vitro embryo production.

P. E. J. Bols (✉)
Department of Veterinary Sciences, Faculty of Pharmaceutical, Biomedical and Veterinary Sciences, University of Antwerp, Wilrijk, Belgium
e-mail: peter.bols@uantwerpen.be

T. A. E. Stout
Department of Equine Sciences, Faculty of Veterinary Medicine, Utrecht University, Utrecht, The Netherlands
e-mail: T.A.E.Stout@uu.nl

© Springer International Publishing AG, part of Springer Nature 2018
H. Niemann, C. Wrenzycki (eds.), *Animal Biotechnology 1*,
https://doi.org/10.1007/978-3-319-92327-7_10

10.1   Introduction

For several decades, puncture and aspiration of bovine (immature) ovarian follicles
has been used to retrieve oocytes for in vitro embryo production (IVP). Several comprehensive reviews on IVP and embryo transfer (ET) in domestic animals have highlighted the availability of 'good'-quality oocytes as the primary prerequisite for
success (Hasler 1998; Galli et al. 2001; Merton et al. 2003; Merton 2014). Cumulus
oocyte complexes (COCs) can be recovered from the ovaries of both slaughtered
cows and living donors. Traditionally, post-mortem oocyte recovery was accomplished by follicle dissection or aspiration with a needle and syringe. However, this
resulted in considerable variation in oocyte number and quality, largely as a result of
differences in recovery techniques (Takagi et al. 1992; Hamano and Kuwayama
1993). The method of oocyte retrieval has an impact on COC morphology and subsequent developmental capacity in vitro, and, in this respect, the importance of an
intact cumulus cell investment for oocyte maturation and in vitro development has
been described extensively (Konishi et al. 1996; Tanghe et al. 2002). Immature
bovine oocytes can be divided into different quality categories based upon light
microscopic evaluation of the compactness of the cumulus investment and the transparency of the cytoplasm (de Loos et al. 1989; Hazeleger et al. 1995). Intimate contact between cumulus cells and the ooplasm is established through cumulus cell
process endings (CCPEs) that extend through channels into the zona pellucida (transzonal processes). In the highest oocyte quality category (category 1), these CCPEs
penetrate the zona pellucida and establish functional gap junctions with the oolemma
(de Loos et al. 1991), which are absent in category 4 oocytes. Following in vitro
maturation, the category 4 oocytes exhibit consistently low developmental capacity.

Understanding the relationship between follicle diameter and the quality of the
enclosed COC during follicle development (Aerts and Bols 2010) is of vital importance for successful follicle and oocyte selection. The follicle constitutes a specific
and defined micro-environment for the oocyte. Growth of the dominant follicle is
associated with an increasing concentration of estradiol-17β in the follicular fluid,
which therefore becomes gradually more estradiol dominated (Assey et al. 1994).
Subordinate follicles either have a lower estradiol-17β/progesterone ratio or are progesterone dominated. Moreover, after ultrasonographically tracking follicle growth
and regression, Price et al. (1995) noted that estradiol-17β concentrations were significantly lower in regressing and histologically atretic compared to non-atretic follicles. With respect to the influence of follicle size on oocyte quality, Arlotto et al.
(1995) reported oocyte growth in all bovine follicle sizes studied, whereas Fair et al.
(1995) demonstrated only a small positive correlation between oocyte diameter and
follicle size. Overall, it appears that the increase in oocyte diameter plateaus at
about 120 μm, when the follicle reaches 3 mm, whereas full meiotic competence is
achieved at an oocyte diameter of 110 μm. Nevertheless, since Lonergan et al.
(1994) obtained more grade 1 COCs (with many layers of cumulus cells) and a
higher number of blastocysts per oocyte from follicles with a diameter >6 mm, it is
probable that full cytoplasmic competence is only reached somewhat later during
follicle growth.

From a practical reproductive perspective, aspiration of immature follicles is particularly interesting when performed on living donors, because the procedure can be repeated and is highly repeatable. In addition, the physiological status of the donor at the time of oocyte recovery can be assessed and manipulated, e.g. by the injection of hormones. This chapter will concentrate on follicle aspiration methods in living donors, with an emphasis on transvaginal ultrasound-guided follicle aspiration, also known as ovum pick-up (OPU), in the cow and to a lesser extent the mare. Following a brief description of the OPU technique per se, we will concentrate on the technical and biological factors that influence the success of OPU.

10.2   Oocyte Retrieval from the Living Donor Cow

The ability to puncture immature follicles within the ovaries of living donors and harvest the oocytes has opened new perspectives in assisted reproduction programs because additional female gametes can be made available for in vitro embryo production (IVP) over an extended time period, which is not the case if the donor animal is slaughtered. In addition, OPU permits hormonal modulation of the donor's ovarian activity prior to oocyte retrieval and thereby an opportunity to influence the quantity and quality of the retrieved COCs. A few important differences exist between post-mortem and in vivo oocyte retrieval. Firstly, transrectal manipulation of the ovary is necessary during oocyte retrieval in the living donor, to facilitate follicle visualization by laparoscopic or ultrasonographic imaging. By contrast, when follicles in the ovaries of slaughtered cows are punctured, a specific follicle can be selected and punctured under direct visual control. Secondly, different mechanical forces play a role when puncturing follicles in vitro, compared to in vivo follicle aspiration with an adjustable aspiration vacuum pressure (Hashimoto et al. 1999).

Different methods have been used to repeatedly collect oocytes from living donor cows; these include puncturing the follicles under laparoscopic guidance (Schellander et al. 1989), which results in high recovery rates but has the disadvantage of being relatively laborious and carries the risk of adhesions developing at the site of puncture. Callesen et al. (1987) were the first to use ultrasonography to collect oocytes from living cattle, using an ultrasonographic transducer equipped with a needle guide via a transcutaneous approach. A transvaginal laparoscopic technique was described by Reichenbach et al. (1994), during which a sterile trocar and cannula were directed into the abdominal cavity through the vaginal wall under rectal guidance; laparoscopy allowed the aspiration of the follicles to be accurately monitored. Pieterse et al. (1988) modified a transvaginal ovum pick-up technique, originally developed for use in human reproduction (Dellenbach et al. 1984), for use in cattle. A big advantage of the transvaginal approach in cattle is that it is possible to both secure and manipulate the ovary per rectum so that it can be moved around the ultrasound transducer and needle, to present the most optimal position for puncture. As a result, a minimally invasive method with high repeatability (Pieterse et al. 1991) for oocyte retrieval from living donor cows became available. Becker et al. (1996) compared transvaginal OPU under ultrasonographic guidance with oocyte retrieval guided by endoscopic instruments. They concluded that the use of ultrasound resulted in better-quality cumulus oocyte complexes, although it is not entirely clear why endoscopic aspiration should cause more damage to the COCs. As a consequence, ultrasound-guided transvaginal oocyte pick-up (abbreviated to 'OPU' for the rest of this chapter) was developed as a successful technique for repeatedly retrieving oocytes from selected heifers and cows of high genetic merit (Kruip et al. 1994), to produce large numbers of calves with known production traits and to shorten the generation interval in cattle breeding programs. Indeed, the ultimate aim was to produce more embryos and pregnancies per donor cow than was possible through multiple ovulation and classical embryo transfer (MOET) programs (Pieterse et al. 1991).

10.2.1 OPU Equipment and Procedure

An OPU system consists of three major components: an ultrasonographic scanner with an appropriate transducer (probe), an aspiration pump, and a needle guidance system connected to an oocyte collecting tube (Figs. 10.1 and 10.2). The transducer and the needle guide are commonly constructed as a single operational unit to enable accurate manipulation of the needle from outside the cow while bringing the transducer into close contact with the ovaries. Mounted alongside the transducer, the puncture needle can be visualized on the ultrasound screen when it is advanced into the sonographic field to enter a follicle; to facilitate visualization, it is helpful to have a biopsy guide on the ultrasound screen and to use needles with a roughened area just behind the tip that is echogenic by dint of trapping air ('echogenic tip'). The needle is in turn connected to a vacuum pump by silicone or Teflon tubing such that follicular contents are aspirated as soon as aspiration pressure is applied via the vacuum pump. The follicular fluid and oocytes are collected into a collection device positioned between the needle and the pump. This oocyte collection device can be a regular embryo filter or a simple Falcon tube sealed with a stopper, into which an afferent tube delivers the follicle aspirate and from which an efferent line is connected to the vacuum pump that applies the aspiration pressure (Figs. 10.1 and 10.2). Although not compulsory, prior to OPU cows can be sedated with detomidine hydrochloride and treated with hyoscine-N-butylbromide to induce relaxation of the intestines. Subsequently, the faeces is removed from the rectum, and epidural anaesthesia is induced using 2% lidocaine to combat excessive straining during the transrectal manipulation. After the tail has been fixed to one side, the vulva and perineum are thoroughly cleaned and disinfected before the OPU device, containing the transducer and the needle guidance system, is inserted into the vagina (Fig. 10.3). While the OPU handle can be manipulated with one hand outside the cow, the head of the ultrasound transducer is positioned cranio-dorsally to the left or right of the cervix, depending on which side oocytes are to be collected. Using the other hand per rectum, the operator fixes the ovary and holds it against the head of the transducer (Fig. 10.4) such that

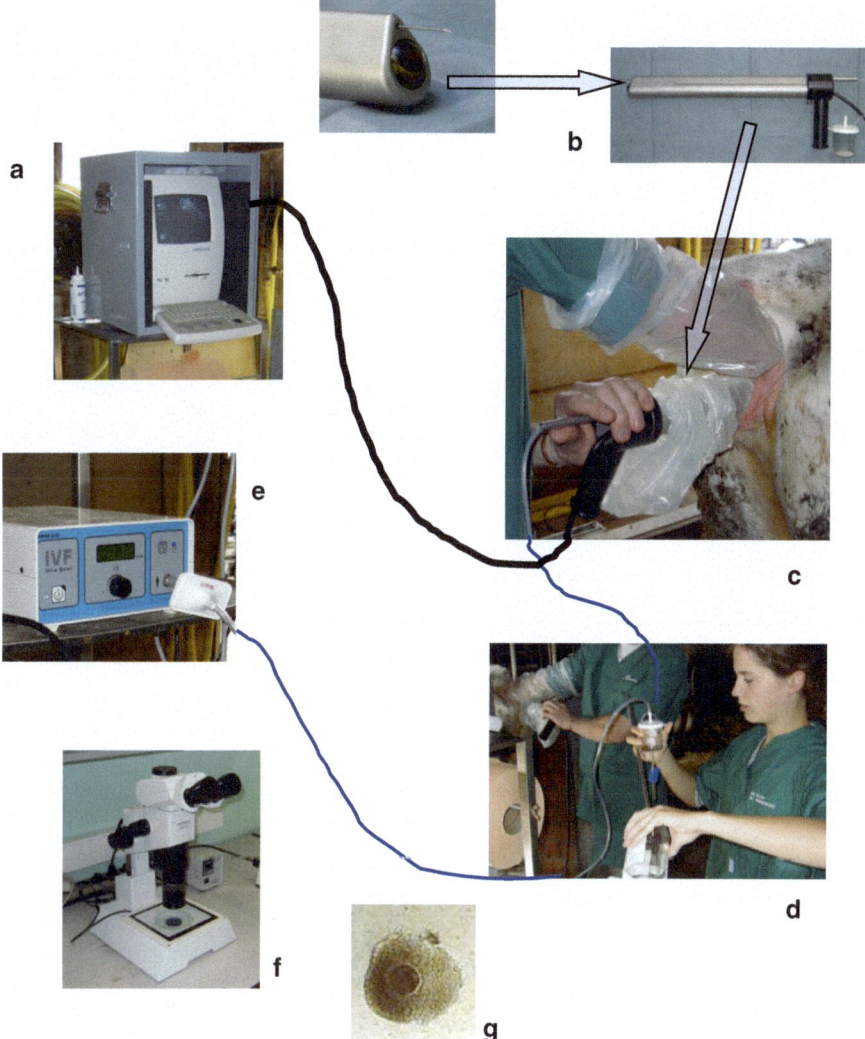

Fig. 10.1 Components of an OPU set up: (**a**) ultrasound scanner (Esaote/Pie Medical, Maastricht, the Netherlands) with (**b**) transducer and needle guidance system, inserted in the vagina of the donor (**c**). The needle is connected to the embryo filter (**d**), which is connected to the aspiration pump (**e**). Cumulus oocyte complexes (**g**) are looked for in the aspirated fluid by means of a stereo microscope (**f**)

the ovary and follicles can be visualized on the ultrasound screen (Fig. 10.5). A biopsy line programmed into the scanner's software is displayed on the screen and indicates where the follicle needs to be positioned for successful puncture. The operator then advances the needle slowly forward until the vaginal wall is pierced and the needle is visualized entering the ultrasound field. By monitoring the needle's position and simultaneously manipulating the ovary per rectum, the needle

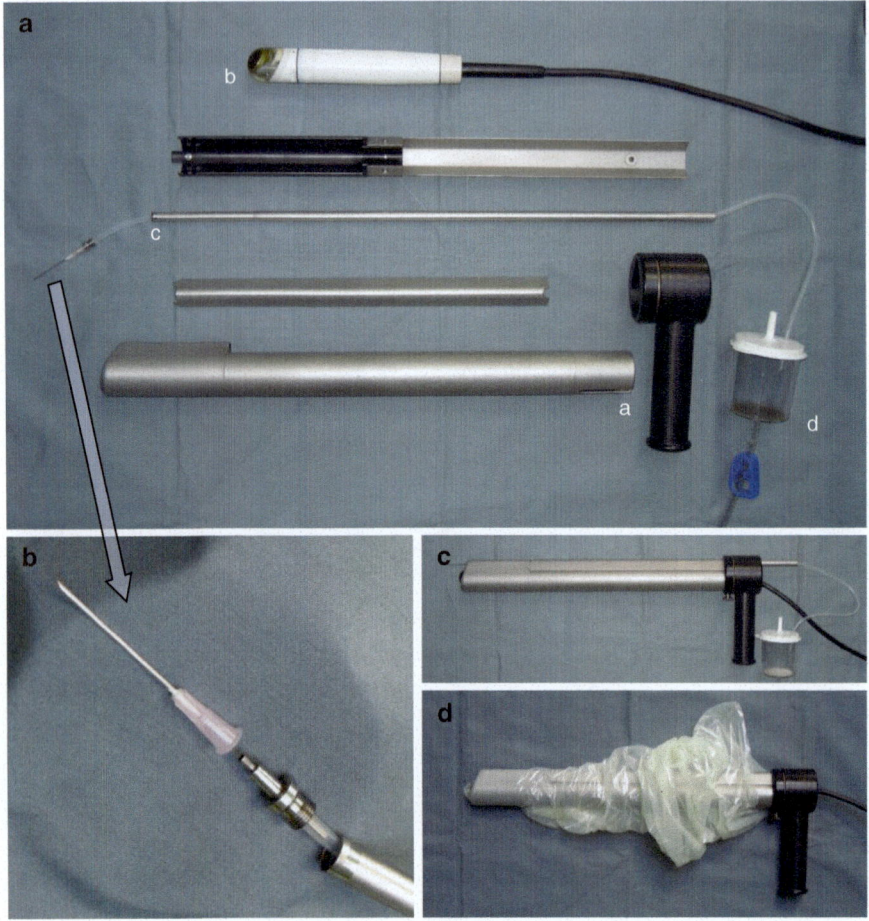

Fig. 10.2 (**a**) OPU device disassembled (**a**) and mounted ready for use (**c** and **d**) with a) intravaginal OPU handle, b) mechanical multiple angle sector transducer – MAP (Esaote/ Pie Medical, Maastricht, the Netherlands), c) needle guidance system and d) oocyte collection filter. (**b**) Detail of puncture needle connected to silicone tubing

can be directed into a follicle. Once the needle enters the follicle, the aspiration pump is activated using the foot pedal and the follicular fluid, and COCs are collected into the embryo filter which contains the oocyte collection medium. Subsequently, the filter contents are washed and transferred to a petri dish, and the oocytes are identified using a stereomicroscope, captured using a glass pipette and placed into maturation medium. After 24 h of maturation, they will be fertilized and cultured for 7 days in vitro to reach the blastocyst stage. The final outcome of OPU, in terms of numbers and quality of retrieved COCs, is influenced by both technical and biological factors (Bols 1997), both of which will be discussed in more detail.

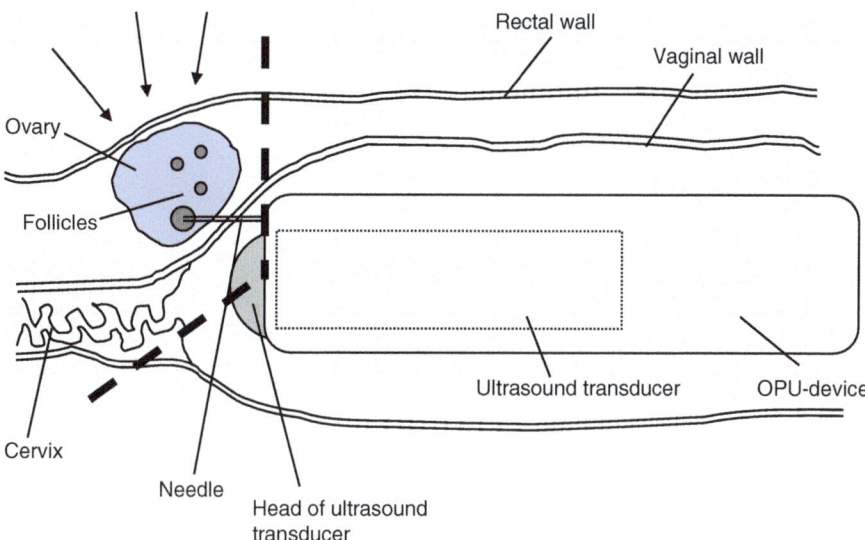

Fig. 10.3 Positioning of the ovary during transrectal palpation. Pressure is exerted on the ovary with the hand positioned intrarectally, from the direction indicated by the arrows. The bold dashed line delineates the scanned area

Fig. 10.4 Positioning of the ovary during transrectal palpation. The puncture needle penetrates the vaginal wall (**a**) when pushed forward. In vivo, the rectal wall lays between the hand of the operator and the ovary (**b**). The white dashed line delineates the position of the intravaginal OPU device

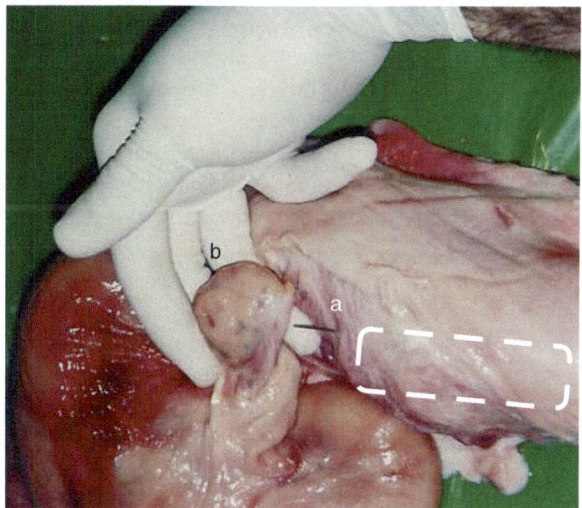

10.3   Technical Factors Influencing OPU Results

Since continuing advances in ultrasound technology have improved image resolution and the accuracy with which ovarian structures can be visualized (Hashimoto et al. 1999; Seneda et al. 2001; Singh et al. 2003; Bols et al. 2004), the 'weakest

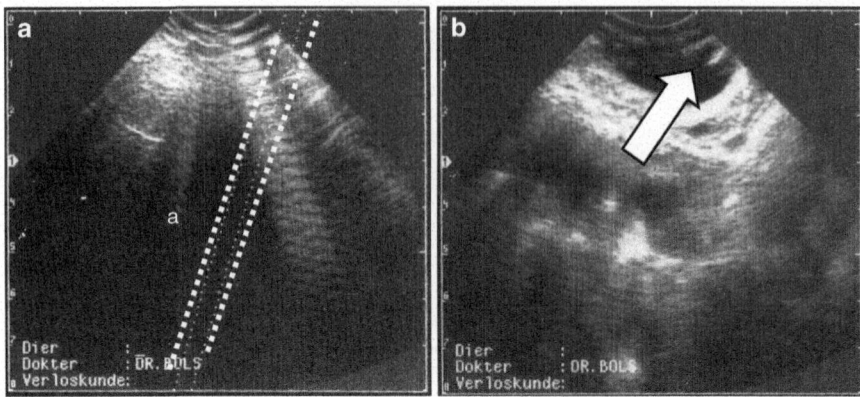

Fig. 10.5 Ultrasound images taken during the OPU procedure. (**a**) The biopsy or puncture line is fixed on the ultrasound screen (a), indicating where the needle will appear within the scanned area; (**b**) needle point penetrates the follicular wall (white arrow)

link' or component of highest concern is now the puncture needle because a sharp needle is a prerequisite for successful OPU (Scott et al. 1994). Traditionally, most operators used 50–60-cm-long needles, with an outer diameter of 1–1.5 mm, which are relatively simple to construct and easy to handle (Looney et al. 1994; Bols 1997). A major disadvantage of these needles is that they become blunt quite quickly and, even with regular resharpening, never regain their original sharpness. In addition, these long, non-disposable needles are relatively expensive and contain a large dead space. Alternative OPU systems have been developed that use disposable 18 gauge epidural needles (Rath 1993) or cheaper, regular hypodermic injection needles (Bols et al. 1995). These needles have the additional advantages of being sterile and available in different diameters and lengths and easy to change.

OPU success rate is quantified firstly in terms of the oocyte recovery rate (RR = number of COCs per 100 follicles punctured), which is influenced by factors including needle diameter, aspiration pressure and operator experience (Bols 1997). As a result, RRs have been reported to vary between 7% and 70% for different OPU teams. Over the years, many different needle diameters and aspiration pressures have been used in either experimental or commercial bovine OPU programs (Bols 1997), which makes it difficult to directly compare recovery rates. In addition, the exact aspiration pressure exerted through the tip of the needle depends on the aspiration device, the length and diameter of the tubing the size and type of collection vessel, as well as on the needle diameter. To make comparisons possible, the aspiration pressure needs to be expressed in terms of the amount of fluid (in ml) that can be aspirated per minute, rather than in mm Hg exerted from the vacuum pump. Indeed, a modest change in needle diameter can triple the rate of fluid aspiration without any change in aspiration pressure (Bols et al. 1996). Given the importance of an intact cumulus cell investment for oocyte maturation and future developmental capacity, any damage to the COC caused by the aspiration procedure has to be assessed for a given system so that preventive measures can be taken. Ideally, the optimal aspiration pressure for a given OPU system should be established by puncturing a substantial number of follicles on ovaries from slaughtered cows. While various vacuum pressures and needle diameters can be tested, COC morphology should be evaluated following aspiration with special attention to the integrity of the cumulus cell investment. In this way a threshold value, or an optimal range, for aspiration pressure can be established that will not result in too much damage to the aspirated COCs but still maintain an acceptable RR. Systems that use simple disposable injection needles allow such an in vitro calibration (Bols et al. 1996). The percentage of retrieved intact COCs usually decreases progressively as the aspiration pressure increases, which is associated primarily with an increase in the number of denuded oocytes, as reported by Ward et al. (2000). As would be expected, higher numbers of good-quality COCs will translate to a higher number of cultured blastocysts produced. Aspirating selected top-quality COCs, which were initially retrieved following slicing of ovaries recovered from slaughtered cows, to assess the net damage that the aspiration procedure can cause, revealed an overall RR of 79% (Bols et al. 1997). In other words, one out of five oocytes was lost during the aspiration process. Fortunately, an average of 82% of the recovered COCs was still surrounded by a compact cumulus investment following aspiration. Thus, on average, around 20% of the initially good-quality COCs were microscopically damaged by the OPU procedure, by (partial) stripping of cumulus cells in a manner likely to impair the oocyte's in vitro developmental potential (Cox et al. 1993). A final very important factor determining OPU outcome is the experience of the operator or the team that is retrieving the oocytes, as evidenced by an in-depth analysis of 7800 OPU sessions performed in a commercial setting by Merton et al. (2003).

10.4 Biological Factors Influencing OPU Results

A substantial body of literature is available on biological factors that might influence the likelihood of blastocyst formation when in vitro embryo production (IVP) is based on COCs recovered via OPU. While there is no doubt that the highest blastocyst rates will be obtained with the best-quality COCs (as stated above), one should bear in mind that the IVP procedure 'as such' is an extremely complex process that critically influences the final blastocyst rate. Since discussing non-OPU factors that affect the success rate of IVP is beyond the scope of this chapter, we will concentrate on a few factors that are directly related to the OPU procedure per se.

10.4.1 Frequency and Timing of Follicle Puncture

The OPU technique has the advantage of being highly repeatable. Pieterse et al. (1991) punctured follicles during different oestrous cycle stages in the same donors, over a 3-month period. However, the presence of a dominant follicle appears to reduce the in vitro developmental competence of oocytes from the subordinate follicles, even at a relatively late stage of dominance (Hendriksen et al. 2004). This is why the dominant follicle is often removed by aspiration prior to a regular oocyte retrieval session 48 h later (DFR). While some studies report no effect of collection frequency on the number of follicles aspirated or the number of COCs collected per session (Garcia and Salaheddine 1998), most researchers agree that a twice-weekly oocyte collection schedule has a positive effect on the number of follicles available for puncture and the number of blastocysts that results (Bols 1997). Indeed, it can be assumed that the developing dominant follicle will be ablated during each session when a cow is punctured twice a week, thereby stimulating an additional wave of smaller follicles to grow (Bergfelt et al. 1994).

10.4.2 Physiological Status and Body Condition of the Donor

In cattle breeding programs, OPU is generally performed on selected healthy heifers with excellent genetic potential for production traits that could in themselves be predictive for oocyte yield and the number of blastocysts produced (Merton et al. 2009). However, OPU can be performed at various stages of a cow's reproductive life; even pregnancy does not exclude OPU, since oocytes can successfully be retrieved during the first 3 months of gestation (Meintjens et al. 1995; Bungartz et al. 1995; Reinders and Van Wagtendonck-de Leeuw 1996). Argov et al. (2004) saw an increase in the number of oocytes recovered when a higher proportion of aspiration sessions were performed in cows in early lactation. On the other hand, undernutrition has a negative effect on the developmental competence of recovered oocytes in vitro, as illustrated by the decreasing percentage of blastocysts associated with decreasing body condition score of the donor (Lopez Ruiz et al. 1996) and an increasing proportion of good-quality oocytes with increasing body condition score (Dominguez 1995).

10.4.3 Breed and Age of the Donor

Early reports suggested that European breeds had significantly more large follicles than zebu or crossbred cows (Dominguez 1995), whereas no differences in the proportion of normal oocytes recovered were apparent. However, over the past 10 years, the use of OPU-IVP has rocketed in Latin-America and in particular in Brazil where the high fecundity of a single breed, the Nelore, has been the foundation for the production of hundreds of thousands of embryos. Indeed, a single OPU session in an average Nelore donor cow can yield up to 50–60 oocytes, resulting in up to 30 in vitro embryos per puncture session (Pontes et al. 2011). Strikingly, these results are obtained without any hormonal stimulation and have led some researchers to conclude that repeated OPU alters follicular dynamics and might increase follicle growth rate in zebu donor cows (Viana et al. 2010). Highly contrasting results have been reported in Belgian Blue donors with impaired fertility, which yielded an average of only 3.1 oocytes and 0.5 embryos per puncture session (Bols et al. 1996).

The use of OPU in young donors is limited by the smaller dimensions of the pelvis. Holstein Friesian heifers can be subjected to OPU from around the age of 6–8 months, depending on the dimensions of the intravaginal handle and transducer used (Rick et al. 1996; Bols et al. 1999). Follicles in calves can also be punctured, but this requires a different approach to access the ovaries (Brogliatti et al. 1995). The major problem with prepubertal donors is the impaired in vitro developmental capacity of the recovered oocytes (Taneja et al. 2000), resulting in a lower overall efficiency of the procedure.

10.4.4 The Role of Hormonal Stimulation to Prepare Donors for OPU

An enormous amount of research has been done on how potential donors can be prepared to maximize oocyte and subsequent embryo yields. An important general remark before describing a few of the possibilities is the fact that long-term, repeated use of OPU in an individual donor cow is possible without any hormonal stimulation (Pieterse et al. 1991). In the long term, the absence of hormonal stimulation offers many advantages because when using hormones to stimulate follicle growth, the blood flow to the ovaries increases enormously, rendering the cows useless for OPU for a few weeks after the initial puncture. Low or suboptimal follicular activity can be remedied in some potential donors, mostly by using FSH-LH combinations or equine chorionic gonadotrophin (eCG = PMSG, pregnant mare serum gonadotrophin). While these hormones have been widely used in ET programs, modifications in the dose and timing of treatments are necessary, because the final aim of stimulation prior to OPU is to generate additional follicles rather than to initiate multiple ovulations. Pieterse et al. (1988) achieved the highest oocyte recovery rates in PMSG-treated donors, which developed larger ovaries and had more follicles than non-stimulated animals. However, a later study (Pieterse et al. 1992) showed that while stimulation resulted in a larger number of aspirated follicles per cycle, it had the opposite effect on oocyte recovery rate (RR), which was lower in stimulated than non-stimulated donors. Positive effects of FSH on the number of follicles with a diameter >6 mm and the number of viable blastocysts have, however, also been reported (Looney et al. 1994; Goodhand et al. 2000). Unfortunately, the increase in the number of follicles, oocytes recovered and embryos produced is often inconsistent and might depend on the cycle stage at which treatment is initiated (Paul et al. 1995). Vos et al. (1994) were able to retrieve five times as many COCs 22 h after, compared to shortly before, the LH surge (in PMSG-treated donors). Stubbings and Walton (1995) found no differences in the mean number of follicles suitable for puncture between non-stimulated cows punctured twice a week and FSH-stimulated cows punctured only once. Subtle changes in FSH dose influenced the sizes, but not the number of follicles, which was mainly a factor of individual donor and OPU session variation (De Roover et al. 2005). Some authors have also used intravaginal progesterone-releasing devices (CIDR) in combination with FSH and LH to prepare oocyte donors, with varying results (Chaubal et al. 2007). It should be noted that

FSH (and probably also other hormonal) treatments might result in asynchrony between the maturation of the oocyte and its surrounding follicle (de Loos et al. 1991) or between nuclear and cytoplasmic maturation (Bousquet et al. 1999), resulting in reduced developmental competence.

As can be expected, hormonal stimulation and OPU puncture frequency together can affect the final embryo yield. De Ruigh et al. (2000) concluded that FSH treatment prior to OPU once every 2 weeks resulted in significantly more COCs and more embryos produced in vitro (expressed per OPU session) than a twice-per-week non-stimulated OPU schedule. However, total embryo production over a 2-week period turned out to be higher with the twice-weekly puncture scheme (four non-stimulated sessions in 2 weeks) than for one FSH-stimulated OPU session every 2 weeks. Goodhand et al. (1999) reported that the puncture of FSH-treated donors once a week produced a similar number of transferable embryos per 'donor week' as aspiration twice a week without FSH treatment. Chaubal et al. (2006) reported that a protocol combining dominant follicle removal and FSH stimulation with a subsequent single OPU per week seemed to be the most productive and cost-effective approach over a 10-week period. When calculating total costs of the procedure, one needs to keep in mind the price of the hormonal treatment, and its administration, which often requires animal handling twice a day for several days.

10.5   OPU-IVP to Treat Bovine Infertility

Compared to ET, where cows can typically be flushed three to four times a year, yielding around five embryos per flush, OPU can be performed as often as twice a week. In healthy donor cows, two embryos per donor per week can be produced, equating to four to five times the average ET yield (Kruip et al. 1994). An important additional advantage of using OPU-IVP is greater flexibility in choice of sire-dam combinations in vitro, i.e. using different bulls on oocytes from the same OPU session, which can accelerate the genetic selection process. In addition, OPU-IVP can be used to produce additional offspring from valuable cows that no longer respond to embryo flushing treatments. The first OPU-IVP calves in Belgium were born in 1995, following oocyte retrieval from Belgian Blue donors with impaired fertility (Bols et al. 1996). Following the transfer of 56 IVP embryos, 12 viable pregnancies were obtained, leading to at least 1 extra calf for 7 out of 12 high genetic merit donors considered to have reached the end of their breeding career. Looney and co-workers (1994) reported OPU in 200 mostly beef cattle donors, of which 50% had a history of good embryo production. An average of 6.3 oocytes were retrieved per session, and 16.4% yielded a blastocyst. Transfer of 813 embryos resulted in 325 pregnancies (40%). Hasler et al. (1995) carried out similar work on 155 infertile dairy cows. An average of 4.1 oocytes suitable for IVF were retrieved per session. Following transfer of 2268 fresh embryos, 1220 pregnancies (53.8%) were obtained. Large data sets like these illustrate that OPU-IVP has evolved to become a routine procedure to produce reliable numbers of embryos in vitro, albeit with a dependency on the breed of cow and the efficacy of the IVP system (Bousquet et al. 1999).

When comparing embryo yields and pregnancy rates between in vivo (classical ET) and in vitro (OPU-IVP) methods using the same donors, the in vitro approach turned out to yield the most embryos (Pontes et al. 2009). Because the ultimate success rate of assisted reproduction is determined by the number of calves produced, a well-synchronized, healthy, recipient herd into which fresh embryos can be transferred is a major prerequisite for success. When fresh transfers cannot keep up with embryo production, reliable embryo cryopreservation methods need to be available, increasing the complexity of the whole operation.

10.6   Donor Health and Repeated OPU

Reports on the impact of the OPU procedure on donor animal health and future reproductive performance are scarce. Pieterse et al. (1991) could not detect any adhesions following OPU, and the procedure did not seem to affect the donor's future fertility. Dairy heifers were closely monitored during two periods of 4–5 weeks while enrolled in a twice-weekly OPU schedule (Petyim et al. 2000). They only occasionally showed signs of oestrus, and corpus luteum-like structures often developed from punctured follicles, which concurred with earlier findings that, based on progesterone profiles, repeated OPU appeared to induce a degree of acyclicity (Bols et al. 1998). At the end of their first OPU period, heifers returned to normal cyclicity (Petyim et al. 2000). Post-mortem findings following the second OPU period included a thickening of the ovarian tunica albuginea and a slight hardening of the ovaries. The authors concluded that OPU did not have major negative effects on ovarian structure or on subsequent ovarian function. Additional research on the effects of OPU revealed a significant rise in FSH levels on the day following puncture (Petyim et al. 2001). In addition, heart rate and cortisol concentrations increased significantly following restraint and epidural injection. However, both parameters returned to normal within 10 min after completion of the OPU procedure.

10.7   Transvaginal Ultrasound-Guided Oocyte Retrieval in the Mare

As with other assisted reproductive technologies, the development and uptake of OPU-IVP in commercial horse breeding has been slower and driven by different primary goals to those that apply to cattle breeding (Galli et al. 2007). While initial reports of transvaginal ultrasound-guided oocyte retrieval in mares (Brück et al. 1992) followed closely behind those in cattle, interest in the technique waned for a number of practical reasons. Most important were the disappointing rates of oocyte recovery from immature follicles (<25% in early studies: see Hinrichs 2012 for review) and the absence of commercially available gonadotrophins capable of stimulating the development of multiple mature follicles from which to harvest in vivo-matured oocytes; taken together this meant that recovering enough high-quality oocytes from living donors to run a viable IVP program appeared an insurmountable challenge. Since conventional in vitro fertilization using equine gametes also proved to be very poorly successful (Hinrichs 2012), commercial interest in equine IVP remained understandably low. However, interest in OPU was rekindled by the development of oocyte transfer (OT) as a tool to examine oocyte developmental competence (Carnevale and Ginther 1995) and to treat severe acquired infertility in mares (Carnevale 2004). Development of OPU was given further impetus by the first reports of intracytoplasmic sperm injection (ICSI) as a technique for successfully producing foals after fertilizing equine oocytes ex vivo (Cochran et al. 1998; McKinnon et al. 2000). Nevertheless, progress remained slow, largely because blastocyst production rates following IVP were much lower (<10% compared to approximately 35%) than those obtained after transfer of sperm-injected oocytes into the oviduct of either synchronized recipient mares (Choi et al. 2004) or progesterone-treated sheep (Tremoleda et al. 2003). The development of DMEM/ Hams F-12-based equine IVP systems capable of supporting blastocyst production rates >35%, at least within an experimental set-up (Choi et al. 2006), was the final breakthrough required for equine IVEP to become a viable clinical technique. Indeed, when Galli et al. (2014) reported producing 0.6 blastocysts per OPU in a commercial OPU-IVP program, it became clear that OPU-IVP could be competitive with commercial embryo transfer, given that embryo recovery rates of 0.3–0.5 per cycle are the norm in commercial sport horse mares inseminated with frozen-thawed or chilled-transported semen (Stout 2006). Most recently, reports of blastocyst production rates of 15–20% per injected oocyte and > 1 per OPU (Hinrichs et al. 2014) even after overnight shipping of oocytes at 20 °C (Galli et al. 2016) have led to a surge in interest in equine OPU-IVP.

10.7.1 Clinical Applications of OPU in the Mare

OPU is the basis for two clinical procedures in horses, oocyte transfer (OT) and in vitro fertilization by intracytoplasmic sperm injection (ICSI) (Fig. 10.6). To date, the main reasons for wanting to use OPU in clinical equine practice has been subfertility. Indeed, OT was developed primarily as a technique for treating subfertility in mares that were not, or only infrequently, able to produce embryos by conventional AI and embryo flushing, due, for example, to repeated failure of normal ovulation or severe pathology of the oviducts, uterus or cervix (Carnevale 2004). OPU-ICSI was similarly introduced initially as a treatment for subfertile mares; however, given its original development as a technique for addressing 'male factor infertility' in human infertility, ICSI also rapidly became an attractive option for addressing stallion subfertility and/or limited availability of semen. Finally, significant improvements in in vitro blastocyst production rates and the realization that OPU-ICSI combined with blastocyst cryopreservation significantly improves the efficiency of recipient mare use have seen OPU-IVP emerge as a desirable method for producing embryos from actively competing sport horse mares (e.g. show jumpers and dressage horse) whose competitive peak overlaps with their most fertile

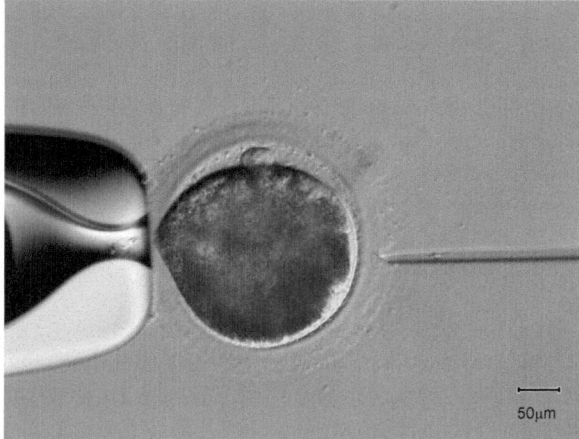

Fig. 10.6 An in vitro-matured MII horse oocyte immediately prior to intracytoplasmic sperm injection. The oocyte is immobilized with a holding pipette with the polar body orientated to 12 o'clock, to minimize the risk of injecting the sperm into the metaphase plate. A sperm is positioned in the tip of a conventional injection needle. The high lipid content of the horse oocyte makes it difficult to visualize cytoplasmic structures

years (Galli et al. 2014). OPU-IVP has the additional advantage over conventional ET that it can be performed as a single outpatient procedure with minimum impact on the training or competition schedule and without the need for any hormonal manipulation of the oestrous cycle; many owners and riders do not like their mares being returned to oestrus since it can negatively affect performance in some mares.

10.8   Oocyte Retrieval from Living Donor Mares

The equipment required for, and procedures involved in, recovering oocytes from living donor mares is essentially the same as those used in cattle, although some modifications are required to account for behavioural and anatomical differences between the species. The most important difference is the fact that immature equine COCs are surrounded by a cumulus investment with fewer cell layers that is attached more firmly to the follicle wall by a broader cumulus cell hillock with projections into an underlying thecal cell pad (Hawley et al. 1995). The practical consequence of this more tenacious attachment of the immature COC to the follicle wall is that simple aspiration of follicular fluid is not sufficient to reliably recover the oocyte. Instead repeated aspiration and flushing of the follicle accompanied by scraping of the follicle wall with the bevel of the aspiration needle is required to achieve a clinically acceptable oocyte recovery rate (Galli et al. 2007). In general, a 60 cm 12 gauge (approx. 2.75 mm outer diameter) double lumen needle is used for equine OPU. Aspiration is performed via the inner stylet which is connected, via a collecting vessel, to the vacuum pump; the vacuum pressure is adjusted to achieve fluid aspiration of roughly 20–25 ml per minute, since higher pressures increase the risk of denuding the already relatively thin equine cumulus cell investment. Once the follicle has been evacuated, it is flushed repeatedly with commercial embryo flushing medium, supplemented with heparin (5–20 i.u. per ml) to prevent clotting of any blood or the gelatinous fluid commonly recovered from large or atretic follicles, and introduced via the outer needle. Using a double lumen needle significantly reduces the risk of an oocyte remaining in the needle's dead space and being repeatedly flushed into and out of a follicle.

10.8.1 Aspirating Immature Follicles

For immature oocyte recovery, follicles from approximately 8–10 mm in diameter are flushed 6–12 times, where larger numbers of flushes are used when few follicles are available for aspiration, to maximize the likelihood of recovering the oocyte. The need to repeatedly flush follicles means that the OPU can be a prolonged procedure (15–45 min) in the mare; epidural anaesthesia using 2% lidocaine is therefore recommended to prevent the mare straining in response to the presence of the ultrasound probe in the vagina and the manipulation of the ovaries via the rectum. In addition, fairly profound sedation with an alpha-2 agonist (e.g. detomidine hydrochloride) potentiated with an opioid analgesic such as butorphanol is recommended to ensure that the mare remains quiet throughout the procedure, while hyoscine-N-butylbromide can be used to further relax the rectum, thereby facilitating manipulation of the ovaries and reducing the risk of damaging the rectum wall. It is also advisable to administer a non-steroidal anti-inflammatory drug (NSAID) to combat pain during and immediately after the OPU procedure and perioperative antibiotics to cover the possibility of contaminants being introduced into the abdominal cavity during OPU. In our experience of >500 OPUs, the procedure is (surprisingly) well tolerated, even in young inexperienced mares, and post-procedure complications have been limited to mild pyrexia and/or abdominal discomfort of short duration (12–36 h) that responds well to NSAIDs. Others have reported occasional rectal bleeding associated either with needle puncture of the rectum wall or as a result of vigorous ovarian manipulation and emphasize the ever-present risk of more serious damage such as a rectal tear or ovarian abscess (Velez et al. 2012); fortunately, the incidence of serious complications appears to be low, and even repeating OPU at 2-week intervals over a period of months appears to have little or no lasting effects on subsequent ovarian structure, cyclicity or fertility (Velez et al. 2012). Recent reports on oocyte recovery rates suggest that, with an established team and system, average RRs from immature follicles of between 50 and 70% can be achieved (Jacobson et al. 2010; Galli et al. 2014, 2016; Hinrichs et al. 2014), although recovery during individual OPU attempts can vary from as little as 20% and up to 100%.

10.8.2 Harvesting In Vivo-Matured Oocytes

The major alternative to harvesting immature oocytes is oocyte recovery from the pre-ovulatory follicle of a donor mare at a set time after hormonal induction of ovulation; indeed, this is the protocol of choice for OT and is also used in some OPU-IVP programs both because oocyte recovery rates from pre-ovulatory follicles are high (>70%: Carnevale et al. 2005; Foss et al. 2013) and because oocytes that undergo in vivo maturation have higher developmental competence, with blastocyst formation rates as high as 40–70% reported albeit on small numbers of oocytes (Jacobson et al. 2010; Foss et al. 2013). OT also aims to utilize the anticipated high developmental competence of in vivo-matured oocytes as a treatment for subfertility of female origin and involves the surgical transfer of a mature (metaphase II) oocyte to the oviduct of an inseminated recipient mare that has had her own oocyte removed by aspiration of the pre-ovulatory follicle (Carnevale 2004). In either situation, oocyte recovery involves aspiration of the single (occasionally 2–3) pre-ovulatory follicle between 20 and 35 h after induction of ovulation using either a long-acting GnRH analogue (e.g. deslorelin acetate), hCG (1500–2500 i.u.) or a combination of the two, in an oestrous mare with a follicle exceeding 35 mm in diameter (Carnevale 2004; Foss et al. 2013). Waiting until 35 h after ovulation induction has the advantage of ensuring that the oocyte has reached MII, i.e. is fully mature, and that the attachment of the COC to the underlying thecal pad has begun to loosen, thereby improving the likelihood of oocyte recovery. On the other hand, a small proportion of mares will ovulate before the 35-h time point and that cycle will therefore be lost. When recovery is performed at 20–24 h after ovulation induction, there is less risk of premature ovulation, but the oocyte will be at approximately the metaphase I stage of maturation and require a further 12–16 h of culture in vitro to complete maturation before transfer into the recipient's oviduct (Carnevale 2004; Galli et al. 2014).

10.8.3 Technical and Biological Factors Influencing OPU Results

As in the cow, the success of OPU-IVP can be divided into two interrelated components, oocyte recovery rate (RR) and blastocyst production rate, where the latter and the pregnancy and foaling rates following transfer of resulting embryos are ultimately most relevant. Historically, RR from immature follicles was poor at around 25% (for review see Hinrichs 2012). However, it is now clear that a RR of >50% can be achieved when aspirating and repeatedly flushing follicles ≥8–10 mm in diameter (Galli et al. 2007; Jacobson et al. 2010; Galli et al. 2014). While this may not quite reach the RR of oocytes from pre-ovulatory follicles (>75%; Carnevale et al. 2005), it is more than compensated by the larger number of oocytes and the fact that in vitro oocyte maturation rates of OPU-derived oocytes is high (>65%: Foss et al. 2013; Galli et al. 2014). One critical technical factor is needle size, with the RR falling when smaller diameter needles are used, e.g. Velez et al. (2012) reported a RR of 38% for a 15 gauge double lumen as compared to 48% for a 12 gauge double lumen needle. While it is not entirely clear exactly why a larger needle is better, it presumably relates either to more rapid flow and greater turbulence during flushing or more effective scraping of the inside of the follicle.

Currently, there is too little data to make firm conclusions about factors influencing the ultimate results of OPU-IVP; indeed, there is very little published data about pregnancy and foaling rates. Nevertheless, the recent upsurge in the use of OPU-IVP is beginning to yield some interesting data. For example, preliminary reports indicate that pregnancy rates exceeding 75% following transfer of fresh (Hinrichs et al. 2014) and exceeding 60% after transfer of cryopreserved (Galli et al. 2007, 2016) OPU-IVP embryos are possible; on the other hand, early pregnancy loss rates appear to be higher than after conventional breeding, AI or ET (>20% versus 5–10%). In addition, mare age, breed, timing of an OPU attempt and time of season all seem to affect aspects of the OPU-IVP process. For example, performing OPU at a fixed interval of 14 days results in a fall in the number of follicles available for puncture (7–9 yielding 3.5–4.5 oocytes; Jacobson et al. 2010; Velez et al. 2012) compared to monitoring mares and delaying the subsequent OPU until follicle numbers have increased. Using the latter approach, Galli et al. (2014) reported aspirating 14–17 follicles during repeated OPU attempts, yielding 9–12 oocytes per OPU. In the clinical program at Utrecht University, the policy is to advise owners to wait until a mare has at least 15 follicles >10 mm, while accepting that some mares will never develop more than 6–10 follicles and need to be aspirated at this point; this policy has resulted in means of 23.5 follicles yielding 12.8 oocytes during 252 commercial OPUs (Claes et al. 2016). With respect to time of season, the autumn and spring transitional periods appear to be optimal for the collection of immature oocytes because mares develop more mid-sized follicles than during the breeding season (e.g., 11.5 versus 6 follicles exceeding 12 mm; Donadeu and Pedersen 2008). Mare age also significantly affects follicle number with mares older than 20 years having significantly fewer follicles during the transitional period than 17–19-year-old mares, which in turn had fewer follicles than 3- to 7-year-olds (Carnevale et al. 1997). These two observations explain why oocyte recovery in a commercial OPU program decreased with increasing mare age and was higher during spring and autumn than in the summer (Claes et al. 2016).

Equine blastocyst production by ICSI is currently a highly operator-dependent process, and, to date, only a handful of laboratories worldwide have been able to generate commercially acceptable embryo production rates (Hinrichs 2012) Even so, it is becoming apparent that there are breed effects on blastocyst production rates with Galli et al. (2014) reporting embryo production rates of 0.84 (11.3%), 0.6 (10%) and 0.29 (4.1%) per OPU (per injected oocyte) for Warmblood, Quarterhorse and Arab mares, respectively. In addition, Claes et al. (2016) recently reported significant effects of antral follicle count (follicles >4 mm at the time of OPU) and donor mare reproductive history on blastocyst production, with fewer blastocysts resulting from mares with lower follicle numbers and with a history of subfertility/fertility using other techniques, irrespective of mare age (Fig. 10.7).

Since OT is a more established technique than OPU-IVP, more information is available about factors affecting pregnancy rates following OT than for OPU-IVP. While operator experience also clearly plays an important role in results, the other principle factor influencing success is age of the donor mare. Indeed, in an experimental setting, OT of oocytes from young mares yielded a 92% pregnancy rate compared to 31% for aged mares (Carnevale and Ginther 1995). Similarly, in a

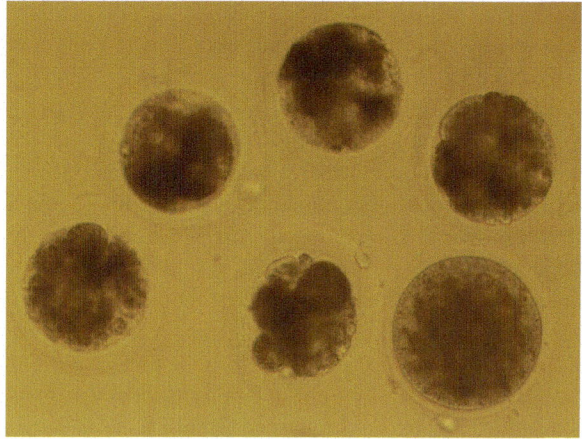

Fig. 10.7 Developing horse embryos 8 days after ICSI. The embryo in the bottom right has developed to the blastocyst stage as evidence by expansion, thinning of the zona pellucida (ZP) and development of a palisading trophoblast layer. In vitro-produced embryos do not produce a confluent blastocyst capsule, which explains the absence of a capsular layer between the trophectoderm and the ZP. The other embryos all underwent cleavage and early cell divisions but are now in various stages of degeneration. It can be challenging to definitively differentiate between blastocysts and degenerate embryos/oocytes

Fig. 10.8 An in vitro-produced horse blastocyst stained with Hoechst 33,342 (Blue) to visualize the nuclei and phalloidin to demonstrate the actin cytoskeleton. Because it takes practice to reliably differentiate viable IVP horse blastocysts from degenerating embryos, staining to demonstrate the presence of numerous cells and formation of a trophoblast layer can be essential to the establishment of an equine IVP program

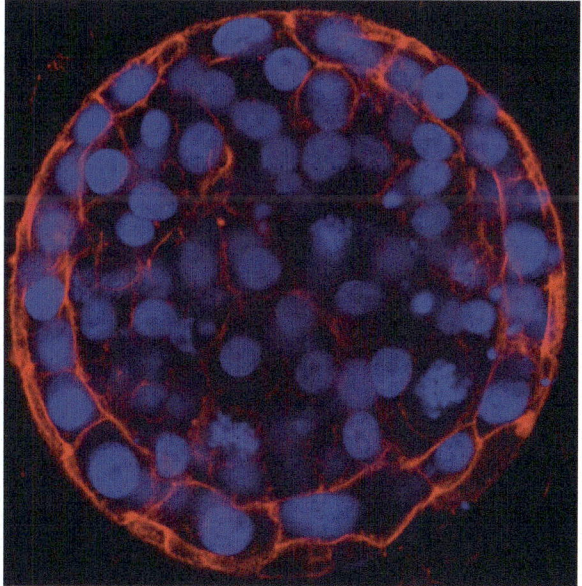

commercial setting, day 15 pregnancy rates averaged 50% for mares <15 years old compared to only 16% for mares >23 years (Carnevale et al. 2005). As for OPU-IVP, pregnancy loss rates in a clinical OT program exceed 20%, presumably reflecting the bias in the donor mare population to aged mares with reduced intrinsic oocyte developmental competence (Hinrichs 2012) (Fig. 10.8).

Conclusion

OPU is now a routine, widely performed procedure in both commercial cattle practice and research into the developmental competence of bovine oocytes. In a commercial setting, the technique offers greater flexibility, in terms of bull use, and is capable of generating more embryos per unit time than conventional multiple ovulation and embryo transfer protocols. In horses, OPU was first introduced into the clinic as a vital component of oocyte transfer, where success is limited by the bias towards aged subfertile mares as donors; nevertheless, OT has allowed production of foals from mares that would otherwise have been considered infertile. Equine OPU-IVP has only very recently become a commercially viable proposition, as a result of significant improvements in immature oocyte recovery and in vitro blastocyst production; nevertheless, OPU-IVP is already proving to be very competitive with AI-ET in terms of numbers of embryos generated per unit time, can be used in cases of both male and female (acquired) infertility and is attracting increasing interest from the owners of competing mares because of its flexibility, availability as an outpatient treatment and lack of any requirement for hormonal manipulation of the oestrous cycle.

References

Aerts JMJ, Bols PEJ (2010) Ovarian follicular dynamics. A review with emphasis on the bovine species. Part II: Antral development, exogenous influence and future prospects. Reprod Dom Anim 45:180–187

Argov N, Arav A, Sklan D (2004) Number of oocytes obtained from cows by OPU in early, but not late lactation increased with plasma insulin and estradiol concentrations and expression of mRNA of the FSH receptor in granulosa cells. Theriogenology 61:947–962

Arlotto T, Schwartz JL, First NL, Leibfried-Rutledge ML (1995) Aspects of follicle and oocyte stage that affect *in vitro* maturation and development of bovine oocytes. Theriogenology 43:943–956

Assey RJ, Hyttel P, Greve T, Purwantara B (1994) Oocyte morphology in dominant and subordinate follicles. Mol Reprod Dev 37:335–344

Becker F, Kanitz W, Nürnberg G, Kurth J, Spitschak M (1996) Comparison of repeated transvaginal ovum pick-up in heifers by ultrasonographic and endoscopic instruments. Theriogenology 46:999–1007

Bergfelt DR, Lightfoot KC, Adams GP (1994) Ovarian dynamics following ultrasound-guided transvaginal follicle ablation in cyclic heifers. Theriogenology 41:161

Bols PEJ 1997 Transvaginal ovum pick-up in the cow: technical and biological modifications. PhD thesis. University of Ghent, Ghent, Belgium

Bols PEJ, Leroy JLMR, Vanholder T, Van Soom A (2004) A comparison of a mechanical sector and a linear array transducer for ultrasound-guided transvaginal oocyte retrieval (OPU) in the cow. Theriogenology 62:906–914

Bols PEJ, Taneja M, Van de Velde A, Riesen J, Schreiber D, Echelard Y, Ziomek C, Yang X (1999) Pregnancies from prepubertal heifers following repeated oocyte collection and IVF between 6 to 12 months of age. Theriogenology 51:298

Bols PEJ, Van Soom A, de Kruif A (1996) Gebruik van de transvaginale Ovum Pick-Up (OPU) techniek: geboorte van de eerste OPU kalveren in België. (Use of transvaginal oocyte pick-up: first OPU calves born in Belgium). Vlaams Diergeneeskundig Tijdschrift 65:86–91

Bols PEJ, Van Soom A, Ysebaert MT, Vandenheede JMM, de Kruif A (1996) Effects of aspiration vacuum and needle diameter on cumulus oocyte complex morphology and developmental capacity of bovine oocytes. Theriogenology 45:1001–1014

Bols PEJ, Vandenheede JMM, Van Soom A, de Kruif A (1995) Transvaginal ovum pick-up (OPU) in the cow: a new disposable needle guidance system. Theriogenology 43:677–687

Bols PEJ, Ysebaert MT, Lein A, Coryn M, Van Soom A, de Kruif A (1998) Effects of long term treatment with bovine somatotropin on follicular dynamics and subsequent oocyte and blastocyst yield during an OPU-IVF program. Theriogenology 49:983–995

Bols PEJ, Ysebaert MT, Van Soom A, de Kruif A (1997) Effects of needle tip bevel and aspiration procedure on the morphology and developmental capacity of bovine compact cumulus oocyte complexes. Theriogenology 47:1221–1236

Bousquet D, Twagiramungu H, Morin N, Brisson C, Carboneau G, Durocher J (1999) *In vitro* embryo production in the cow: an effective alternative to the conventional embryo production approach. Theriogenology 51:59–70

Brogliatti GM, Swan CD, Adams GP (1995) Transvaginal ultrasound-guided oocyte collection in 10 to 16 weeks of age calves. Theriogenology 43:177

Brück I, Raun K, Synnestvedt B, Greve T (1992) Follicle aspiration in the mare using a transvaginal ultrasound-guided technique. Equine Vet J 24:58–59

Bungartz L, Lucas-Hahn A, Rath D, Niemann H (1995) Collection of oocytes from cattle via follicular aspiration aided by ultrasound with or without gonadotropin pretreatment and in different reproductive stages. Theriogenology 43:667–676

Callesen H, Greve T, Christensen F (1987) Ultrasonically guided aspiration of bovine follicular oocytes. Theriogenology 27:217

Carnevale EM (2004) Oocyte transfer and gamete intrafallopian transfer in the mare. Anim Reprod Sci 82-83:617–624

Carnevale EM, Coutinho da Silva MA, Panzani D, Stokes JE, Squires EL (2005) Factors affecting the success of oocyte transfer in a clinical program for subfertile mares. Theriogenology 64:519–527

Carnevale EM, Ginther OJ (1995) Defective oocytes as a cause of subfertility in old mares. Biol Reprod Monogr 1:209–214

Carnevale EM, Hermenet MJ, Ginther OJ (1997) Age and pasture effects on vernal transition in mares. Theriogenology 47:1009–1018

Chaubal SA, Ferre LB, Molina JA, Faber DC, Bols PEJ, Rezamand P, Tian X, Yang X (2007) Hormonal treatments for increasing the oocyte and embryo production in an OPU-IVP system. Theriogenology 67:719–728

Chaubal SA, Molina JA, Ohlrichs CA, Ferre LB, Faber DC, Bols PEJ, Riesen JW, Tian X, Yang X (2006) Comparison of different transvaginal ovum pick-up protocols to optimise oocyte retrieval and embryo production over a 10-week period in cows. Theriogenology 65:1631–1648

Choi YH, Love LB, Varner DD, Hinrichs K (2006) Holding immature equine oocytes in the absence of meiotic inhibitors: effect on germinal vesicle chromatin and blastocyst development after intracytoplasmic sperm injection. Theriogenology 66:955–963

Choi YH, Roasa LM, Love CC, Varner DD, Brinsko SP, Hinrichs K (2004) Blastocyst formation rates *in vivo* and *in vitro* of in vitro-matured equine oocytes fertilized by intracytoplasmic sperm injection. Biol Reprod 70:1231–1238

Claes A, Galli C, Colleoni S, Necchi D, Lazzari G, Deelen C, Beitsma M, Stout T (2016) Factors influencing oocyte recovery and *in vitro* production of equine embryos in a commercial OPU/ICSI program. J Equine Vet Sci 41:68

Cochran R, Meintjes M, Reggio B, Hylan D, Carter J, Pinto C, Paccamonti D, Godke RA (1998) Live foals produced from sperm-injected oocytes derived from pregnant mares. J Equine Vet Sci 18:736–740

Cox JF, Hormazabal J, Santa Maria A (1993) Effect of the cumulus on *in vitro* fertilization of bovine matured oocytes. Theriogenology 40:1259–1267

de Loos FAM, Bevers MM, Dieleman SJ, Kruip TAM (1991) Morphology of preovulatory bovine follicles as related to oocyte maturation. Theriogenology 35:527–535

de Loos F, Van Vliet C, Van Maurik P, Kruip TAM (1989) Morphology of immature bovine oocytes. Gamete Res 24:197–204

De Roover R, Genicot G, Leonard S, Bols P, Dessy F (2005) Ovum pick-up and *in vitro* embryo production in cows superstimulated with an individually adapted superstimulation protocol. Anim Reprod Sci 86:13–25

De Ruigh L, Mullaart E, van Wagtendonk-de Leeuw AM (2000) The effect of FSH stimulation prior to ovum pick-up on oocyte and embryo yield. Theriogenology 53:349

Dellenbach P, Nisand I, Moreau L, Feger B, Plumere C, Gerlinger P, Brun B, Rumpler Y (1984) Transvaginal sonographically controlled ovarian follicle puncture for egg retrieval. Lancet 1(8392):1467

Dominguez MM (1995) Effects of body condition, reproductive status and breed on follicular population and oocyte quality in cows. Theriogenology 43:1405–1418

Donadeu FX, Pedersen HJ (2008) Follicle development in mares. Reprod Dom Anim 43(Suppl. 2):224–231

Fair T, Hyttel P, Greve T (1995) Bovine oocyte diameter in relation to maturational competence and transcriptional activity. Mol Reprod Dev 42:437–442

Foss R, Ortis H, Hinrichs K (2013) Effect of potential oocyte transport protocols on blastocyst rates after intracytoplasmic sperm injection in the horse. Equine Vet J 45(suppl):39–43

Galli C, Colleoni S, Claes A, Beitsma M, Deelen C, Necchi D, Duchi R, Lazzari G, Stout T (2016) Overnight shipping of equine oocytes from remote locations to an ART laboratory enables access to the flexibility of ovum pick up-ICSI and embryo cryopreservation technologies. J Equine Vet Sci 41:82

Galli C, Colleoni S, Duchi R, Lagutina I, Lazzari G (2007) Developmental competence of equine oocytes and embryos obtained by *in vitro* procedures ranging from *in vitro* maturation and ICSI to embryo culture, cryopreservation and somatic cell nuclear transfer. Anim Reprod Sci 98:39–55

Galli C, Crotti G, Notari C, Turini P, Duchi R, Lazzari G (2001) Embryo production by ovum pick up from live donors. Theriogenology 55:1341–1357

Galli C, Duchi R, Colleoni S, Lagutina I, Lazzari G (2014) Ovum pick up, intracytoplasmic sperm injection and somatic cell nuclear transfer in cattle, buffalo and horses: from the research laboratory to clinical practice. Theriogenology 81:138–151

Garcia A, Salaheddine M (1998) Effects of repeated ultrasound-guided transvaginal follicular aspiration on bovine oocyte recovery and subsequent follicular development. Theriogenology 50:575–585

Goodhand KL, Staines ME, Hutchinson JSM, Broadbent PJ (2000) *In vivo* oocyte recovery and in vitro embryo production from bovine oocyte donors treated with progestagen, oestradiol and FSH. Anim Reprod Sci 63:145–158

Goodhand KL, Watt RG, Staines ME, Hutchinson JSM, Broadbent PJ (1999) *In vivo* oocyte recovery and *in vitro* embryo production from bovine donors aspirated at different frequencies or following FSH treatment. Theriogenology 51:951–961

Hamano S, Kuwayama M (1993) *In vitro* fertilization and development of bovine oocytes recovered from the ovaries of individual donors: a comparison between the cutting and aspiration method. Theriogenology 39:703–712

Hashimoto S, Takakura R, Kishi M, Sudo T, Minami N, Yamada M (1999) Ultrasound-guided follicle aspiration: the collection of bovine cumulus-oocyte complexes from ovaries of slaughtered or live cows. Theriogenology 51:757–765

Hashimoto S, Takakura R, Minami N, Yamada M (1999) Ultrasound-guided follicle aspiration: effect of the frequency of a linear transvaginal probe on the collection of bovine oocytes. Theriogenology 52:131–138

Hasler JF (1998) The current status of oocyte recovery, *in vitro* embryo production and embryo transfer in domestic animals, with an emphasis on the bovine. J Anim Sci 76(3 Suppl):52–74

Hasler JF, Henderson WB, Hurtgen PJ, Jin ZO, McCauly AD, Mower SA, Neely B, Shuey LS, Stokes JE, Trimmer SA (1995) Production, freezing and transfer of bovine IVF embryos and subsequent calving results. Theriogenology 43:141–152

Hawley LR, Enders AC, Hinrichs K (1995) Comparison of equine and bovine oocyte-cumulus morphology within the ovarian follicle. Biol Reprod Monogr 1:243–252

Hazeleger NL, Hill DJ, Stubbings RB, Walton JS (1995) Relationship of morphology and follicular fluid environment of bovine oocytes to their developmental potential *in vitro*. Theriogenology 43:509–522

Hendriksen PJM, Steenweg WNM, Harkema JC, Merton JS, Bevers MM, Vos PLAM, Dieleman SJ (2004) Effect of different stages of the follicular wave on *in vitro* developmental competence of bovine oocytes. Theriogenology 61:909–920

Hinrichs K (2012) Assisted reproduction techniques in the horse. Reprod Fertil Dev 25:80–93

Hinrichs K, Choi YH, Love CC, Spacek S (2014) Use of in vitro maturation of oocytes , intracytoplasmic sperm injection and in vitro culture to the blastocyst stage in a commercial equine assisted reproduction program. J Equine Vet Sci 34:176

Jacobson CC, Choi YH, Hayden SS, Hinrichs K (2010) Recovery of mare oocytes on a fixed biweekly schedule, and resulting blastocyst formation after intracytoplasmic sperm injection. Theriogenology 73:1116–1126

Konishi M, Aoyagi Y, Takedomi T, Itakura H, Itoh T, Yazawa S (1996) Presence of granulosa cells during oocyte maturation improved development of IVM-IVF bovine oocytes that were collected by ultrasound-guided transvaginal aspiration. Theriogenology 45:573–581

Kruip TAM, Boni R, Wurth YA, Roelofsen MWM, Pieterse MC (1994) Potential use of ovum pick-up for embryo production and breeding in cattle. Theriogenology 42:675–684

Lonergan P, Monaghan P, Rizos D, Boland MP, Gordon I (1994) Effect of follicle size on bovine oocyte quality and developmental competence following maturation, fertilization and culture *in vitro*. Mol Reprod Dev 37:48–53

Looney CR, Lindsey BR, Gonseth CL, Johnson DL (1994) Commercial aspects of oocyte retrieval and *in vitro* fertilization (IVF) for embryo production in problem cows. Theriogenology 41:67–72

Lopez Ruiz L, Alvarez N, Nunez I, Montes I, Solano R, Fuentes D, Pedroso R, Palma GA, Brem G (1996) Effect of body condition on the developmental competence of IVM/IVF bovine oocytes. Theriogenology 45:292

McKinnon AO, Lacham-Kaplan O, Trounson AO (2000) Pregnancies produced from fertile and infertile stallions by intracytoplasmic sperm injection (ICSI) of single frozen–thawed spermatozoa into in vivo matured mare oocytes. J Reprod Fertil 56:513–517

Meintjens M, Bellow MS, Broussard JR, Paul JB, Godke RA (1995) Transvaginal aspiration of oocytes from hormone-treated pregnant beef cattle for *in vitro* fertilization. J Anim Sci 73:967–974

Merton S 2014 Factors affecting the outcome of in vitro bovine embryo production using ovum pick-up derived cumulus oocyte complexes. PhD Thesis, Faculty of Veterinary Medicine, University of Utrecht, the Netherlands

Merton JS, Ask B, Onkundi DC, Mullaart E, Colenbrander B, Nielen M (2009) Genetic parameters for oocyte number and embryo production within a bovine ovum pick-up *in vitro* production embryo production program. Theriogenology 72:885–893

Merton JS, de Roos APW, Mullaart E, de Ruigh L, Kaal L, Vos PLAM, Dieleman SJ (2003) Factors affecting oocyte quality and quantity n commercial application of embryo technologies in the cattle breeding industry. Theriogenology 59:651–674

Paul JB, Looney CR, Lindsay BR, Godke RA (1995) Gonadotropin stimulation of cattle donors at estrus for transvaginal oocyte collection. Theriogenology 43:294

Petyim S, Båge R, Forsberg M, Rodriguez-Martinez H, Larsson B (2000) The effect of repeated follicular puncture on ovarian function in dairy heifers. J Vet Med A 47:627–640

Petyim S, Båge R, Forsberg M, Rodriguez-Martinez H, Larsson B (2001) Effects of repeated follicular punctures on ovarian morphology and endocrine parameters in dairy heifers. J Vet Med A 48:449–463

Pieterse MC, Kappen KA, Kruip TAM, Taverne MAM (1988) Aspiration of bovine oocytes during transvaginal ultrasound scanning of the ovaries. Theriogenology 30:751–762

Pieterse MC, Vos PLAM, Kruip TAM, Willemse AH, Taverne MAM (1991) Characteristics of bovine estrous cycles during repeated transvaginal, ultrasound-guided puncturing of follicles for ovum pick-up. Theriogenology 35:401–413

Pieterse MC, Vos PLAM, Kruip TAM, Wurth YA, van Beneden TH, Willemse AH, Taveme MAM (1991) Transvaginal ultrasound guided follicular aspiration of bovine oocytes. Theriogenology 35:19–24

Pieterse MC, Vos PLAM, Kruip TAM, Wurth YA, van Beneden TH, Willemse AH, Taverne MAM (1992) Repeated transvaginal ultrasound-guided ovum pick-up in ECG-treated cows. Theriogenology 37:273

Pontes JHF, Melo Sterza FA, Basso AC, Ferreira CR, Sanches BV, Rubin KCP, Seneda MM (2011) Ovum pick-up, *in vitro* embryo production, and pregnancy rates from a large-scale commercial program using Nelore cattle (Bos indicus) donors. Theriogenology 75:1640–1646

Pontes JHF, Nonato-Junior I, Sanches BV, Ereno-Junior JC, Uvo S, Barreiros TRR, Oliveira JA, Hasler JF, Seneda MM (2009) Comparison of embryo yield and pregnancy rate between *in vivo* and *in vitro* methods in the same Nelore (Bos indicus) donor cows. Theriogenology 71:690–697

Price CA, Carrière PD, Bhatia B, Groome NP (1995) Comparison of hormonal and histological changes during follicular growth, as measured by ultrasonography, in cattle. J Reprod Fertil 103:63–68

Rath D (1993) Current status of ultrasound-guided retrieval of bovine oocytes. Embryo Transfer Newsl 11:10–15

Reichenbach MD, Wiebke NH, Mödl J, Zhu J, Brem G (1994) Laparoscopy through the vaginal fornix of cows for the repeated aspiration of follicular oocytes. Vet Rec 135:353–356

Reinders JMC, Van Wagtendonck-de Leeuw AM (1996) Improvement of a MOET program by addition of *in vitro* production of embryos after ovum pick-up from pregnant donor heifers. Theriogenology 45:354

Rick G, Hadeler KG, Lemme E, Lucas-Hahn A, Rath D, Schindler L, Niemann H (1996) Long-term ultrasound guided ovum pick-up in heifers from 6 to 15 months of age. Theriogenology 45:356

Schellander K, Fayrer-Hosken R, Keefer C, Brown L, Malter H, Mcbride C, Brackett B (1989) *In vitro* fertilisation of bovine follicular oocytes recovered by laparoscopy. Theriogenology 31:927–933

Scott CA, Robertson L, de Moura RTD, Paterson C, Boyd JS (1994) Technical aspects of trans-vaginal ultrasound-guided follicular aspiration in cows. Vet Rec 134:440–443

Seneda MM, Esper CS, Garcia JM, de Oliveira JA, Vantini R (2001) Relationship between follicle size and ultrasound-guided transvaginal oocyte recovery. Anim Reprod Sci 67:37–43

Singh J, Adams GP, Pierson RA (2003) Promise of new imaging technologies for assessing ovarian function. Anim Rep Sci 78:371–399

Stout TA (2006) Equine embryo transfer: review of developing potential. Equine Vet J 38:467–478

Stubbings RB, Walton JS (1995) Effect of ultrasonically-guided follicle aspiration on estrous cycle and follicular dynamics in Holstein cows. Theriogenology 43:705–712

Takagi Y, Mori K, Takahashi T, Sugawara S, Masaki J (1992) Differences in development of bovine oocytes recovered by aspiration or by mincing. J Anim Sci 70:1923–1927

Taneja M, Bols PEJ, Van de Velde A, Ju J-C, Schreiber D, Tripp MW, Levine H, Echelard Y, Riesen J, Yang X (2000) Developmental competence of juvenile calf oocytes *in vitro* and *in vivo*: influence of donor animal variation and repeated gonadotropin stimulation. Biol Reprod 62:206–213

Tanghe S, Van Soom A, Nauwynck H, Corijn M, de Kruif A (2002) Minireview: functions of the cumulus oophorus during oocyte maturation, fertilization and ovulation. Mol Reprod Dev 61:414–424

Tremoleda JL, Stout TA, Lagutina I, Lazzari G, Bevers MM, Colenbrander B, Galli C (2003) Effects of *in vitro* production on horse embryo morphology, cytoskeletal characteristics, and blastocyst capsule formation. Biol Reprod 69:1895–1906

Velez IC, Arnold C, Jacobson CC, Norris JD, Choi YM, Edwards JF, Hayden SS, Hinrichs K (2012) Effects of repeated transvaginal aspiration of immature follicles on mare health and ovarian status. Equine Vet J 44:78–83

Viana JHM, Palhao MP, Siqueira LGB, Fonseca JF, Camargo LSA (2010) Ovarian follicular dynamics, follicle deviation, and oocyte yield in Gyr breed (*Bos indicus*) cows undergoing repeated ovum pick-up. Theriogenology 73:966–972

Vos PLAM, de Loos FAM, Pieterse MC, Bevers MM, Taverne MAM, Dieleman SJ (1994) Evaluation of transvaginal ultrasound-guided follicle puncture to collect oocytes and follicular fluids at consecutive times relative to the preovulatory LH surge in eCG/PG treated cows. Theriogenology 41:829–840

Ward FA, Lonergan P, Enright BP, Boland MP (2000) Factors affecting recovery and quality of oocytes for bovine embryo production *in vitro* using ovum pick-up technology. Theriogenology 54:433–446

Preservation of Gametes and Embryos

11

Amir Arav and Joseph Saragusty

Abstract

Cryopreservation is the practical implementation of the scientific field of cryobiology. It was developed particularly over the last two centuries, having major milestones in the field of animal reproduction. Technologies such as directional freezing of sperm and sperm desiccation, as well as oocyte and embryo freezing and vitrification, are discussed and described in this chapter. Hereinafter, we describe the major breakthroughs of the past two centuries and our foresight for the near future.

11.1 History of Cryopreservation

Cryopreservation refers to preservation of biological samples, such as proteins, cells, tissues, organs and even whole animals, at low temperatures. Currently its most popular meaning is the preservation of such cells and tissues at very low sub-zero temperatures.

In the late nineteenth century, researchers such as Ferdinand Cohn (1871), Hermann Kunisch (1880) or Hermann Müller-Thurgau (1886) made observations of plant tissue while freezing it by taking the microscope, specimens and themselves out during cold winter days. It was, however, Hans Molisch who developed the first cryomicroscope in 1897 (Molisch 1982). Molisch, then in Austria, packaged the

A. Arav (✉)
FertileSafe Ltd., Ness Ziona, Israel
e-mail: amir.arav@mail.huji.ac.il

J. Saragusty
Laboratory of Embryology, Faculty of Veterinary Medicine, University of Teramo,
Teramo, Italy
e-mail: saragusty@izw-berlin.de

© Springer International Publishing AG, part of Springer Nature 2018   235
H. Niemann, C. Wrenzycki (eds.), *Animal Biotechnology 1*,
https://doi.org/10.1007/978-3-319-92327-7_11

Fig. 11.1 An illustration
of the first cryomicroscope
(Photo: A. Arav)

microscope into a box with ice and other coolants, allowing limited control from outside the box (illustrated photo in Fig. 11.1). He observed that plant cells are damaged due to the growth of ice crystals. He stated that "…the direct observation of the freezing cell is the best means of obtaining information on the causes of death by freezing" (Molisch 1982).

11.1.1 Cryopreservation: The Beginning

The nineteenth century was the century at which modern science became professionalised and institutionalised with many breakthroughs in modern scientific fields such as chemistry, physics, biology (evolution) and the scientific methodology. One of the great scientists of that time was Joseph Louis Gay-Lussac; one of his discoveries was the effect of small volume of water droplets on supercooling. B. J. Luyet and P.M Gehenio wrote in their 1940 book titled *Life and Death at Low Temperatures* about the observation of Gay-Lussac that "Some of the oldest investigations on subcooling were made by Gay Lussac (1836) who observed that water can be subcooled to −12 °C when it is enclosed in small tubes" (Luyet and Gehenio 1940).

In 1858 Johann Rudolf Albert Mousson sprayed droplets of water, less than 0.5 mm in diameter, on a dry surface and observed that the smaller the drops, the longer they stayed subcooled (Mousson 1858). Not only the volume was important to achieve supercooling but also the cooling velocity and the concentration of the supercooled or supersaturated solutions, among other factors that might have an influence on inducing crystallisation as was mentioned by Luyet and Gehenio (Luyet and Gehenio 1940). They wrote "To avoid freezing, the temperature should drop at a rate of some hundred degrees per second, within the objects themselves" and also "The only method of vitrifying a substance is to take it in the liquid or gas state and cool it rapidly so as to skip over the zone of crystallization temperatures in less time than is necessary for the material to freeze. ... It is evident that when the crystals grow faster one must traverse the crystallization zone more rapidly if one wants to avoid crystallization" (Luyet and Gehenio 1940). These early experiments on supercooling were the basis for vitrification. In 1938 Luyet and Hodapp published the first report on successful cryopreservation of spermatozoa, done by vitrification (Luyet and Hodapp 1938). A decade later, in 1949, Christopher Polge, Audrey Smith and Alan Parkes (Polge et al. 1949), when trying to duplicate Luyet's and Hodapp's results with fowl spermatozoa, accidently discovered the cryoprotective property of glycerol and so opened the field of slow freezing.

Still today, the two methods for gametes' cryopreservation are slow freezing and vitrification. Slow freezing has the advantage of using low concentrations of cryoprotectants (CPs), which are associated with chemical toxicity and osmotic shock. Vitrification, on the other hand, is a very rapid method that causes the sample to turn into a glassy amorphous state instead of creating ice crystals. It has the benefit of reducing injuries caused to the cells by chilling and crystallisation. Sherman and Lin (Sherman and Lin 1958) showed that mouse oocytes need 8–10 min for equilibration in a freezing solution containing 5% glycerol at 37 °C. Such oocytes survived freezing to −10 °C and maintenance at that temperature for 3.5 h. In addition, they demonstrated that mouse oocytes survive supercooling to −20 °C after slow cooling at 0.6 °C/min; however, oocytes that were cooled faster or to lower temperatures were damaged due to intracellular crystallisation.

In the early 1970s, two groups were competing on achieving the first success of slow freezing of embryos. One group included Whittingham, Leibo and Mazur and the other Wilmut and Polge. Whittingham had reported partial success in freezing embryos to −79 °C for 30 min using polyvinylpyrrolidone (PVP) (Whittingham 1971); however, this experiment could not be duplicated, but instead both groups published in 1972 their reports, showing the first survival of mouse embryos after slow freezing (Whittingham et al. 1972; Wilmut 1972) and live offspring (Whittingham et al. 1972). The technique included cooling at a slow rate in the presence of 1 mol/L DMSO, which most likely was the ingredient that enabled this success. In 1976, sheep embryos were slow frozen by Willadsen using 1.5 mol/L DMSO and a cooling rate of 0.3 °C/min (Willadsen et al. 1976). However, the first farm animal to be born after transfer of frozen/thawed embryos was a calf. This was published by Wilmut and Rowson in 1973 (Wilmut and Rowson 1973). Since then, dozens of species have been successfully cryopreserved by slow freezing (for a review, see (Saragusty and Arav 2011)).

11.2   Vitrification of Oocytes and Embryos

For many years, slow freezing, and not vitrification, was the method of choice for embryo cryopreservation. In 1985, the first successful vitrification of mouse embryos, using a relatively large volume sample, was achieved (Rall and Fahy 1985). Rall and Fahy vitrified mouse embryos with a mixture of DMSO, acetamide, and polyethylene glycol and in a relatively large volume inside a 0.25-mL straw that was plunged into liquid nitrogen (LN). As stated above, vitrification is the process in which a sample solidifies without the formation of ice crystals, thus resulting in a glassy amorphous state. The main factors that influence the probability for vitrification to occur are (1) sample's volume, the lower the volume, the greater the chances for vitrification to occur; (2) cooling rate, as cooling rate increases, the chances for ice nuclei to form or to grow into ice crystals decrease;. and (3) sample's viscosity, the higher the viscosity of the sample, the higher the chances to avoid crystallisation [see Arav equation].

$$\text{Probability of vitrification} = \frac{\text{Cooling / warming rate} \times \text{Viscosity}}{\text{Volume}}$$

In 1989 the "minimum drop size" method (MDS) was developed by A. Arav (Arav 1989; Arav 1992). The volume used for vitrification was in the range of 0.07 μL (70 nL), and the concentration of the vitrification solution (VS) was about 50% lower than of the VS used for large-volume vitrification. The method was named "minimum drop size" because this was the minimal size that maintained oocytes or embryos without damage owing to desiccation. Vitrification of oocytes, on the other hand, although initially attempted in the late 1980s, had not been applied clinically until recently. Vitrification is currently producing very satisfactory outcomes by means of methodologies that use a minimal volume (Cobo et al. 2008; Kuwayama 2007; Vajta et al. 1998). Despite very favourable results, success in vitrification varies between laboratories and individuals, partially because the transfer of oocytes or embryos between solutions requires much experience. This operator dependency is a source of inconsistency. Currently, there is no methodology that is compatible with full automation of the vitrification process. Here we report a method that, by virtue of a fully automated, operator-independent process, including the immersion into LN, will simplify and standardise the vitrification procedures worldwide. For this purpose we have developed a fully automated vitrification device (Sarah, FertileSafe, Ness Ziona, Israel; Figs. 11.2 and 11.3). Using this device, we vitrified mouse oocytes and embryos at the blastocyst stage. Results showed that 95% (19/20) of the MII oocytes regained isotonic volumes, and all (100%) of the surviving oocytes were viable according to the live/dead stains (Arav et al. 2016). Rewarmed embryos had 95% (38/40) blastulation rate (day 4) and 75% (30/40) hatching rate (day 5). The fresh embryos (controls) had a similar (two-tailed Z-test, $Z = 0.4317$, $P = 0.67$) hatching rate of 80% (16/20).

Fig. 11.2 Sarah automatic
vitrification and warming
device

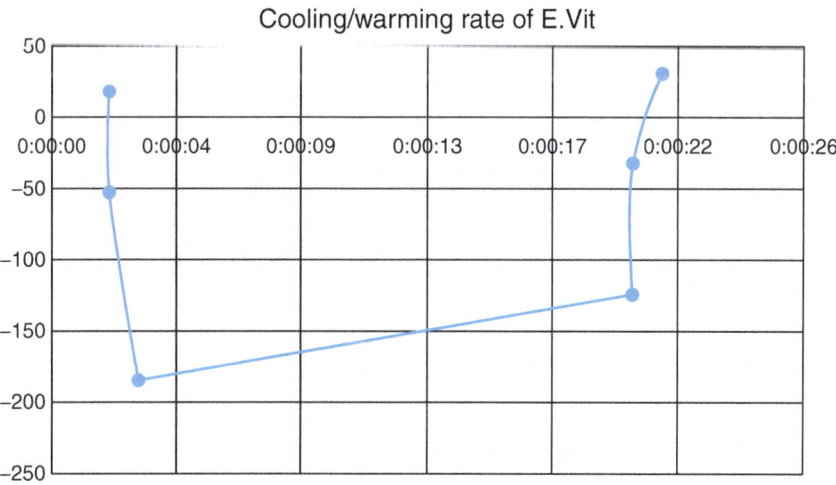

Fig. 11.3 A typical cooling and warming curve when vitrifying in the Sarah automatic vitrification
and warming device, using the E.Vit (fertileSafe, Ness Ziona, Israel) vitrification straws that were
specially designed to allow fast transfer of the oocytes or embryos between solutions and high cooling and warming rates. Here the sample was cooled from room temperature to −181 °C then gradually warmed up to −123 °C and then warmed at very high warming rate up to about 31 °C

11.2.1 Safe Vitrification Cooling and Storage

Cryopreservation of biological samples by direct exposure to liquid nitrogen (LN) and their storage in standard LN tanks is problematic due to the risk posed by potential contamination with viruses, bacteria, fungi and spores that may survive in LN (Bielanski et al. 2003; Bielanski 2012; Vajta et al. 2015). The risk of contamination might be due to contaminants already residing in the LN when supplied to the laboratory or, alternatively, due to cross-contamination between samples where the LN acts as transmitting medium. This risk of microbial infection becomes a particularly serious threat when the biological samples are intended to be transferred into recipients, as is done in IVF procedures, ovarian cortex transplantation and future stem cell therapies. Therefore, the use of safe cryopreservation protocols is important and highly desired. At present, heat-sealing is considered the best and safest method for firm closure of storage carriers. Other methods such as cotton plugs, beads and PVP powder (commonly used for sperm), waterproof mechanical closures such as screwed caps of cryovials or protective caps of several vitrification devices do not provide appropriate protection. The drawback of these sealing methods, particularly of heat-sealing, is that they slow the cooling rate and increase sample volume, thus reducing the chances for successful vitrification.

To overcome these potential risks of contamination posed by both closed and open systems, we developed two devices. The first is a bench-top device that produces clean liquid air from air filtered through 0.22-µm filter (CLAir, FertileSafe, Ness Ziona, Israel). Liquid air (LA) has the same temperature and properties as LN. The second is a sterile storage device, which enables storing samples in a sterile manner in standard LN tanks. This device preserves the sample at LN temperature while insulating it from being in contact with the surrounding LN (Esther, FertileSafe, Ness Ziona, Israel).

Example of Vitrification Protocol for Bovine Blastocysts

For Cooling

1. Vial 1 is filled with 2 mL of 25% of equilibration solution (ES), which is 7.5% ethylene glycol or EG, 7.5% DMSO and 10% foetal calf serum or FCS in PBS (i.e. 1.875% EG, 1.875% DMSO, 10% FCS in PBS).
2. Vial 2 is filled with 2 mL of 50% ES (i.e. 3.75% EG, 3.75% DMSO and 10% FCS in PBS).
3. Vial 3 is filled with 2 mL of 100% ES.
4. Vial 4 is filled with 2 mL of 100% vitrification solution or VS (16% EG + 16% DMSO + 0.5 M sucrose + 10% FCS in PBS).
5. Vials are placed into their respective slots on the automatic vitrification and warming device (Sarah, FertileSafe Ltd., Ness Ziona, Israel).
6. Embryos are loaded into 0.25-mL straws with holding medium (i.e. PBS + 10% FCS).
7. Straws are loaded into their holders and the start button is pressed.
8. At the end of the process, the straws with the vitrified embryos are transferred into liquid nitrogen for storage.

For Warming
1. Vial 1 is filled with 2 mL of 1.0 M sucrose + 10% FCS in PBS.
2. Vial 2 is filled with 2 mL of 0.5 M sucrose + 10% FCS in PBS.
3. Vial 3 is filled with 2 mL of 0.25 M sucrose + 10% FCS in PBS.
4. Vial 4 is filled with holding medium (i.e. PBS + 10% FCS).
5. Vials are placed in their respective slots on the Sarah.
6. Vitrified straws are loaded and the start button is pushed.
7. At the end of warming, the tip of the straw is cut (leaving the cotton wool end intact), and the straw with the embryos is ready for transfer into the cow's uterus (there is no need for a microscope for the warming process).

11.3 Directional Freezing

Sperm cryopreservation has been attempted already in the eighteenth century (Spallanzani 1776) and, to some extent, practised during the first half of the twentieth century (Luyet and Hodapp 1938; Phillips and Lardy 1940). The science of cryopreservation, however, really started with the seminal work of Father Basile J. Luyet (Luyet 1937; Luyet and Hodapp 1938) who realised that ice formation can cause damage to the cryopreserved cells and should therefore be avoided. In the absence of thickeners and other cryoprotective agents, the only way to achieve that was by kinetic vitrification, i.e. vitrification of very small volumes at very high cooling rates. In the decade that followed, several attempts were made to cryopreserve spermatozoa with varying rates of success (Hoagland and Pincus 1942; Jahnel 1938; Schaffner 1942). The era of sperm freezing was launched when Polge and colleagues discovered the cryoprotective properties of glycerol in 1949 (Polge et al. 1949), a discovery that was made in parallel also in Russia by I. V. Smirnov (cited in Katkov et al. 2012). The developments in sperm cryopreservation that followed are described elsewhere in this book (see Waberski, Chapter 4). We wish to concentrate here on one specific development, the directional freezing technique.

The idea of directional freezing may have arisen from the casting industry. Casting proceeds in such a way that molten feed metal coming through the sprue is continuously available for the solid-liquid boundary. The metal shrinks when it cools and solidifies, and, in the absence of continuous feed metal, defects will form in it. This process is known as directional solidification. When water solidifies, it expands rather than shrinks so the physical forces acting on the sample when confined to the boundaries of its container are different (Saragusty et al. 2009c). It turned out, however, that by guiding the freezing process in a unidirectional manner, it is possible to control many aspects of the freezing process so one could study it. The first steps of directional freezing were therefore on the stage of the microscope when Körber (Körber et al. 1983), Rubinsky and Ikeda (Rubinsky and Ikeda 1985) and others constructed their directional cryomicroscopes in the early 1980s (Figs. 11.4 and 11.5). Subsequently, work with the directional freezing technology pursued two parallel, and mutually complementing, lines of investigation. On the

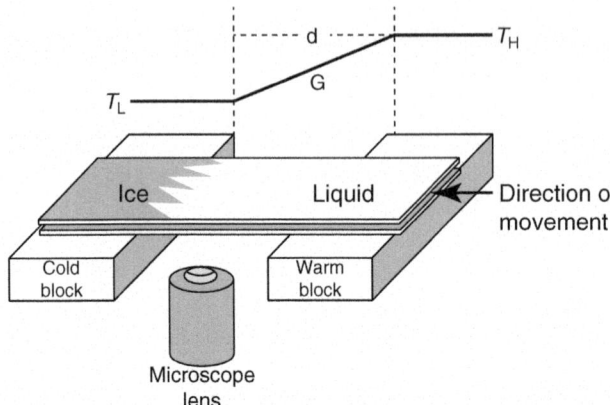

Fig. 11.4 Schematic diagram of the directional freezing technology as demonstrated here on a microscope stage. The apparatus is composed of a warm (suprazero temperature, e.g., +5 °C; T_H) and a cold (very low temperature, e.g., −50 °C; T_L) blocks and a pushing mechanism that pushes the sample (mounted glass slide, glass tube, etc.) at a constant rate from the warm block to the cold block. Ice crystals will grow in a direction opposite to the direction of movement of the sample. The cooling rate (B) is a function of the temperature gradient between the blocks (G) and the velocity (v). The gradient (G) is associated with the temperature difference between the warm (T_H) and cold (T_L) blocks and the distance (d) between them. The microscope may be inverted, as illustrated here, or upright

one hand, research continued using this technology to advance our basic understanding of processes occurring during cryopreservation. On the other hand, and especially since a freezing device relying on this technology has been developed (Arav 1999), researchers utilised this technology for applied cryobiological studies. With this device, heat is removed from the samples to the surrounding highly thermal conductive cold metal block as well as to the unfrozen part of the sample that is still in the warm block. At the same time, the controlled movement of the sample from the warm to the cold blocks at a constant velocity ensures ideal morphology of the ice crystals and their growth in a direction opposite to the direction of movement. These characteristics help in achieving better protection of the cells or tissues in the sample throughout the freezing process.

Behaviour of the ice front and its effect on survival of cells in the solution were studied by Brower and colleagues (Brower et al. 1981) showing that cells can survive if they are pushed ahead of the ice front during initial freezing to allow some dehydration and equilibration and are then engulfed by the advancing ice front. This was followed by studies on the behaviour of binary water-salt solutions during directional freezing with the aim of better understanding what are the stresses experienced by the cells at the ice front. Körber and colleagues showed that the concentration of the solution ahead of the advancing ice front was changing in agreement with phase diagrams and theoretical predictions (Körber et al. 1983; Körber and Scheiwe 1983), a topic that was also studied at the time by Rubinsky (Rubinsky 1983). These authors showed that with time or velocity, the concentration of solutes in the unfrozen media increased and demonstrated the advantage of the directional

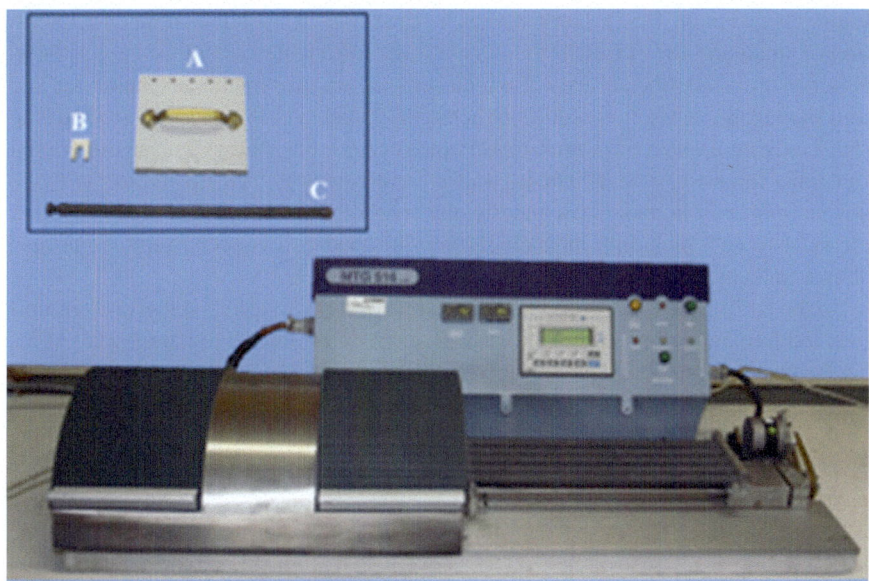

Fig. 11.5 The directional freezing machine. (**a**) The cassette into which up to five HollowTubes are loaded. The cassette acts as the warm block and goes into its slot under the right green cover on the machine. (**b**) Holder for a pushing rod. Five holders keep the five rods in place. (**c**) The engine, seen on the right side of the machine, moves five such pushing rods at a constant velocity. The rods push the HollowTubes from the warm block into and through the cold block and then out at the other end into the collection chamber, which is under the green cover on the left. The figure was reproduced from Methods in Cryopreservation and Freeze-Drying, Directional Freezing for Large Volume Cryopreservation, 2015, Vol. 1257, pp. 381–397, Saragusty J with permission of Springer (Saragusty 2015)

freezing microscope as a cryobiological investigation tool. The microscope made it possible to visualise the morphology of the advancing ice and, in cases of cells suspensions, the interaction between the cells and the ice front. It was also possible to perform densitometric measurements while being able to control single parameters such as the velocity or the temperature gradient through which the sample is propagated. It was then shown that particles in the solution may be repulsed by the ice front or engulfed by it, depending on the velocity in which the ice front advances and the size of the particles (Körber et al. 1985) and in case of gas bubbles also the saturation level of the solution (Lipp et al. 1987). These studies were thoroughly reviewed by Körber (Körber 1988). In later studies, during the 1990s, the applicability of the general cooling rate eq. $B = G \times v$ to the directional freezing was demonstrated (Beckmann et al. 1990). In this eq. B is the cooling rate (change in temperatures divided by change in time or $\Delta T/\Delta t$), G is the temperature gradient (change in temperature divided by the distance or $\Delta T/d$) and v is the velocity (distance divided by the change in time or $d/\Delta t$). The whole topic of cryoinjury was then studied, using directional freezing apparatuses. Hubel and colleagues demonstrated the association between cooling rate and ice-cell association as a source of injury (Hubel et al. 1992). Later studies by the same group demonstrated that addition of small amount of trehalose (10 mM) was sufficient to alter this ice-cell interaction and push up the velocity at which cells are trapped by the advancing ice and thus damaged (Hubel et al. 2007). Other agents, such as PBS, glycerol or OptiPrep (60% iodixanol), were all shown to affect ice crystal morphology during directional freezing (Saragusty et al. 2009a, b), thus affecting the way the cells interact with the advancing ice front and, consequently, the level of cryoinjury. The effects of other physical forces, including mechanical pressure and cell concentration, were also shown to be involved in cryoinjury (Saragusty et al. 2009c).

In the early 1980s, cryobiologists started working with ice crystal growth inhibitors and freezing temperature depressors known as antifreeze glycoproteins (AFP). These proteins are found in various insects (Duman 1982), fishes (Knight et al. 1984) and plants (Griffith et al. 1992) but can also be produced synthetically (Bar et al. 2006; Matsumoto et al. 2006). At first, directional freezing studies showed the positive effects of these proteins on various biological systems including oocytes and embryos (Arav et al. 1993; Arav et al. 1994; Rubinsky et al. 1991; Rubinsky et al. 1992). The effect of these proteins on ice crystal morphology and the kinetic supercooling at the ice-solution interface were also studied (Furukawa et al. 2005) and recently reviewed (Bar Dolev et al. 2016). In parallel, other studies, using the same directional freezing technology, showed that the addition of AFP could also lead to cell injury due to intracellular ice formation (Ishiguro and Rubinsky 1994; Koushafar and Rubinsky 1997). In the following years, the directional freezing technology was also utilised to study the relationship between the cooling rate and the 3D ice crystal morphology and their interaction with the suspended cells, using confocal laser scanning microscope (Ishiguro and Koike 1998). More studies on cryoinjury (Arav et al. 2002; Li et al. 2010; Li et al. 2013) demonstrating how interactions between the cells and extracellular ice crystals could lead to intracellular ice formation in a directional manner.

Directional freezing has been used successfully to cryopreserve a variety of cell and tissue types, including oocytes and embryos (Arav 1989; Rubinsky et al. 1991), whole ovaries (Arav 2003; Arav et al. 2005; Arav et al. 2007; Gavish et al. 2004a, b; Maffei et al. 2013b; Maffei et al. 2013a; Revel et al. 2001; Revel et al. 2004) and ovarian tissue (Maffei et al. 2013a), whole livers (Gavish et al. 2008) and hearts (Elami et al. 2008), cartilage tissue (Arav 2012) and a variety of cells, including granulosa cells (Loi et al. 2008), umbilical cord mononuclear cells (Natan et al. 2009), red blood cells (Arav and Natan 2011; Arav and Natan 2012) and more. However, cryopreservation of spermatozoa is by far the most widespread application of the directional freezing technology (Table 11.1).

Since their conception, the directional freezing machines went through many changes. Some models were designed to freeze straws, but the majority of the machines in use these days freeze in larger-volume HollowTube®. In the early days, freezing was done in normal glass tubes of 12 mL. Some experiments were even performed with 50-mL glass tubes, but this was dropped after a while. Early models were also capable of rotating the tubes with the samples while pushing them forward. The assumption was that if not rotated, the sperm in the sample will sink to

Table 11.1 Sperm cryopreservation by the directional freezing technology

Species	Study's findings	Reference
African elephant	First successful AI with frozen-thawed semen in elephants	Hildebrandt et al. (2012)
African elephant	DF as a successful method for cryopreservation in this species	Hermes et al. (2013)
Asian elephant	DF as a successful method for cryopreservation in this species	Hermes et al. (2003)
Asian elephant	Report on a successful protocol for freezing sperm in this species	Saragusty et al. (2009a)
Asian elephant	Sperm stored up to 12 h post DF and thawing maintain acceptable in vitro values	(O'Brien et al. 2012)
Bottlenose dolphin	Directional freezing superior to straw freezing. AI with frozen-thawed sorted sperm produced an offspring	O'Brien and Robeck (2006)
Bottlenose dolphin	Double freezing with sorting can be done with DF being better than straw freezing	Montano et al. (2012)
Bottlenose dolphin	Mid-horn AI with sexed spermatozoa frozen by DF result in good conception rate	Robeck et al. (2013)
Beluga	Trehalose is better than glycerol and DF is superior to straw freezing	O'Brien and Robeck (2010)
Beluga	First successful AI with offspring with frozen-thawed sperm in this species	Robeck et al. (2010)
Cattle	Similar pregnancy rate following single freezing in straw and double freezing by DF	Arav et al. (2002)
Cattle	Better membrane and acrosome integrity in DF compared to freezing in straws	Hayakawa et al. (2007)
Cattle	Iodixanol as a protective agent during DF	Saragusty et al. (2009d)
Cattle	Large-scale field study on double freezing: DF then straw vs. only straw with similar conception rates	Saragusty et al. (2009b)
Cattle	Demonstration that the controlled ice nucleation stage is not needed during DF	Saragusty et al. (2016)
Common hippopotamus	Post-castration epididymal sperm freezing with clear differences between males	Saragusty et al. (2010)
Donkey	DF superior to straw freezing in all parameters evaluated and with high conception rate when used for AI at the correct timing	Saragusty et al. (2017)
European brown hare	High fertility rates in captive hares following AI with frozen-thawed wild hare sperm	Hildebrandt et al. (2009)
Dorcas and mountain gazelles	Roadkill epididymal sperm rescue. Successful freezing with between-male differences	Saragusty et al. (2006)
Goat	Acceptable post-thaw and pregnancy rates	Gacitua and Arav (2005)
Horse	Slower freezing rate and larger temperature gradient provide better post-thaw values	Sieme et al. (2001)
Horse	DF superior to pellet freezing	Rubei et al. (2004)

(continued)

Table 11.1 (continued)

Species	Study's findings	Reference
Horse	No difference in post-thaw in vitro parameters between DF and straw freezing	Zirkler et al. (2005)
Horse	DF superior to straw freezing	Gacitua et al. (2006)
Horse	DF better than straw freezing in all parameters evaluated	Saragusty et al. (2007)
Human	DF results in quality similar to commercial standard freezing equipment	Gao et al. (2002)
Killer whale	Lower glycerol concentration and slower cooling rates better for straw freezing. DF superior to straw freezing in all evaluations	Robeck et al. (2011)
King penguin	Chilled preservation at 5 °C better than 21 °C. DF better than straw freezing in membrane integrity	O'Brien and Robeck (2014)
Onager	Optimal results for epididymal sperm obtained when testicles shipped at 4 °C, frozen by DF and maintained at 22 °C post-thaw	Prieto Pablos et al. (2015)
Yunan Diannan miniature pig	60-s induced ice nucleation and 1.5-mm/s velocity resulted in best post-thaw parameters and acceptable fertilisation rate	Zheng et al. (2010)
Rabbit	Double freezing result in acceptable fertility and kindling rates	Si et al. (2006)
Rhesus macaque	DF superior to straw freezing, with high in vitro fertilisation and blastocyst rates	Si et al. (2010)
Rhinoceros	SMI highest in DF when frozen with K^+/EDTA-supplemented extender	Reid et al. (2006)
Rhinoceros	DF superior to straw freezing in various in vitro evaluations and in maintaining sperm head volume	Reid et al. (2009)
Rhinoceros	First successful AI with frozen-thawed semen in rhinoceroses	Hermes et al. (2009)
Sheep	Successful cryopreservation with DF	Arav et al. (2000)

DF directional freezing, *AI* artificial insemination

the bottom where they might get damaged once in contact with the cold block of the machine. This was later abandoned when experiments showed that freezing without the rotation worked well. The glass tubes were also replaced by the HollowTubes® which are basically two tubes of different diameters that are placed one inside the other and fused together at the bottom. The sample is placed in the space between these tubes. When using these tubes, cooling is done both through the inner tube by convection and through the outer wall by convection and conduction. These tubes come in several sizes, ranging from 2 to 8 mL. This brings us to one of the main advantages of the directional freezing technology for sperm cryopreservation, namely, the ability to cryopreserve large volumes. Economically, freezing in large volume is of great advantage to any breeding program that relies on progeny testing. For example, in the dairy cattle industry, semen is collected in large quantities over a long time period from a large number of bulls, but in the end semen from only about 10% of the bulls is used in the breeding program, while everything else is discarded as unsuitable. If one calculates the costs involved in freezing and discarding such a huge number of straws, it is clear that freezing in large and reusable glass tubes can save enormously large amount of money. The glass tubes also seem to occupy smaller space in the LN storage tanks (one 8-mL tube occupies less space compared to 32 straws of 0.25 mL), saving storage space and LN costs. It was based on such calculations that we proposed the double freezing solution (Arav et al. 2002; Saragusty et al. 2009b). Based on this concept, bovine semen can be cryopreserved from a large number of bulls in 8-mL tubes till the progeny tests are completed. The tubes from those bulls that will not be used can be thawed, the contents discarded and the tubes cleaned, sterilised and reused. Tubes with semen from bulls that are to be used in the breeding program can be thawed, repackaged in 0.25-mL straws (insemination doses) and refrozen to be ready for use. The idea of double freezing can also be utilised in conjunction with sperm sex sorting. Samples can be frozen in large volumes, shipped to the sorting centre where they are thawed, sorted and then refrozen in insemination doses to be shipped to where they are to be used.

A second notable advantage of the directional freezing technology is the fact that in experiments spanning a large number of species, the directional freezing proved to be superior to conventional freezing in straws. These studies include, for example, rhinoceroses (Reid et al. 2009), bottlenose dolphins (O'Brien and Robeck 2006), beluga (O'Brien and Robeck 2010), king penguin (O'Brien and Robeck 2014), onager (Prieto Pablos et al. 2015), killer whale (Robeck et al. 2011), horse (Gacitua et al. 2006; Rubei et al. 2004; Saragusty et al. 2007), rhesus macaque (Si et al. 2010), cattle (Hayakawa et al. 2007) and donkey (Saragusty et al. 2017). This superiority also explains why Cogent, probably the world's largest cattle sperm sex sorting company, elected to use this technology to preserve its products (Cogent 2013). Another interesting aspect is the fact that while in conventional freezing different species require different freezing protocols, when using the directional freezing, experience shows that one freezing protocol basically fits all species. Yet, noting all these advantages, one might wonder how come the directional freezing did not become a widely used technology. Many speculations exist, but probably the most plausible one is related to marketing. We think that if there had been proper and more targeted marketing, this technology would have been today widely in use. In the absence of this, directional freezing of sperm remained almost entirely where it was when it was conceived more than 15 years ago—in limited use in scientific investigations and esoteric applications in wildlife.

Example of a Sperm Freezing Protocol

After collection and evaluation of the sample, it is either washed to remove the seminal fluid or not, depending on the species, diluted in cryopreservation media and chilled to ~5 °C. Samples are then loaded into prechilled HollowTubes (IMT Ltd., Ness Ziona, Israel), leaving a space of about 1 cm between the level of the fluid and the labelled stopper. On the machine the warm block is set to +5 °C, the cold block to −50 °C and the collection chamber to −100 °C. A controlled ice nucleation stage is set into the program although a recent study showed that this stage may be eliminated (Saragusty et al. 2016). This ice nucleation position should be adjusted within the program, depending on the size of the HollowTube. Velocity of the machine is set to the desired speed, which, in most studies, was shown to be 1.0–1.5 mm/s. Tubes with the samples are tilted to resuspend the spermatozoa homogenously in the media and then loaded into the warm block. The engine is activated, and freezing proceeds till the tubes reach the collection chamber from which they are transferred directly into liquid nitrogen. The machine automatically returns to the starting position to allow freezing of the next batch.

Thawing is done by first holding the tube at room temperature for 90 s and then placing it for 60 s in a dedicated thawing device (Harmony CryoCare, Chester, UK) in a 37 °C water bath, in which water flows through and around the tube while the tube is rotating.

For more elaborated explanation, please refer to Saragusty (2015).

11.4   Desiccation

Cryopreservation works fairly well for gametes of both sexes as well as embryos of many domestic and wildlife species. Of course, as discussed above, various species have their unique aspects, sensitivities and limitations, but, as a whole, germplasm can be cryopreserved, stored and eventually used in assisted reproductive programs. This effective preservation method, however, comes with a heavy price tag. Maintaining cryopreserved samples in storage under liquid nitrogen (cryostorage) has high maintenance costs and requires dedicated specialised facilities, trained staff and guaranteed and continuous liquid nitrogen supply. Cryostorage is energy-dependent, has safety concerns and presents a risk of pathogen transmission between samples, all serious issues in clinical practice. Due to liquid nitrogen (LN) nature, shipping samples between locations are complicated and costly. Besides these intrinsic problems, the industrial production and distribution of LN and the energy demands of the dedicated storage facilities have a serious environmental impact, leaving a massive carbon footprint. For all these reasons, seeking an alternative that will help overcome these limitations is highly desirable.

As in many other instances, we would be right to check how extended preservation is done in Nature. To achieve long-term storage, Nature reduces metabolism, reduces chemical reactions and protects against temperature fluctuations and radiation. To achieve all these, Nature's method is simply to get rid of the water, the component without which chemical and biological reactions are brought to a halt. See, for example, the following account written by Antonie van Leeuwenhoek, the renowned Dutch microscopist, in 1702 (van Leeuwenhoek 1800):

...In order more fully to satisfy myself in this respect, on the third of September, about seven in the morning, I took some of this dry sediment, which I had taken out of the leaden gutter and had stood almost two days in my study, and put a little of it into two separate glass tubes, wherein I poured some rain water which had been boiled and afterwards cooled... As soon as I had poured on the water... I examined it, and perceived some of the Animalcules lying closely heaped together. In a short time afterwards they began to extend their bodies, and in half an hour at least an hundred of them were swimming about the glass....

The process is known as anhydrobiosis or life without water. By definition, anhydrobiosis is an extremely dehydrated state in which organisms show no detectable metabolism but retain the ability to resume biological activity after rehydration (Sakurai et al. 2008). Preservation in the dry state is very common in plants (seeds) and many prokaryotes, but it can also be found in some eukaryotes (Hand et al. 2011), including rotifers, tardigrades, nematodes, crustaceans, insects and more. What unifies them all is that they are relatively small, they have little or no control over the loss of water from their bodies, and they are generally inhabitants of ephemerally wet habitats. They desiccate at various developmental stages. In the absence of water, there can be no biochemical reactions, metabolism declines beyond detectable levels, and there is no water to freeze or boil and no active cell processes to be disrupted so they can withstand various environmental extremes. Anhydrobiosis allows animals to survive long periods without water, effectively extending their lifespan and facilitating reproduction or development at the most suitable conditions. Loss of water is gradual and slow, allowing the accumulation of a host of membranes, proteins and nucleus protective agents to as much as 50% of their dry weight. These protective agents include disaccharides, primarily trehalose (Crowe et al. 1984), late embryogenesis abundant proteins (LEAp) (Hand et al. 2011), anhydrin (Goyal et al. 2005), heat-shock proteins (Clark et al. 2007) and more. Drying has been used for hundreds of years as a food preservation technique. Today it is widely used for pharmaceutical, bacterial, viral, fungal and yeast preparations as well as in the food industry (instant coffee, milk and egg powder, dried yeast and more).

Attempts to desiccate eukaryotic cells centred over the years almost entirely on spermatozoa, with some work on other cell types such as blood cells (Arav and Natan 2011; Crowe et al. 2005; Goodrich et al. 1992), fibroblasts (Das et al. 2010; Guo et al. 2000; Zhang et al. 2017) and other cell types (Li et al. 2012; Loi et al. 2008; Natan et al. 2009). Probably two main reasons stand behind the selection of spermatozoa as the main target. First, spermatozoa are relatively small in size with very little water in them, and second, thanks to the protamines, the nucleus in spermatozoa is highly condensed and as such is more stable and less prone to DNA damages (Perreault et al. 1988; Yanagida et al. 1991). The famous paper by Polge and colleagues reporting the use of glycerol as cryoprotectant (Polge et al. 1949) is hailed as the starting point of the era of modern cryopreservation but not everybody remembers that in the same paper Polge et al. also reported on poultry sperm freeze-drying experiments. And impressively, they reported the recovery of as many as

50% of motile spermatozoa after rehydration. With 15% final glycerol concentration and only 3 h of drying, much moisture must have remained in the samples. The fact that the dried samples did not survive even 2 hours at room temperature further supports the notion that sufficient drying probably was not achieved. In the years that followed, several other attempts at freeze-drying were reported for bovine (Bialy and Smith 1957; Saacke and Almquist 1961; Sherman 1957) and human (Sherman 1954; Sherman 1963) spermatozoa, and although some reports claimed to be successful in obtaining surviving motile spermatozoa (Larson and Graham 1976; Meryman and Kafig 1959) and even offspring (Yushchenko 1957), these results could not be reproduced (Nei and Nagase 1961; Saacke and Almquist 1961), even by the same authors (Meryman and Kafig 1963).

Injecting sperm into oocytes was done at least since the early 1960s (Hiramoto 1962), yet it was only after the intracytoplasmic sperm injection (ICSI) technique it was shown to lead to live offspring in the early 1990s (Palermo et al. 1992) that the utilisation of immotile spermatozoa became a viable option. This led to the first success with freeze-dried spermatozoa (Wakayama and Yanagimachi 1998), demonstrating that "dead" spermatozoa do not mean dead DNA. Following their success with mice, a large body of researchers have attempted to dry spermatozoa. These reports include other studies on mice as well as on various other species (Table 11.2).

To freeze-dry successfully, several conditions need to be met:

1. Cellular membranes need to be stabilised to minimise damages occurring during the different lyophilisation steps:
   (a) During freezing, mechanical damage is caused primarily by ice crystal formation. Increase in ice crystals' size during slow freezing will increase mechanical damage (Saragusty et al. 2009c) and osmotic stress due to the increased concentration of solutes (Mazur et al. 1972) and lipid phase transition as lipids in the cell membranes turn from a liquid form into a gel form (Drobnis et al. 1993).
   (b) During the drying process, damage is induced by free radicals, peroxides, browning reactions, cross-linking of proteins and membrane deformation and fusion (Crowe et al. 1994; Loomis et al. 1979).
   (c) During rehydration there is a lyotropic phase transition as the membranes change back from gel into liquid (van Bilsen et al. 1994). Lysis of cellular membranes may also result from faster-than-desired flow of water back into the cells.
2. The matrix/solution in which the cells are being lyophilised should have high glass transition temperature (Tg) to support and protect the biological materials for long periods of time at elevated temperatures.

In a way, drying can be viewed as the extension of cryopreservation. When freezing cells in suspension, ice forms in the solution, leading to ever-increasing osmotic pressure and thus to dehydration of the cells. This process proceeds till the temperature is low enough and the extra- and intracellular compartments reach high enough

Table 11.2 Sperm desiccation

Species	Drying technique	Study endpoint	Reference
Boar	Freeze-drying	Blastocyst formation	Kwon et al. (2004), Men et al. (2013), Meng et al. (2010), Nakai et al. (2007), Olaciregui et al. (2017a)
Boar	Freeze-drying	MPN formation	García et al. (2014)
Boar	Freeze-drying	Oocyte gene activation	Men et al. (2016)
Boar	Freeze-drying	Ultrastructural analysis	Pfaller et al. (1976)
Boar	Evaporative-drying	Blastocyst formation	Li et al. (2017)
Bovine	Freeze-drying	Blastocyst formation	Hara et al. (2013), Keskintepe et al. (2002), Martins et al. (2007)
Bovine	Freeze-drying	Zygote, methylation	Abdalla et al. (2009b)
Bovine	Freeze-drying	Fertilisation[a]	Meryman and Kafig (1959)
Bovine	Freeze-drying	Sperm asters and microtubule formation	Hara et al. (2011)
Bovine	Freeze-drying	Meiosis resumption	Abdalla et al. (2009a)
Bovine	Freeze-drying	Sperm motility	Bialy and Smith (1957), Meryman and Kafig (1959), Saacke and Almquist (1961), Sherman (1957)
Bovine	Freeze-drying	Ultrastructural analysis	Pfaller et al. (1976)
Bovine	Heat drying	Blastocyst formation	Lee and Niwa (2006)
Bovine	Convective drying	Sperm motility, membrane integrity	Sitaula et al. (2009)
Buffalo	Freeze-drying	DNA integrity	Shahba et al. (2016)
Cat	Freeze-drying	Blastocyst formation	Moisan et al. (2005)
Cat	Freeze-drying	Cleavage rate	Ringleb et al. (2011)
Cat	Freeze-drying	DNA integrity	Magalhães et al. (2012)
Cat	Air-drying	Blastocyst formation	Moisan et al. (2005)
Cat	Microwave	Blastocyst formation	Patrick et al. (2017)
Chimpanzee	Freeze-drying	MPN development in mouse oocyte	Kaneko et al. (2014)
Dog	Freeze-drying	MPN formation	Watanabe et al. (2009)
Dog	Freeze-drying	DNA integrity	Olaciregui et al. (2015)
Fat-tailed Dunnart	Freeze-drying	Motility, viability, acrosome and DNA integrity	Czarny et al. (2009)

(continued)

Table 11.2 (continued)

Species	Drying technique	Study endpoint	Reference
Giraffe	Freeze-drying	MPN development in mouse oocyte	Kaneko et al. (2014)
Hamster	Freeze-drying	Live offspring	Muneto and Horiuchi (2011)
Hamster	Freeze-drying	MPN formation	Katayose et al. (1992)
Horse	Freeze-drying	Live offspring	Choi et al. (2011)
Horse	Freeze-drying	DNA integrity	Olaciregui et al. (2016), Oldenhof et al. (2017)
Human	Freeze-drying	Sperm survival	Sherman (1954), Sherman (1963)
Human	Freeze-drying	MPN formation	Katayose et al. (1992)
Human	Freeze-drying	Chromosome integrity	Kusakabe et al. (2008)
Human	Freeze-drying	DNA integrity	Arav and Saragusty (2016), Gianaroli et al. (2012)
Human	Freeze-drying	Ultrastructural analysis	Zhu et al. (2016)
Jaguar	Freeze-drying	MPN development in mouse oocyte	Kaneko et al. (2014)
Long-haired rat	Freeze-drying	MPN development in mouse oocyte	Kaneko et al. (2014)
Mouse	Freeze-drying	Live offspring	Kaneko and Nakagata (2005), Kaneko and Nakagata (2006), Kaneko and Serikawa (2012a), Wakayama and Yanagimachi (1998), Wakayama et al. (2017), Ward et al. (2003)
Mouse	Freeze-drying	Day ≥14 foetuses	Bhowmick et al. (2003), Kaneko et al. (2003), Kawase et al. (2007a), Kusakabe et al. (2001), Kusakabe et al. (2008)
Mouse	Freeze-drying	Blastocyst formation	Kawase et al. (2005), Kawase et al. (2007b)
Mouse	Freeze-drying	Chromosome integrity	Kusakabe and Tateno (2011), Kusakabe and Tateno (2017)
Mouse	Partial convective drying	Live offspring	Li et al. (2007), McGinnis et al. (2005)
Mouse	Convective drying	Live offspring	Liu et al. (2012), Liu et al. (2014)
Mouse	Convective drying	Day 15 foetuses	Bhowmick et al. (2003)
Nile tilapia	Freeze-drying	ICSI could not be performed	Poleo et al. (2005)
Poultry	Freeze-drying	Sperm motility	Polge et al. (1949)
Rabbit	Freeze-drying	Live offspring	Liu et al. (2004)
Ram	Freeze-drying	Blastocyst formation	Olaciregui et al. (2017b), Palazzese et al. (2017)

Table 11.2 (continued)

Species	Drying technique	Study endpoint	Reference
Rat	Heat drying	Live offspring	Lee et al. (2013)
Rat	Freeze-drying	Live offspring	Hirabayashi et al. (2005), Hochi et al. (2008), Kaneko and Serikawa (2012b)
Rhesus macaque	Vacuum drying	Blastocyst formation	Klooster et al. (2011), Meyers et al. (2009)
Rhesus macaque	Freeze-drying	Sperm asters, MPN formation	Sánchez-Partida et al. (2008)
Weasel	Freeze-drying	MPN development in mouse oocyte	Kaneko et al. (2014)

[a]The authors and others later failed to duplicate these results (Meryman and Kafig 1963; Saacke and Almquist 1961)

Abbreviations: *MPN* male pronuclei, *ICSI* intracytoplasmic sperm injection

viscosity to form glass. Drying follows basically the same idea only instead of removing the water from the system in the form of ice as happens during freezing, water is removed through sublimation (freeze-drying, also known as lyophilisation) or evaporation (air drying, vacuum drying, heat drying, etc.). The vast majority of studies used lyophilisation as the drying technique; however, some success was also reported when using other drying methods such as air drying (Alonso et al. 2015), drying with inert gas at ambient temperatures (Bhowmick et al. 2003; Li et al. 2007; Liu et al. 2012; Liu et al. 2014; McGinnis et al. 2005), heat drying (Lee and Niwa 2006; Lee et al. 2013), spin drying (Chakraborty et al. 2011), microwave drying (Patrick et al. 2017) or vacuum drying (Meyers 2006). In the presence of sugars (disaccharides), such as trehalose or sucrose, the lipid phase transition and the glass transition temperatures considerably increase, allowing the formation of glass (vitrification) at above-zero temperatures (Crowe et al. 1998). Another role suggested for these sugars is the replacement of water at the polar head group region of the membranes and in protein structures, thereby enhancing their stability (Crowe et al. 1992; Leslie et al. 1995).

Despite many studies on the subject, advancement in this field is still largely by trial and error, inherently a very time-consuming, expensive and inefficient scientific investigation. Although desiccated spermatozoa still do not resume motility upon rehydration, several interesting and rather uniform aspects emerged. First, drying seems to be damaging to the cellular membrane, including the tail, with the neck and end-piece being especially sensitive. It is clear, thus, that better membrane stabilisation or possibly altered membrane fluidity may improve survival. Second, the DNA seems to be fairly stable and survive well such insults. Differences between species and even between males within the same species suggest varying levels of sensitivity. Third, although a variety of solutions have been used, our impression is that, as a group, they are mostly not very different from each other and it is possible that a universal drying solution is out there waiting to be discovered. Fourth, composition of the desiccation solution, termed here xeroprotective media, is different from the cryopreservation media specifically designed for the species in question.

Still, despite a decrease in sperm motility after freezing, relatively high motility can often be preserved if freezing is done properly. This raises the question when exactly do the spermatozoa lose their motility and membrane integrity? Does it happen during the drying process or is it an outcome of rehydration? Fifth, researchers seem to be divided into two schools with respect to how the samples are frozen. One group seems to assume that motility/viability a priori cannot be preserved, so samples are frozen with minimal cryoprotection by snap freezing. The other group of thought still strives to regain some motility after rehydration, so proper cryopreservation techniques are used to freeze the samples, and alterations to the xeroprotective media are introduced with the aim of preserving motility.

Another interesting aspect of desiccation is the fact that while drying of bacteria or yeast is successful and the culture "returns to life" after rehydration, desiccated spermatozoa are generally considered dead (no motility, loss of cellular membrane integrity). When it comes to yeast or gram-negative bacteria, they both have cellular membrane and cell wall. Could it be that the additional layer, which is missing in eukaryotic cells, provides the needed protection? Possible, but if this was the case, drying would not be successful in gram-positive bacteria, which is not the case. An alternative explanation suggests that the secret is the ability to multiply. Bacteria, both gram-positive and gram-negative, and yeast possess the ability to multiply, usually with relatively short generation time. This is not the case for the majority of eukaryotic cells. Experiments in drying eukaryotic cells show that under some conditions, a few viable cells can be found. When this happens with bacteria or yeast, these few surviving organisms can start multiplying themselves to form new populations in culture. Following this line of thought one would expect that eukaryotic cells that possess the ability to multiply themselves will also do so after rehydration. This is exactly what happens when the human hepatoma cell line, HepG2 or hematopoietic stem cells were lyophilised (Buchanan et al. 2010; Li et al. 2012; Natan et al. 2009). Drying of other cell types such as sheep granulosa cells (Loi et al. 2008) or porcine foetal fibroblasts (Das et al. 2010) and, of course, spermatozoa from a number of species may have resulted in cellular survival, but these cell types do not naturally multiply.

Oocyte drying, like oocyte freezing, is a big challenge due to their size and high water content. In an attempt to overcome this limiting factor, drying was attempted on germinal vesicles extracted from cat oocytes (Graves-Herring et al. 2013). These cannot survive on their own, with or without drying. The DNA of these germinal vesicles, however, remained viable and was able to direct embryonic development once inside a fresh ooplast/oocyte.

As an alternative to desiccation of gametes, the possibility to desiccate somatic cells has been explored primarily over the past 20 years or so. Somatic cells have the advantage of carrying both the maternal and the paternal genetic material; they are highly available, easy to obtain and, in many cases, easy to grow and multiply in vitro. These cells, when used for somatic cell nuclear transfer (SCNT), can lead to normal embryonic development to term, as was demonstrated in 1997 by the birth of Dolly the sheep (Wilmut et al. 1997) and a large number of other species since then. In 2008 it was shown that freeze-dried somatic cells can also lead to embryonic development in vitro (Loi et al. 2008). As of writing these lines, an off-spring derived from SCNT with desiccated somatic cell is still to be born. We esti-mate that this would not take very long to happen. With recent advancements in assisted reproductive techniques, such desiccated somatic cells can be utilised in yet another way. A transformative technology, published in 2006, made it possible to transform somatic cells into pluripotent cells using a cocktail of four transcription factors – *Pou5f1*, *Sox2*, *Klf4* and *Myc* (Takahashi and Yamanaka 2006). These cells, known as "induced pluripotent stem cells" or iPSCs, have the potential to develop into any of the animal's tissue types, germ cells included. A number of recent stud-ies have demonstrated that by complex and extended in vitro culture, such iPSCs can be directed to develop into functional male or female gamete-like cells known as artificial gametes (for a recent review, see Hendriks et al. (2015)). As a proof of principle, birth of healthy and fertile offspring from such artificial gametes has been demonstrated in mice (Hayashi et al. 2011; Hayashi and Saitou 2013; Hikabe et al. 2016). Although, at present, this technique has only been demonstrated in mice and only with iPSCs derived from fresh somatic cells, we believe that in the not-too-distant future, we shall see applications of this technology to other species and pos-sibly also with desiccated somatic cells.

Sperm Freeze-Drying Protocol (Arav and Saragusty 2016)
After semen evaluation, the sample is washed to remove the seminal plasma and then diluted at a ratio of 1:1 or more in lyophilisation solution (LyoS) to a concentration of about 1×10^6 spermatozoa/mL. The LyoS is composed of 0.25-M sucrose, 0.25-M trehalose and 0.6% w/v human (or bovine) serum albumin in α-MEM Eagle medium. Droplets of 10 µL are dropped directly into sterile liquid air generated by CLAir® (FertileSafe, Ness Ziona, Israel), a bench-top device. A few pellets are thawed in HEPES medium at 37 °C for post-thaw evaluation. Other pellets are dried in a lyophilisation device (Darya®, FertileSafe) that was sterilised in a laboratory autoclave before use. Drying is performed at a pressure of 10 mTorr (0.0133 mbar) with shelf tem-perature set at −35 °C and condenser temperature set at −110 °C. Drying proceeds for 48 h. Samples are then sealed under vacuum.

For rehydration, pellets are exposed to 10 µL/pellet of either α-MEM Eagle medium or LyoS previously warmed to 37 °C

References

Abdalla H, Hirabayashi M, Hochi S (2009a) The ability of freeze-dried bull spermatozoa to induce calcium oscillations and resumption of meiosis. Theriogenology 71:543–552. https://doi.org/10.1016/j.theriogenology.2008.08.021

Abdalla H, Hirabayashi M, Hochi S (2009b) Demethylation dynamics of the paternal genome in pronuclear-stage bovine zygotes produced by *in vitro* fertilization and ooplasmic injection of freeze-thawed or freeze-dried spermatozoa. J Reprod Dev 55:433–439. https://doi.org/10.1262/jrd.20229

Alonso A, Baca Castex C, Ferrante A, Pinto M, Castañeira C, Trasorras V, Gambarotta MC, Losinno L, Miragaya M (2015) *In vitro* equine embryo production using air-dried spermatozoa, with different activation protocols and culture systems. Andrologia 47:387–394. https://doi.org/10.1111/and.12273

Arav A (1989) Cryopreservation of oocytes and embryos. DVM Thesis, University of Bologna, Bologna, Italy

Arav A (1992) Vitrification of oocytes and embryos. In: Lauria A, Gandolfi F (eds) New trends in embryo transfer. Portland Press, Cambridge, UK, pp 255–264

Arav A (1999) Device and methods for multigradient directional cooling and warming of biological samples. United States, patent # US 5,873,254 (Assigned to: Interface Multigrad Technology)

Arav A (2003) Large tissue freezing. J Assist Reprod Genet 20:351. https://doi.org/10.1023/A:1025436228056

Arav A (2012) Directional freezing of reproductive cells, tissues, and organs. In: Nagy ZP, Varghese AC, Agarwal A (eds) Practical manual of in vitro fertilization. Springer, New York, pp 547–550

Arav A, Gacitua H, Zenou A, Malmakov N, Gootwine E, and Bor A (2000) Cryopreservation of ram semen using various freezing extenders and new freezing device. In: Small ruminant reproduction, satellite of the 14th International Congress on Animal Reproduction, June 30, 2000, Sundens, Norway

Arav A, Gavish Z, Elami A, Silber S, Patrizio P (2007) Ovarian survival 6 years after whole organ cryopreservation and transplantation. Fertil Steril 88:S352 (abstract). https://doi.org/10.1016/j.fertnstert.2005.02.006

Arav A, Natan D (2011) Freeze drying (lyophilization) of red blood cells. J Trauma Acute Care Surg 70:S61–S64. https://doi.org/10.1097/TA.0b013e31821a6083

Arav A, Natan D (2012) Freeze drying of red blood cells: the use of directional freezing and a new radio frequency lyophilization device. Biopreserv Biobank 10:386–394. https://doi.org/10.1089/bio.2012.0021

Arav A, Natan Y, Levi Setti PE, Leong M, and Patrizio P (2016) Vitrification of oocytes and blastocysts using a fully automated cryopreservation device. In: 32nd Annual Meeting of the European society of human reproduction and embryology, 3–6 July, 2016, Helsinki, Finland. pp. i226–i227

Arav A, Revel A, Nathan Y, Bor A, Gacitua H, Yavin S, Gavish Z, Uri M, Elami A (2005) Oocyte recovery, embryo development and ovarian function after cryopreservation and transplantation of whole sheep ovary. Hum Reprod 20:3554–3559. https://doi.org/10.1016/j.fertnstert.2005.02.006

Arav A, Rubinsky B, Fletcher G, Seren E (1993) Cryogenic protection of oocytes with antifreeze proteins. Mol Reprod Dev 36:488–493. https://doi.org/10.1002/mrd.1080360413

Arav A, Rubinsky B, Seren E, Roche JF, Boland MP (1994) The role of thermal hysteresis proteins during cryopreservation of oocytes and embryos. Theriogenology 41:107–112. https://doi.org/10.1016/S0093-691X(05)80055-X

Arav A, Saragusty J (2016) Directional freezing of sperm and associated derived technologies. Anim Reprod Sci 169:6–13. https://doi.org/10.1016/j.anireprosci.2016.02.007

Arav A, Zeron Y, Shturman H, Gacitua H (2002) Successful pregnancies in cows following double freezing of a large volume of semen. Reprod Nutr Dev 42:583–586. https://doi.org/10.1051/rnd:2002044

Bar M, Bar-Ziv R, Scherf T, Fass D (2006) Efficient production of a folded and functional, highly disulfide-bonded β-helix antifreeze protein in bacteria. Protein Express Purif 48:243–252. https://doi.org/10.1016/j.pep.2006.01.025

Bar Dolev M, Braslavsky I, Davies PL (2016) Ice-binding proteins and their function. Annu Rev Biochem 85:515–542. https://doi.org/10.1146/annurev-biochem-060815-014546

Beckmann J, Korber C, Rau G, Hubel A, Cravalho EG (1990) Redefining cooling rate in terms of ice front velocity and thermal gradient: first evidence of relevance to freezing injury of lymphocytes. Cryobiology 27:279–287. https://doi.org/10.1016/0011-2240(90)90027-2

Bhowmick S, Zhu L, McGinnis L, Lawitts J, Nath BD, Toner M, Biggers J (2003) Desiccation tolerance of spermatozoa dried at ambient temperature: production of fetal mice. Biol Reprod 68:1779–1786. https://doi.org/10.1095/biolreprod.102.009407

Bialy G, Smith VR (1957) Freeze-drying of bovine spermatozoa. J Dairy Sci 40:739–745. https://doi.org/10.3168/jds.S0022-0302(57)94548-4

Bielanski A (2012) A review of the risk of contamination of semen and embryos during cryopreservation and measures to limit cross-contamination during banking to prevent disease transmission in ET practices. Theriogenology 77:467–482. https://doi.org/10.1016/j.theriogenology.2011.07.043

Bielanski A, Bergeron H, Lau PC, Devenish J (2003) Microbial contamination of embryos and semen during long term banking in liquid nitrogen. Cryobiology 46:146–152. https://doi.org/10.1016/S0011-2240(03)00020-8

Brower WE, Freund MJ, Baudino MD, Ringwald C (1981) An hypothesis for survival of spermatozoa via encapsulation during plane front freezing. Cryobiology 18:277–291. https://doi.org/10.1016/0011-2240(81)90099-7

Buchanan SS, Pyatt DW, Carpenter JF (2010) Preservation of differentiation and clonogenic potential of human hematopoietic stem and progenitor cells during lyophilization and ambient storage. PLoS One 5:e12518. https://doi.org/10.1371/journal.pone.0012518

Chakraborty N, Chang A, Elmoazzen H, Menze M, Hand S, Toner M (2011) A spin-drying technique for lyopreservation of mammalian cells. Ann Biomed Eng 39:1582–1591. https://doi.org/10.1007/s10439-011-0253-1

Choi YH, Varner DD, Love CC, Hartman DL, Hinrichs K (2011) Production of live foals via intracytoplasmic injection of lyophilized sperm and sperm extract in the horse. Reproduction 142:529–538. https://doi.org/10.1530/rep-11-0145

Clark MS, Thorne MAS, Purać J, Grubor-Lajšić G, Kube M, Reinhardt R, Worland MR (2007) Surviving extreme polar winters by desiccation: clues from Arctic springtail (*Onychiurus arcticus*) EST libraries. BMC Genomics 8:1–12. https://doi.org/10.1186/1471-2164-8-475

Cobo A, Domingo J, Pérez S, Crespo J, Remohí J, Pellicer A (2008) Vitrification: an effective new approach to oocyte banking and preserving fertility in cancer patients. Clin Transl Oncol 10:268–273. https://doi.org/10.1007/s12094-008-0196-7

Cogent (2013) 'The science.' Available at http://www.cogentuk.com/other-services/sexed-semen/the-science/ [Verified 11 December 2017]

Crowe JH, Carpenter JF, Crowe LM (1998) The role of vitrification in anhydrobiosis. Annu Rev Physiol 60:73–103. https://doi.org/10.1146/annurev.physiol.60.1.73

Crowe JH, Crowe LM, Chapman D (1984) Preservation of membranes in anhydrobiotic organisms: the role of trehalose. Science 223:701–703. https://doi.org/10.1126/science.223.4637.701

Crowe JH, Crowe LM, Wolkers WF, Oliver AE, Ma X, Auh J-H, Tang M, Zhu S, Norris J, Tablin F (2005) Stabilization of dry mammalian cells: lessons from nature. Integr Comp Biol 45:810–820. https://doi.org/10.1093/icb/45.5.810

Crowe JH, Hoekstra FA, Crowe LM (1992) Anhydrobiosis. Annu Rev Physiol 54:579–599. https://doi.org/10.1146/annurev.ph.54.030192.003051

Crowe JH, Leslie SB, Crowe LM (1994) Is vitrification sufficient to preserve liposomes during freeze-drying? Cryobiology 31:355–366. https://doi.org/10.1006/cryo.1994.1043

Czarny NA, Harris MS, De Iuliis GN, Rodger JC (2009) Acrosomal integrity, viability, and DNA damage of sperm from dasyurid marsupials after freezing or freeze drying. Theriogenology 72:817–825. https://doi.org/10.1016/j.theriogenology.2009.05.018

Das ZC, Gupta MK, Uhm SJ, Lee HT (2010) Lyophilized somatic cells direct embryonic development after whole cell intracytoplasmic injection into pig oocytes. Cryobiology 61:220–224. https://doi.org/10.1016/j.cryobiol.2010.07.007

Drobnis EZ, Crowe LM, Berger T, Anchordoguy TJ, Overstreet JW, Crowe JH (1993) Cold shock damage is due to lipid phase transitions in cell membranes: a demonstration using sperm as a model. J Exp Zool 265:432–437. https://doi.org/10.1002/jez.1402650413

Duman JG (1982) Insect antifreezes and ice-nucleating agents. Cryobiology 19:613–627. https://doi.org/10.1016/0011-2240(82)90191-2

Elami A, Gavish Z, Korach A, Houminer E, Schneider A, Schwalb H, Arav A (2008) Successful restoration of function of frozen and thawed isolated rat hearts. J Thorac Cardiovasc Surg 135:666–672.e1. https://doi.org/10.1016/j.jtcvs.2007.08.056

Furukawa Y, Inohara N, Yokoyama E (2005) Growth patterns and interfacial kinetic supercooling at ice/water interfaces at which anti-freeze glycoprotein molecules are adsorbed. J Cryst Growth 275:167–174. https://doi.org/10.1016/j.jcrysgro.2004.10.085

Gacitua H, Arav A (2005) Successful pregnancies with directional freezing of large volume buck semen. Theriogenology 63:931–938. https://doi.org/10.1016/j.theriogenology.2004.05.012

Gacitua H, Pettit MT, Saragusty J, Arav A (2006) Directional freezing of large volume equine semen. Reprod Domest Anim 41:318 (abstract). https://doi.org/10.1111/j.1439-0531.2006.00728_3.x

Gao D, Watson PF, He L, Yu J, Critser J (2002) Development of a directional solidification device for cell cryopreservation. Cell Preserv Technol 1:231–238. https://doi.org/10.1089/15383440260682062

García A, Gil L, Malo C, Martínez F, Kershaw-Young C, de Blas I (2014) Effect of different disaccharides on the integrity and fertilising ability of freeze-dried boar spermatozoa: a preliminary study. Cryo Lett 35:277–285

Gavish Z, Ben-Haim M, Arav A (2008) Cryopreservation of whole murine and porcine livers. Rejuv Res 11:765–772. https://doi.org/10.1089/rej.2008.0706

Gavish Z, Dekel I, Shneorson O, Gacitua H, and Arav A (2004a) Cryopreservation of sperm and whole ovaries 24 hours post-mortem. In: Israel Fertility Association Congress, 12–13 May, 2004, Tel Aviv, Israel, p. 64 (abstract)

Gavish Z, Dekel I, Shneorson O, Gacitua H, and Arav A (2004b) Directional freezing of wild gazelle sperm and whole ovaries. In: Forty-First Annual Meeting of the Society for Cryobiology in association with the Japanese Societies & Associations for Cryobiology, Cryopreservation and Cryomedicine, and the Chinese Cryobiological Society', Beijing, China, p. 314 (abstract)

Gianaroli L, Magli MC, Stanghellini I, Crippa A, Crivello AM, Pescatori ES, Ferraretti AP (2012) DNA integrity is maintained after freeze-drying of human spermatozoa. Fertil Steril 97:1067–1073. https://doi.org/10.1016/j.fertnstert.2012.02.014

Goodrich RP, Sowemimo-Coker SO, Zerez CR, Tanaka KR (1992) Preservation of metabolic activity in lyophilized human erythrocytes. Proc Nat Acad Sci USA 89:967–971. https://doi.org/10.1073/pnas.89.3.967

Goyal K, Walton LJ, Browne JA, Burnell AM, Tunnacliffe A (2005) Molecular anhydrobiology: identifying molecules implicated in invertebrate anhydrobiosis. Integr Comp Biol 45:702–709. https://doi.org/10.1093/icb/45.5.702

Graves-Herring JE, Wildt DE, Comizzoli P (2013) Retention of structure and function of the cat germinal vesicle after air-drying and storage at supra-zero temperature. Biol Reprod 88(139):1–7. https://doi.org/10.1095/biolreprod.113.108472

Griffith M, Ala P, Yang DSC, Hon W-C, Moffatt BA (1992) Antifreeze protein produced endogenously in winter rye leaves. Plant Physiol 100:593–596

Guo N, Puhlev I, Brown DR, Mansbridge J, Levine F (2000) Trehalose expression confers desiccation tolerance on human cells. Nat Biotechnol 18:168–171. https://doi.org/10.1038/72616

Hand SC, Menze MA, Toner M, Boswell L, Moore D (2011) LEA proteins during water stress: not just for plants anymore. Annu Rev Physiol 73:115–134. https://doi.org/10.1146/annurev-physiol-012110-142203

Hara H, Abdalla H, Morita H, Kuwayama M, Hirabayashi M, Hochi S (2011) Procedure for bovine ICSI, not sperm freeze-drying, impairs the function of the microtubule-organizing center. J Reprod Dev 57:428–432. https://doi.org/10.1262/jrd.10-167N

Hara H, Tagiri M, Hirabayashi M, Hochi S (2013) Effect of cake collapse on the integrity of freeze-dried bull spermatozoa. Reprod Fertil Dev 26:144 (Abstract). https://doi.org/10.1071/RDv26n1Ab60

Hayakawa H, Yamazaki T, Oshi M, Hoshino M, Dochi O, Koyama H (2007) Cryopreservation of conventional and sex-sorted bull sperm using a directional freezing method. Reprod Fertil Dev 19:176–177 (abstract). https://doi.org/10.1071/RDv19n1Ab118

Hayashi K, Ohta H, Kurimoto K, Aramaki S, Saitou M (2011) Reconstitution of the mouse germ cell specification pathway in culture by pluripotent stem cells. Cell 146:519–532. https://doi.org/10.1016/j.cell.2011.06.052

Hayashi K, Saitou M (2013) Generation of eggs from mouse embryonic stem cells and induced pluripotent stem cells. Nat Protoc 8:1513–1524. https://doi.org/10.1038/nprot.2013.090

Hendriks S, Dancet EAF, van Pelt AMM, Hamer G, Repping S (2015) Artificial gametes: a systematic review of biological progress towards clinical application. Hum Reprod Update 21:285–296. https://doi.org/10.1093/humupd/dmv001

Hermes R, Arav A, Saragusty J, Göritz F, Pettit M, Blottner S, Flach E, Eshkar G, Boardman W, and Hildebrandt TB (2003) Cryopreservation of Asian elephant spermatozoa using directional freezing. In: Annual meeting of the American Association of Zoo Veterinarians, 4.10.-10.10.2003, Minneapolis, MN, USA. (Ed. C Kirk-Baer), p. 264 (abstract). (Yulee, FL, USA)

Hermes R, Göritz F, Saragusty J, Sos E, Molnar V, Reid CE, Schwarzenberger F, Hildebrandt TB (2009) First successful artificial insemination with frozen-thawed semen in rhinoceros. Theriogenology 71:393–399. https://doi.org/10.1016/j.theriogenology.2008.10.008

Hermes R, Saragusty J, Göritz F, Bartels P, Potier R, Baker B, Streich WJ, Hildebrandt TB (2013) Freezing African elephant semen as a new population management tool. PLoS One 8:e57616. https://doi.org/10.1371/journal.pone.0057616

Hikabe O, Hamazaki N, Nagamatsu G, Obata Y, Hirao Y, Hamada N, Shimamoto S, Imamura T, Nakashima K, Saitou M, Hayashi K (2016) Reconstitution *in vitro* of the entire cycle of the mouse female germ line. Nature 539(7628):299–303 In Press. https://doi.org/10.1038/nature20104

Hildebrandt TB, Hermes R, Saragusty J, Potier R, Schwammer HM, Balfanz F, Vielgrader HD, Baker B, Bartels P, Göritz F (2012) Enriching the captive elephant population genetic pool through artificial insemination with frozen-thawed semen collected in the wild. Theriogenology 78:1398–1404. https://doi.org/10.1016/j.theriogenology.2012.06.014

Hildebrandt TB, Roellig K, Goeritz F, Fassbender M, Krieg R, Blottner S, Behr B, Hermes R (2009) Artificial insemination of captive European brown hares (*Lepus europaeus* PALLAS, 1778) with fresh and cryopreserved semen derived from free-ranging males. Theriogenology 72:1065–1072. https://doi.org/10.1016/j.theriogenology.2009.06.026

Hirabayashi M, Kato M, Ito J, Hochi S (2005) Viable rat offspring derived from oocytes intracytoplasmically injected with freeze-dried sperm heads. Zygote 13:79–85. https://doi.org/10.1017/S096719940500300X

Hiramoto Y (1962) Microinjection of the live spermatozoa into sea urchin eggs. Exp Cell Res 27:416–426. https://doi.org/10.1016/0014-4827(62)90006-X

Hoagland E, Pincus GG (1942) Revival of mammalian sperm after immersion in liquid nitrogen. J Genet Physiol 25:337–344

Hochi S, Watanabe K, Kato M, Hirabayashi M (2008) Live rats resulting from injection of oocytes with spermatozoa freeze-dried and stored for one year. Mol Reprod Dev 75:890–894. https://doi.org/10.1002/mrd.20825

Hubel A, Cravalho EG, Nunner B, Korber C (1992) Survival of directionally solidified B-lymphoblasts under various crystal growth conditions. Cryobiology 29:183–198. https://doi.org/10.1016/0011-2240(92)90019-X

Hubel A, Darr TB, Chang A, Dantzig J (2007) Cell partitioning during the directional solidification of trehalose solutions. Cryobiology 55:182–188. https://doi.org/10.1016/j.cryobiol.2007.07.002

Ishiguro H, Koike K (1998) Three-dimensional behavior of ice crystals and biological cells during freezing of cell suspensions. Ann N Y Acad Sci 858:235–244

Ishiguro H, Rubinsky B (1994) Mechanical interactions between ice crystals and red blood cells during directional solidification. Cryobiology 31:483–500. https://doi.org/10.1006/cryo.1994.1059

Jahnel F (1938) Resistance of human spermatozoa to deep cold. Klin Wochenschr 17:1273–1274

Kaneko T, Ito H, Sakamoto H, Onuma M, Inoue-Murayama M (2014) Sperm preservation by freeze-drying for the conservation of wild animals. PLoS One 9:e113381. https://doi.org/10.1371/journal.pone.0113381

Kaneko T, Nakagata N (2005) Relation between storage temperature and fertilizing ability of freeze-dried mouse spermatozoa. Comp Med 55:140–144

Kaneko T, Nakagata N (2006) Improvement in the long-term stability of freeze-dried mouse spermatozoa by adding of a chelating agent. Cryobiology 53:279–282. https://doi.org/10.1016/j.cryobiol.2006.06.004

Kaneko T, Serikawa T (2012a) Long-term preservation of freeze-dried mouse spermatozoa. Cryobiology 64:211–214. https://doi.org/10.1016/j.cryobiol.2012.01.010

Kaneko T, Serikawa T (2012b) Successful long-term preservation of rat sperm by freeze-drying. PLoS One 7:e35043. https://doi.org/10.1371/journal.pone.0035043

Kaneko T, Whittingham DG, Yanagimachi R (2003) Effect of pH value of freeze-drying solution on the chromosome integrity and developmental ability of mouse spermatozoa. Biol Reprod 68:136–139. https://doi.org/10.1095/biolreprod.102.008706

Katayose H, Matsuda J, Yanagimachi R (1992) The ability of dehydrated hamster and human sperm nuclei to develop into pronuclei. Biol Reprod 47:277–284. https://doi.org/10.1095/biolreprod47.2.277

Katkov II, Bolyukh VF, Chernetsov OA, Dudin PI, Grigoriev AY, Isachenko V, Isachenko E, Lulat AG-M, Moskovtsev SI, Petrushko MP, Pinyaev VI, Sokol KM, Sokol YI, Sushko AB, Yakhnenko I (2012) Kinetic vitrification of spermatozoa of vertebrates: what can we learn from nature. In: Katkov II (ed) Current frontiers in cryobiology, vol 1. InTech, Rijeka, Croatia, pp 3–40

Kawase Y, Araya H, Kamada N, Jishage K, Suzuki H (2005) Possibility of long-term preservation of freeze-dried mouse spermatozoa. Biol Reprod 72:568–573. https://doi.org/10.1095/biolreprod.104.035279

Kawase Y, Hani T, Kamada N, Jishage K-i, Suzuki H (2007b) Effect of pressure at primary drying of freeze-drying mouse sperm reproduction ability and preservation potential. Reproduction 133:841–846. https://doi.org/10.1530/rep-06-0170

Kawase Y, Tachibe T, Jishage K, Suzuki H (2007a) Transportation of freeze-dried mouse spermatozoa under different preservation conditions. J Reprod Dev 53:1169–1174

Keskintepe L, Pacholczyk G, Machnicka A, Norris K, Curuk MA, Khan I, Brackett BG (2002) Bovine blastocyst development from oocytes injected with freeze-dried spermatozoa. Biol Reprod 67:409–415. https://doi.org/10.1095/biolreprod67.2.409

Klooster KL, Burruel VR, Meyers SA (2011) Loss of fertilization potential of desiccated rhesus macaque spermatozoa following prolonged storage. Cryobiology 62:161–166. https://doi.org/10.1016/j.cryobiol.2011.02.002

Knight CA, De Vries AL, Oolman LD (1984) Fish antifreeze protein and the freezing and recrystallization of ice. Nature 308:295–296. https://doi.org/10.1038/308295a0

Körber C (1988) Phenomena at the advancing ice–liquid interface: solutes, particles and biological cells. Q Rev Biophys 21:229–298. https://doi.org/10.1017/S0033583500004303

Körber C, Rau G, Cosman MD, Cravalho EG (1985) Interaction of particles and a moving ice-liquid interface. J Cryst Growth 72:649–662. https://doi.org/10.1016/0022-0248(85)90217-9

Körber C, Scheiwe MW (1983) Observations of the non-planar freezing of aqueous salt solutions. J Cryst Growth 61:307–316. https://doi.org/10.1016/0022-0248(83)90367-6

Körber C, Scheiwe MW, Wollhöver K (1983) Solute polarization during planar freezing of aqueous salt solutions. Int J Heat Mass Transf 26:1241–1253. https://doi.org/10.1016/S0017-9310(83)80179-3

Koushafar H, Rubinsky B (1997) Effect of antifreeze proteins on frozen primary prostatic adenocarcinoma cells. Urology 49:421–425. https://doi.org/10.1016/S0090-4295(96)00572-9

Kusakabe H, Szczygiel MA, Whittingham DG, Yanagimachi R (2001) Maintenance of genetic integrity in frozen and freeze-dried mouse spermatozoa. Proc Nat Acad Sci USA 98:13501–13506. https://doi.org/10.1073/pnas.241517598

Kusakabe H, Tateno H (2011) Characterization of chromosomal damage accumulated in freeze-dried mouse spermatozoa preserved under ambient and heat stress conditions. Mutagenesis 26:447–453. https://doi.org/10.1093/mutage/ger003

Kusakabe H, Tateno H (2017) Prevention of high-temperature-induced chromosome damage in mouse spermatozoa freeze-dried using Ca2+ chelator-containing buffer alkalinized with NaOH or KOH. Cryobiology 79:71–77. https://doi.org/10.1016/j.cryobiol.2017.08.007

Kusakabe H, Yanagimachi R, Kamiguchi Y (2008) Mouse and human spermatozoa can be freeze-dried without damaging their chromosomes. Hum Reprod 23:233–239. https://doi.org/10.1093/humrep/dem252

Kuwayama M (2007) Highly efficient vitrification for cryopreservation of human oocytes and embryos: the Cryotop method. Theriogenology 67:73–80. https://doi.org/10.1016/j.theriogenology.2006.09.014

Kwon IK, Park KE, Niwa K (2004) Activation, pronuclear formation, and development *in vitro* of pig oocytes following intracytoplasmic injection of freeze-dried spermatozoa. Biol Reprod 71:1430–1436. https://doi.org/10.1095/biolreprod.104.031260

Larson EV, Graham EF (1976) Freeze-drying of spermatozoa. Dev Biol Stand 36:343–348

Lee K-B, Niwa K (2006) Fertilization and development *in vitro* of bovine oocytes following intracytoplasmic injection of heat-dried sperm heads. Biol Reprod 74:146–152. https://doi.org/10.1095/biolreprod.105.044743

Lee K-B, Park K-E, Kwon I-K, Tripurani SK, Kim KJ, Lee JH, Niwa K, Kim MK (2013) Develop to term rat oocytes injected with heat-dried sperm heads. PLoS One 8:e78260. https://doi.org/10.1371/journal.pone.0078260

Leslie SB, Israeli E, Lighthart B, Crowe JH, Crowe LM (1995) Trehalose and sucrose protect both membranes and proteins in intact bacteria during drying. Appl Environ Microbiol 61:3592–3597

Li MW, Biggers JD, Elmoazzen HY, Toner M, McGinnis L, Lloyd KCK (2007) Long-term storage of mouse spermatozoa after evaporative drying. Reproduction 133:919–929. https://doi.org/10.1530/rep-06-0096

Li S, Chakraborty N, Borcar A, Menze MA, Toner M, Hand SC (2012) Late embryogenesis abundant proteins protect human hepatoma cells during acute desiccation. Proc Nat Acad Sci USA 109:20859–20864. https://doi.org/10.1073/pnas.1214893109

Li X-X, Diao Y-F, Wei H-J, Wang S-Y, Cao X-Y, Zhang Y-F, Chang T, Li D-L, Kim MK, Xu B (2017) Tauroursodeoxycholic acid enhances the development of porcine embryos derived from *in vitro*-matured oocytes and evaporatively dried spermatozoa. Sci Rep 7:6773. https://doi.org/10.1038/s41598-017-07185-w

Li Y, Wang H, Tingrui P (2013) Intracellular ice formation (IIF) during freeze-thaw repetitions. Int J Heat Mass Transf 64:436–443. https://doi.org/10.1016/j.ijheatmasstransfer.2013.04.036

Li Y, Wang F, Wang H (2010) Cell death along single microfluidic channel after freeze-thaw treatments. Biomicrofluidics 4:1–10. https://doi.org/10.1063/1.3324869

Lipp G, Korber C, Englich S, Hartmann U, Rau G (1987) Investigation of the behavior of dissolved gases during freezing. Cryobiology 24:489–503. https://doi.org/10.1016/0011-2240(87)90053-8

Liu JL, Kusakabe H, Chang CC, Suzuki H, Schmidt DW, Julian M, Pfeffer R, Bormann CL, Tian XC, Yanagimachi R, Yang X (2004) Freeze-dried sperm fertilization leads to full-term development in rabbits. Biol Reprod 70:1776–1781. https://doi.org/10.1095/biolreprod.103.025957

Liu J, Lee GY, Lawitts JA, Toner M, Biggers JD (2012) Preservation of mouse sperm by convective drying and storing in 3-O-methyl-D-glucose. PLoS One 7:e29924. https://doi.org/10.1371/journal.pone.0029924

Liu J, Lee GY, Lawitts JA, Toner M, Biggers JD (2014) Live pups from evaporatively dried mouse sperm stored at ambient temperature for up to 2 years. PLoS One 9:e99809. https://doi.org/10.1371/journal.pone.0099809

Loi P, Matsukawa K, Ptak G, Clinton M, Fulka J Jr, Natan Y, Arav A (2008) Freeze-dried somatic cells direct embryonic development after nuclear transfer. PLoS One 3:e2978. https://doi.org/10.1371/journal.pone.0002978

Loomis SH, O'Dell SJ, Crowe JH (1979) Anhydrobiosis in nematodes: inhibition of the browning reaction of reducing sugars with dry proteins. J Exp Zool 208:355–360. https://doi.org/10.1002/jez.1402080312

Luyet B (1937) The vitrification of organic colloids and protoplasm. Biodynamica 1:1–14

Luyet BJ, Gehenio PM (1940) Life and death at low temperatures. Biodynamica, Normandy, MO, p 335

Luyet BJ, Hodapp EL (1938) Revival of frog's spermatozoa vitrified in liquid air. Exp Biol Med 39:433–434. https://doi.org/10.3181/00379727-39-10229p

Maffei S, Hanenberg M, Pennarossa G, Silva JRV, Brevini TAL, Arav A, Gandolfi F (2013b) Direct comparative analysis of conventional and directional freezing for the cryopreservation of whole ovaries. Fertil Steril 100:1122–1131. https://doi.org/10.1016/j.fertnstert.2013.06.003

Maffei S, Pennarossa G, Brevini TAL, Arav A, Gandolfi F (2013a) Beneficial effect of directional freezing on *in vitro* viability of cryopreserved sheep whole ovaries and ovarian cortical slices. Hum Reprod 29:114–124. https://doi.org/10.1093/humrep/det377

Magalhães LCO, Melo-Oña CM, Sudano MJ, Paschoal DM, Crocomo LF, Ackermann CL, Villaverde AISB, Landim-Alvarenga FC, Lopes MD (2012) An easy-to-perform method to assess viability of feline freeze-dried sperm. Reprod Fertil Dev 25:182 (abstract). https://doi.org/10.1071/RDv25n1Ab69

Martins CF, Bao SN, Dode MN, Correa GA, Rumpf R (2007) Effects of freeze-drying on cytology, ultrastructure, DNA fragmentation, and fertilizing ability of bovine sperm. Theriogenology 67:1307–1315. https://doi.org/10.1016/j.theriogenology.2007.01.015

Matsumoto S, Matsusita M, Morita T, Kamachi H, Tsukiyama S, Furukawa Y, Koshida S, Tachibana Y, Nishimura S-i, Todo S (2006) Effects of synthetic antifreeze glycoprotein analogue on islet cell survival and function during cryopreservation. Cryobiology 52:90–98. https://doi.org/10.1016/j.cryobiol.2005.10.010

Mazur P, Leibo SP, Chu EH (1972) A two-factor hypothesis of freezing injury. Evidence from Chinese hamster tissue-culture cells. Exp Cell Res 71:345–355. https://doi.org/10.1016/0014-4827(72)90303-5

McGinnis LK, Zhu L, Lawitts JA, Bhowmick S, Toner M, Biggers JD (2005) Mouse sperm desiccated and stored in trehalose medium without freezing. Biol Reprod 73:627–633. https://doi.org/10.1095/biolreprod.105.042291

Men NT, Kikuchi K, Furusawa T, Dang-Nguyen TQ, Nakai M, Fukuda A, Noguchi J, Kaneko H, Viet Linh N, Xuan Nguyen B, Tajima A (2016) Expression of DNA repair genes in porcine oocytes before and after fertilization by ICSI using freeze-dried sperm. Anim Sci J 87:1325–1333. https://doi.org/10.1111/asj.12554

Men NT, Kikuchi K, Nakai M, Fukuda A, Tanihara F, Noguchi J, Kaneko H, Linh NV, Nguyen BX, Nagai T, Tajima A (2013) Effect of trehalose on DNA integrity of freeze-dried boar sperm, fertilization, and embryo development after intracytoplasmic sperm injection. Theriogenology 80:1033–1044. https://doi.org/10.1016/j.theriogenology.2013.08.001

Meng X, Gu X, Wu C, Dai J, Zhang T, Xie Y, Wu Z, Liu L, Ma H, Zhang D (2010) Effect of trehalose on the freeze-dried boar spermatozoa. Sheng Wu Gong Cheng Xue Bao 26:1143–1149 (abstract). [In Chinese]

Meryman HT, Kafig E (1959) Survival of spermatozoa following drying. Nature 184:470–471. https://doi.org/10.1038/184470a0

Meryman HT, Kafig E (1963) Special article: freeze-drying of bovine spermatozoa. J Reprod Fertil 5:87–94. https://doi.org/10.1530/jrf.0.0050087

Meyers SA (2006) Dry storage of sperm: applications in primates and domestic animals. Reprod Fertil Dev 18:1–5. https://doi.org/10.1071/RD05116

Meyers SA, Li MW, Enders AC, Overstreet JW (2009) Rhesus macaque blastocysts resulting from intracytoplasmic sperm injection of vacuum-dried spermatozoa. J Med Primatol 38:310–317. https://doi.org/10.1111/j.1600-0684.2009.00352.x

Moisan AE, Leibo SP, Lynn JW, Gómez MC, Pope CE, Dresser BL, Godke RA (2005) Embryonic development of felid oocytes injected with freeze-dried or air-dried spermatozoa. Cryobiology 51:373 (abstract). https://doi.org/10.1016/j.cryobiol.2005.10.001

Molisch H (1982) Investigations into the freezing death of plants. Cryo Lett 3:328–391

Montano GA, Kraemer D, Love CC, Robeck T, O'Brien J (2012) Evaluation of motility, membrane status and DNA integrity of frozen-thawed bottlenose dolphin (*Tursiops truncatus*) spermatozoa after sex-sorting and recryopreservation. Reproduction 143:799–813. https://doi.org/10.1530/rep-11-0490

Mousson A (1858) Einige Thatsachen betreffend das Schmelzen und Gefrieren des Wassers. Annalen der Physik und Chemie 181:161–174. https://doi.org/10.1002/andp.18581811002 [In German]

Muneto T, Horiuchi T (2011) Full-term development of hamster embryos produced by injecting freeze-dried spermatozoa into oocytes. J Mamm Ova Res 28:32–39. https://doi.org/10.1274/jmor.28.32

Nakai M, Kashiwazaki N, Takizawa A, Maedomari N, Ozawa M, Noguchi J, Kaneko H, Shino M, Kikuchi K (2007) Effects of chelating agents during freeze-drying of boar spermatozoa on DNA fragmentation and on developmental ability *in vitro* and *in vivo* after intracytoplasmic sperm head injection. Zygote 15:15–24. https://doi.org/10.1017/S0967199406003935

Natan D, Nagler A, Arav A (2009) Freeze-drying of mononuclear cells derived from umbilical cord blood followed by colony formation. PLoS One 4:e5240. https://doi.org/10.1371/journal.pone.0005240

Nei T, Nagase H (1961) Attempts to freeze-dry bull spermatozoa. Low Temp Sci B 19:107–115

O'Brien JK, Robeck TR (2006) Development of sperm sexing and associated assisted reproductive technology for sex preselection of captive bottlenose dolphins (*Tursiops truncatus*). Reprod Fertil Dev 18:319–329. https://doi.org/10.1071/RD05108

O'Brien JK, Robeck TR (2010) Preservation of beluga (*Delphinapterus leucas*) spermatozoa using a trehalose-based cryodiluent and directional freezing technology. Reprod Fertil Dev 22:653–663. https://doi.org/10.1071/RD09176

O'Brien JK, Robeck TR (2014) Semen characterization, seasonality of production, and *in vitro* sperm quality after chilled storage and cryopreservation in the king penguin (*Aptenodytes patagonicus*). Zoo Biol 33:99–109. https://doi.org/10.1002/zoo.21111

O'Brien JK, Steinman KJ, Montano GA, Love CC, Saiers RL, Robeck TR (2012) Characteristics of high-quality Asian elephant (*Elephas maximus*) ejaculates and *in vitro* sperm quality after prolonged chilled storage and directional freezing. Reprod Fertil Dev 25:790–797. https://doi.org/10.1071/RD12129

Olaciregui M, Luño V, Domingo P, González N, Gil L (2017b) *In vitro* developmental ability of ovine oocytes following intracytoplasmic injection with freeze-dried spermatozoa. Sci Rep 7:1096. https://doi.org/10.1038/s41598-017-00583-0

Olaciregui M, Luño V, Gonzalez N, De Blas I, Gil L (2015) Freeze-dried dog sperm: dynamics of DNA integrity. Cryobiology 71:286–290. https://doi.org/10.1016/j.cryobiol.2015.08.001

Olaciregui M, Luño V, González N, Domingo P, de Blas I, Gil L (2017a) Chelating agents in combination with rosmarinic acid for boar sperm freeze-drying. Reprod Biol 17:193–198. https://doi.org/10.1016/j.repbio.2017.05.001

Olaciregui M, Luño V, Martí JI, Aramayona J, Gil L (2016) Freeze-dried stallion spermatozoa: evaluation of two chelating agents and comparative analysis of three sperm DNA damage assays. Andrologia 48:900–906. https://doi.org/10.1111/and.12530

Oldenhof H, Zhang M, Narten K, Bigalk J, Sydykov B, Wolkers WF, Sieme H (2017) Freezing-induced uptake of disaccharides for preservation of chromatin in freeze-dried stallion sperm during accelerated aging. Biol Reprod 97:892–901. https://doi.org/10.1093/biolre/iox142

Palazzese L, Anzalone DA, Gosálvez J, Loi P, Saragusty J (2017) DNA fragmentation of epididymal freeze-dried ram spermatozoa impairs embryo development. Reprod Fertil Dev 30:162 (Abstract). https://doi.org/10.1071/RDv30n1Ab45

Palermo G, Joris H, Devroey P, Van Steirteghem AC (1992) Pregnancies after intracytoplasmic injection of single spermatozoon into an oocyte. Lancet 340:17–18. https://doi.org/10.1016/0140-6736(92)92425-F

Patrick JL, Elliott GD, Comizzoli P (2017) Structural integrity and developmental potential of spermatozoa following microwave-assisted drying in the domestic cat model. Theriogenology 103:36–43. https://doi.org/10.1016/j.theriogenology.2017.07.037

Perreault SD, Barbee RR, Slott VL (1988) Importance of glutathione in the acquisition and maintenance of sperm nuclear decondensing activity in maturing hamster oocytes. Dev Biol 125:181–
186. https://doi.org/10.1016/0012-1606(88)90070-X

Pfaller W, Rovan E, Mairbäurl H (1976) A comparison of the ultrastructure of spray-frozen and
freeze-etched or freeze-dried bull and boar spermatozoa with that after chemical fixation. J
Reprod Fertil 48:285–290. https://doi.org/10.1530/jrf.0.0480285

Phillips PH, Lardy HA (1940) A yolk-buffer pabulum for the preservation of bull semen. J Dairy
Sci 23:399–404. https://doi.org/10.3168/jds.S0022-0302(40)95541-2

Poleo GA, Godke RR, Tiersch TR (2005) Intracytoplasmic sperm injection using cryopreserved,
fixed, and freeze-dried sperm in eggs of Nile tilapia. Mar Biotechnol 7:104–111. https://doi.
org/10.1007/s10126-004-0162-5

Polge C, Smith AU, Parkes AS (1949) Revival of spermatozoa after vitrification and dehydration
at low temperatures. Nature 164:666. https://doi.org/10.1038/164666a0

Prieto Pablos MT, Saragusty J, Santiago-Moreno J, Stagegaard J, Göritz F, Hildebrandt TB,
Hermes R (2015) Cryopreservation of onager (*Equus hemionus onager*) epididymal spermatozoa. J Zoo Wildl Med 46:517–525. https://doi.org/10.1638/2014-0243.1

Rall WF, Fahy GM (1985) Ice-free cryopreservation of mouse embryos at −196°C by vitrification.
Nature 313:573–575. https://doi.org/10.1038/313573a0

Reid CE, Blottner S, Hildebrandt TB, Göritz F, Hermes R (2006) The effect of potassium, EDTA,
cytochalsin D, and xanthurinic acid on the post-freeze/thaw parameters of rhinoceros spermatozoa using an egg-yolk based extender. Reprod Domest Anim 41:28 (abstract). https://doi.
org/10.1111/j.1439-0531.2006.00663.x

Reid CE, Hermes R, Blottner S, Goeritz F, Wibbelt G, Walzer C, Bryant BR, Portas TJ, Streich
WJ, Hildebrandt TB (2009) Split-sample comparison of directional and liquid nitrogen
vapour freezing method on post-thaw semen quality in white rhinoceroses (*Ceratotherium
simum simum* and *Ceratotherium simum cottoni*). Theriogenology 71:275–291. https://doi.
org/10.1016/j.theriogenology.2008.07.009

Revel A, Elami A, Bor A, Yavin S, Natan Y, Arav A (2001) Intact sheep ovary cryopreservation and
transplantation. Fertil Steril 76:S42–S43 (abstract)

Revel A, Elami A, Bor A, Yavin S, Natan Y, Arav A (2004) Whole sheep ovary cryopreservation and
transplantation. Fertil Steril 82:1714–1715. https://doi.org/10.1016/j.fertnstert.2004.06.046

Ringleb J, Waurich R, Wibbelt G, Streich WJ, Jewgenow K (2011) Prolonged storage of epi-
didymal spermatozoa does not affect their capacity to fertilise *in vitro*-matured domestic cat
(*Felis catus*) oocytes when using ICSI. Reprod Fertil Dev 23:818–825. https://doi.org/10.1071/
RD10192

Robeck TR, Gearhart SA, Steinman KJ, Katsumata E, Loureiro JD, O'Brien JK (2011) *In vitro*
sperm characterization and development of a sperm cryopreservation method using directional
solidification in the killer whale (*Orcinus orca*). Theriogenology 76:267–279. https://doi.
org/10.1016/j.theriogenology.2011.02.003

Robeck TR, Montano GA, Steinman KJ, Smolensky P, Sweeney J, Osborn S, O'Brien JK (2013)
Development and evaluation of deep intra-uterine artificial insemination using cryopreserved
sexed spermatozoa in bottlenose dolphins (*Tursiops truncatus*). Anim Reprod Sci 139:168–
181. https://doi.org/10.1016/j.anireprosci.2013.04.004

Robeck TR, Steinman KJ, Montano GA, Katsumata E, Osborn S, Dalton L, Dunn JL, Schmitt T,
Reidarson T, O'Brien JK (2010) Deep intra-uterine artificial inseminations using cryopreserved
spermatozoa in beluga (*Delphinapterus leucas*). Theriogenology 74:989–1001. https://doi.
org/10.1016/j.theriogenology.2010.04.028

Rubei M, Degl'Innocenti S, De Vries PJ, Catone G, and Morini G (2004) Directional freezing
(Harmony Cryocare–Multi Thermal Gradient 516): A new tool for equine semen cryopreserva-
tion. In: 15th International Congress on Animal Reproduction', 8–12 August, Porto Seguro,
Brazil. (Eds. LR França, HP Godinho, M Henry and MIV Melo), p. 503 (abstract)

Rubinsky B (1983) Solidification processes in saline solutions. J Cryst Growth 62:513–522

Rubinsky B, Arav A, DeVries AL (1991) Cryopreservation of oocytes using directional cooling
and antifreeze glycoproteins. Cryo Lett 12:93–106

Rubinsky B, Arav A, Devries AL (1992) The cryoprotective effect of antifreeze glycopeptides from antarctic fishes. Cryobiology 29:69–79. https://doi.org/10.1016/0011-2240(92)90006-N

Rubinsky B, Ikeda M (1985) A cryomicroscope using directional solidification for the controlled freezing of biological material. Cryobiology 22:55–68. https://doi.org/10.1016/0011-2240(85)90008-2

Saacke RG, Almquist JO (1961) Freeze-drying of bovine spermatozoa. Nature 192:995–996. https://doi.org/10.1038/192995a0

Sakurai M, Furuki T, Akao KI, Tanaka D, Nakahara Y, Kikawada T, Watanabe M, Okuda T (2008) Vitrification is essential for anhydrobiosis in an African chironomid Polypedilum vanderplanki. Proc Nat Acad Sci U.S.A. 105:5093–5098. https://doi.org/10.1073/pnas.0706197105

Sánchez-Partida LG, Simerly CR, Ramalho-Santos J (2008) Freeze-dried primate sperm retains early reproductive potential after intracytoplasmic sperm injection. Fertil Steril 89:742–745. https://doi.org/10.1016/j.fertnstert.2007.02.066

Saragusty J (2015) Directional freezing for large volume cryopreservation. In: Wolkers WF, Oldenhof H (eds) Methods in cryopreservation and freeze-drying. Springer Verlarg, New York, pp 381–397

Saragusty J, Arav A (2011) Current progress in oocyte and embryo cryopreservation by slow freezing and vitrification. Reproduction 141:1–19. https://doi.org/10.1530/rep-10-0236

Saragusty J, Gacitua H, King R, Arav A (2006) Post-mortem semen cryopreservation and characterization in two different endangered gazelle species (Gazella gazella and Gazella dorcas) and one subspecies (Gazella gazelle acaiae). Theriogenology 66:775–784. https://doi.org/10.1016/j.theriogenology.2006.01.055

Saragusty J, Gacitua H, Pettit MT, Arav A (2007) Directional freezing of equine semen in large volumes. Reprod Domest Anim 42:610–615. https://doi.org/10.1111/j.1439-0531.2006.00831.x

Saragusty J, Gacitua H, Rozenboim I, Arav A (2009c) Do physical forces contribute to cryodamage? Biotechnol Bioeng 104:719–728. https://doi.org/10.1002/bit.22435

Saragusty J, Gacitua H, Rozenboim I, Arav A (2009d) Protective effects of iodixanol during bovine sperm cryopreservation. Theriogenology 71:1425–1432. https://doi.org/10.1016/j.theriogenology.2009.01.019

Saragusty J, Gacitua H, Zeron Y, Rozenboim I, Arav A (2009b) Double freezing of bovine semen. Anim Reprod Sci 115:10–17. https://doi.org/10.1016/j.anireprosci.2008.11.005

Saragusty J, Hildebrandt TB, Behr B, Knieriem A, Kruse J, Hermes R (2009a) Successful cryopreservation of Asian elephant (Elephas maximus) spermatozoa. Anim Reprod Sci 115:255–266. https://doi.org/10.1016/j.anireprosci.2008.11.010

Saragusty J, Lemma A, Hildebrandt TB, Göritz F (2017) Follicular size predicts success in artificial insemination with frozen-thawed sperm in donkeys. PLoS One 12:e0175637. https://doi.org/10.1371/journal.pone.0175637

Saragusty J, Osmers J-H, Hildebrandt TB (2016) Controlled ice nucleation - is it really needed for large-volume sperm cryopreservation? Theriogenology 85:1328–1333. https://doi.org/10.1016/j.theriogenology.2015.12.019

Saragusty J, Walzer C, Petit T, Stalder G, Horowitz I, Hermes R (2010) Cooling and freezing of epididymal sperm in the common hippopotamus (Hippopotamus amphibius). Theriogenology 74:1256–1263. https://doi.org/10.1016/j.theriogenology.2010.05.031

Schaffner CS (1942) Longivity of fowl spermatozoa in frozen condition. Science 96:337. https://doi.org/10.1126/science.96.2493.337

Shahba MI, El-Sheshtawy RI, El-Azab A-SI, Abdel-Ghaffar AE, Ziada MS, Zaky AA (2016) The effect of freeze-drying media and storage temperature on ultrastructure and DNA of freeze-dried buffalo bull spermatozoa. Asia Pac J Reprod 5:524–535. https://doi.org/10.1016/j.apjr.2016.11.002

Sherman JK (1954) Freezing and freeze-drying of human spermatozoa. Fertil Steril 5:357–371. https://doi.org/10.1016/S0015-0282(16)31685-5

Sherman JK (1957) Freezing and freeze-drying of bull spermatozoa. Am J Phys 190:281–286

Sherman JK (1963) Improved methods of preservation of human spermatozoa by freezing and freeze-drying. Fertil Steril 14:49–64. https://doi.org/10.1016/S0015-0282(16)34746-X

Sherman JK, Lin TP (1958) Survival of unfertilized mouse eggs during freezing and thawing. Exp Biol Med 98:902–905. https://doi.org/10.3181/00379727-98-24224

Si W, Hildebrandt TB, Reid C, Krieg R, Ji W, Fassbender M, Hermes R (2006) The successful double cryopreservation of rabbit (*Oryctolagus cuniculus*) semen in large volume using the directional freezing technique with reduced concentration of cryoprotectant. Theriogenology 65:788–798. https://doi.org/10.1016/j.theriogenology.2005.06.010

Si W, Lu Y, He X, Ji S, Niu Y, Tan T, Ji W (2010) Directional freezing as an alternative method for cryopreserving rhesus macaque (*Macaca mulatta*) sperm. Theriogenology 74:1431–1438. https://doi.org/10.1016/j.theriogenology.2010.06.015

Sieme H, Arav A, Klus N, and Klug E (2001) Cryopreservation of stallion spermatozoa using a directional freezing technique. In: Proceedings of the Second Meeting of the European Equine Gamete Group', 26–29 September, Loosdrecht, The Netherlands. (Eds. TAE Stout and JF Wade) pp. 6–8. (Newmarket, Suffolk, UK)

Sitaula R, Elmoazzen H, Toner M, Bhowmick S (2009) Desiccation tolerance in bovine sperm: a study of the effect of intracellular sugars and the supplemental roles of an antioxidant and a chelator. Cryobiology 58:322–330. https://doi.org/10.1016/j.cryobiol.2009.03.002

Spallanzani L (1776) Opuscoli di fisica animale e vigitabile. In: Opuscolo II. Osservazioni e sperienze intorno ai vermicelli spermaici dell' homo e degli animali. (Presso la Societa Tipografica: Modena, Italy)

Takahashi K, Yamanaka S (2006) Induction of pluripotent stem cells from mouse embryonic and adult fibroblast cultures by defined factors. Cell 126:663–676. https://doi.org/10.1016/j.cell.2006.07.024

Vajta G, Holm P, Kuwayama M, Booth PJ, Jacobsen H, Greve T, Callesen H (1998) Open pulled straw (OPS) vitrification: a new way to reduce cryoinjuries of bovine ova and embryos. Mol Reprod Dev 51:53–58. https://doi.org/10.1002/(SICI)1098-2795(199809)51:1<53::AID-MRD6>3.0.CO;2-V

Vajta G, Rienzi L, Ubaldi FM (2015) Open versus closed systems for vitrification of human oocytes and embryos. Reprod Biomed Online 30:325–333. https://doi.org/10.1016/j.rbmo.2014.12.012

van Bilsen DGJL, Hoekstra FA, Crowe LM, Crowe JH (1994) Altered phase behavior in membranes of aging dry pollen may cause imbibitional leakage. Plant Physiol 104:1193–1199. https://doi.org/10.1104/pp.104.4.1193

van Leeuwenhoek A (1800) On certain Animalcules found in the sediment in gutters on the roofs of houses. In: The selected works of Antony Van Leeuwenhoek. Vol. 2. pp. 207–214. (G. Sidney: London)

Wakayama S, Kamada Y, Yamanaka K, Kohda T, Suzuki H, Shimazu T, Tada MN, Osada I, Nagamatsu A, Kamimura S, Nagatomo H, Mizutani E, Ishino F, Yano S, Wakayama T (2017) Healthy offspring from freeze-dried mouse spermatozoa held on the international Space Station for 9 months. Proc Nat Acad Sci USA 114:5988–5993. https://doi.org/10.1073/pnas.1701425114

Wakayama T, Yanagimachi R (1998) Development of normal mice from oocytes injected with freeze-dried spermatozoa. Nat Biotechnol 16:639–641. https://doi.org/10.1038/nbt0798-639

Ward MA, Kaneko T, Kusakabe H, Biggers JD, Whittingham DG, Yanagimachi R (2003) Long-term preservation of mouse spermatozoa after freeze-drying and freezing without cryoprotection. Biol Reprod 69:2100–2108. https://doi.org/10.1095/biolreprod.103.020529

Watanabe H, Asano T, Abe Y, Fukui Y, Suzuki H (2009) Pronuclear formation of freeze-dried canine spermatozoa microinjected into mouse oocytes. J Assist Reprod Genet 26:531–536. https://doi.org/10.1007/s10815-009-9358-y

Whittingham DG (1971) Survival of mouse embryos after freezing and thawing. Nature 233:125–126. https://doi.org/10.1038/233125a0

Whittingham DG, Leibo SP, Mazur P (1972) Survival of mouse embryos frozen to −196° and −269°C. Science 178:411–414. https://doi.org/10.1126/science.178.4059.411

Willadsen SM, Polge C, Rowson LEA, Moor RM (1976) Deep freezing of sheep embryos. J Reprod Fertil 46:151–154. https://doi.org/10.1530/jrf.0.0460151

Wilmut I (1972) The effect of cooling rate, warming rate, cryoprotective agent and stage of development of survival of mouse embryos during freezing and thawing. Life Sci 11:1071–1079. https://doi.org/10.1016/0024-3205(72)90215-9

Wilmut I, Rowson LE (1973) Experiments on the low-temperature preservation of cow embryos. Vet Rec 92:686–690

Wilmut I, Schnieke AE, McWhir J, Kind AJ, Campbell KH (1997) Viable offspring derived from fetal and adult mammalian cells. Nature 385:810–813. https://doi.org/10.1038/385810a0

Yanagida K, Yanagimachi R, Perreault SD, Kleinfeld RG (1991) Thermostability of sperm nuclei assessed by microinjection into hamster oocytes. Biol Reprod 44:440–447. https://doi.org/10.1095/biolreprod44.3.440

Yushchenko NP (1957) Proof of the possibility of preserving mammalian spermatozoa in a dried state. Proc Lenin Acad Agric Sci 22:37–40

Zhang M, Oldenhof H, Sydykov B, Bigalk J, Sieme H, Wolkers WF (2017) Freeze-drying of mammalian cells using trehalose: preservation of DNA integrity. Sci Rep 7:6198. https://doi.org/10.1038/s41598-017-06542-z

Zheng H, Li B, Yang S-H, Chen L-L, He B-L, Jiao J-L (2010) Cryopresevation of sperms with directional freezing method in Yunnan Diannan miniature pig. Chin J Comp Med 2010:06 (Abstract)

Zhu WJ, Li J, Xiao LJ (2016) Changes on membrane integrity and ultrastructure of human sperm after freeze-drying. J Reprod Contracept 27:76–81. https://doi.org/10.7669/j.issn.1001-7844.2016.02.0076

Zirkler H, Gerbes K, Klug E, Sieme H (2005) Cryopreservation of stallion semen collected from good and poor freezers using a directional freezing device (harmony CryoCare - multi thermal gradient 516). Anim Reprod Sci 89:291–294 (abstract). https://doi.org/10.1016/j.anireprosci.2005.07.004

In Vitro Production of (Farm) Animal Embryos

12

Christine Wrenzycki

Abstract

Over the past decades, in vitro production (IVP) of bovine embryos has been significantly improved, and in particular bovine IVP is now widely applied under field conditions. This in vitro technique provides new opportunities for cattle producers, particularly in the dairy industry, to overcome infertility and to increase dissemination of animals with high genetic merit. Improvements in OPU/IVP resulted in large-scale international commercialization. More than half a million IVP embryos are generated on the yearly basis demonstrating the enormous potential of this technology. These advances and the fact that bovine and human early development is remarkably similar have prompted the use of bovine embryos as a model system to study early mammalian embryogenesis including humans. In horses, OPU/IVP is also an established procedure for breeding infertile and sports mares throughout the year. It requires ICSI because conventional IVF does not work in this species. In small ruminants, application of IVP on the commercial and research basis is low compared to other livestock species.

Despite all the improvements, embryos generated in vitro still differ from their in vivo-derived counterparts. Embryos must adjust to multiple microenvironments at preimplantation stages. Consequently, maintaining or mimicking the in vivo situation in vitro will aid to improving the quality and developmental competence of the resulting embryo.

The successful clinical application of the techniques in reproductive biotechnology requires both species-specific clinical skills and extensive laboratory experience.

C. Wrenzycki
Clinic for Veterinary Obstetrics, Gynecology and Andrology of Large and Small Animals;
Chair of Molecular Reproductive Medicine, Faculty of Veterinary Medicine,
Justus-Liebig-University Giessen, Giessen, Germany
e-mail: Christine.Wrenzycki@vetmed.uni-giessen.de

© Springer International Publishing AG, part of Springer Nature 2018
H. Niemann, C. Wrenzycki (eds.), *Animal Biotechnology 1*,
https://doi.org/10.1007/978-3-319-92327-7_12

12.1 Introduction

The birth of the first IVF calf derived from an in vivo matured oocyte in 1981 (Brackett et al. 1982) and the discovery of heparin as capacitating agent for bull sperm in 1986 (Parrish et al. 1986) were two key events, ultimately resulting in efficient IVP systems for bovine preimplantation embryos, including in vitro maturation (IVM) of the immature oocyte to the matured metaphase II stage, in vitro fertilization (IVF), and subsequent in vitro culture (IVC) of embryos to the desired stage of development, preferably the blastocyst stage. The first calves produced entirely from IVM-IVF-IVC were born in 1987 (Fukuda et al. 1990).

While artificial insemination (AI) is an effective way to disseminate the genetics of valuable sires, with the implementation of embryo biotechnologies, female genetics can also be distributed worldwide. In the last decades, major advances were made in multiple ovulation and embryo transfer (MOET), ovum pickup (Pieterse et al. 1991; Kruip et al. 1991) combined with in vitro production of embryos (OPU/IVP) and cryopreservation (Colleau 1991).

The current technology of OPU/IVP harvesting immature oocytes from living cows can routinely be performed twice a week for an extended period of time without any detrimental effects on the donor's cow fertility (Chastant-Maillard et al. 2003). At present, the application of IVP combined with ovum pickup (OPU) from valuable donors is increasing (again) due to new breeding strategies based on genomic selection using SNP (single nucleotide polymorphism) chips. Depending on the chip used, thousands of these SNPs can be analyzed even in a biopsy taken from an embryo. This technology is now reaching routine usage for genomic selection (GS) in cattle (Ponsart et al. 2014).

With regard to IVP efficiency (Table 12.1), approximately 80–90% of immature bovine oocytes undergo nuclear maturation in vitro, about 80% undergo fertilization, 30–40% develop to the blastocyst stage, and around 50% of the transferred embryos establish and maintain a pregnancy (Galli et al. 2014; Lonergan et al. 2016; Wrenzycki et al. 2007).

According to the International Embryo Technology Society (IETS) statistics (Fig. 12.1), the number of embryos produced in vitro and transferred into recipients has increased more than 10 times in the last century and is now approaching the numbers of embryos produced in vivo by superovulation. This indicates that OPU and IVP are considered a reliable and cost-effective technique and have acquired a significant role in cattle breeding. Detailed data stemming from the year 2016 are shown in Tables 12.2a and 12.2b.

Table 12.1 Efficiencies of bovine IVP

Maturation rate[a]	90%
Fertilization rate[a]	75–80%
Cleavage rate[a]	60–70%
Blastocyst rate[a]	30–40%
Pregnancy rate[b]	50–60%
Calves born[c]	70–90%

[a]calculated on the number of immature oocytes
[b]calculated per embryo transfer
[c]calculated per pregnancy

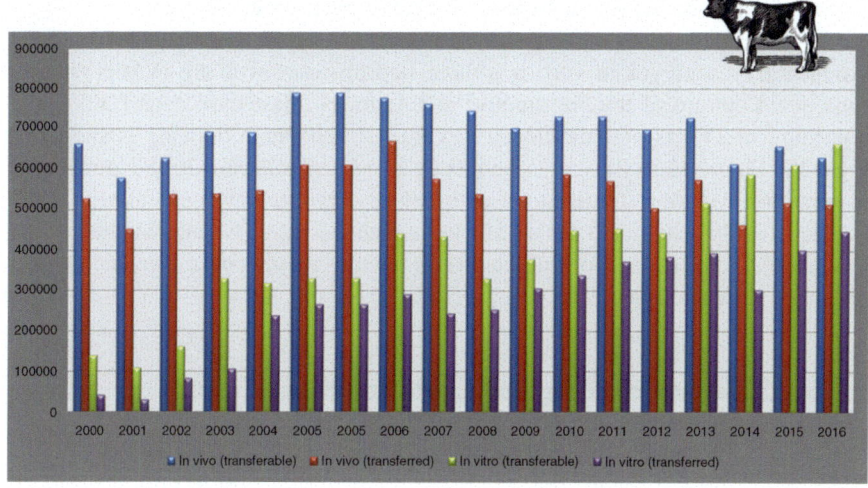

Fig. 12.1 Bovine embryo transfer statistics published annually by the International Embryo Technology Society (IETS)

Table 12.2a Collection and transfer of bovine OPU/IVP embryos by regions (year 2016)

	Collection			Transfers	
Region	Donors	Oocytes	Embryos	Fresh embryos	Frozen embryos
Africa	619	19,062	2167	379	246
Asia[a]	3177	59,224	9438	3250	1164
Europe	10,651	94,407	18,879	10,424	3635
North America	45,918	805,072	260,574	80,825	50,672
Oceania	2241	21,587	6304	4732	1702
South America	49,739	1,138,302	378,291	230,263	65,235
Grand Total	112,345	2,137,654	675,653	329,873	122,654

[a]Data from 2015

Table 12.2b Collection and transfer of bovine abattoir-derived IVP embryos by regions (year 2016)

	Collection			Transfers	
Region	Donors	Oocytes	Embryos	Fresh embryos	Frozen embryos
Africa[a]	156	2033	235	0	0
Asia[a]	35,335	714,783	56,740	10,685	7831
Europe	256	18,317	1095	40	133
North America[a]	5	9117	1037	273	418
Oceania	4	60	16	8	4
South America	1345	5017	1511	0	0
Grand Total	37,101	749,327	60,634	11,006	8386

[a]Data from 2015

12.2   General Steps of In Vitro Production of Embryos

Production of embryos in vitro is a three-step process involving IVM, IVF, and subsequent culture of the presumptive zygote to the blastocyst stage (IVC). The procedure of IVP is best developed in cattle. A schematic drawing is shown in Fig. 12.2. Figure 12.3a illustrates the processes occurring during oocyte growth and maturation, ovulation, fertilization, and early embryonic development within the oviduct, finally leading to the blastocyst which has already entered the uterus. Representative pictures of oocytes and early bovine embryos of in vivo and in vitro origin are shown in Fig. 12.3b.

12.2.1 Collection of Cumulus-Oocyte Complexes (COC)

COC can be collected from ovaries of slaughtered or euthanized animals or from those of live animals. With slaughterhouse ovaries, COC are usually isolated by aspiration or slicing, less often via isolation and dissection of follicles. With these methodologies, cumulus-enclosed immature oocytes at the germinal vesicle (GV) stage can be harvested. Immature oocytes are also collected from antral follicles of living animals via the OPU technology. This is a noninvasive and repeatable technique which can also be used to collect MII oocytes shortly before ovulation with or without hormonal stimulation similar as in human-assisted reproduction.

OPU without hormonal pre-stimulation can be routinely done twice a week. Twice a week OPU can be done for an extended period of time without detrimental

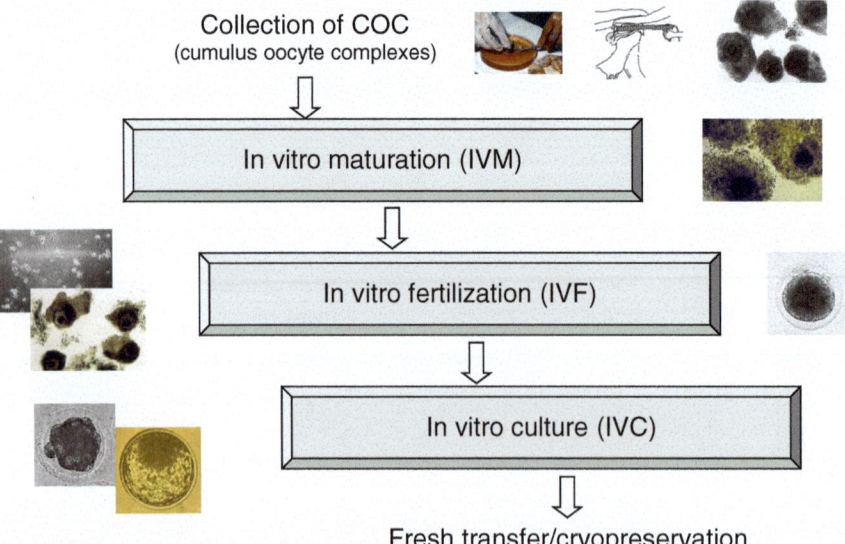

Fig. 12.2 In vitro production of embryos (for further information, see text)

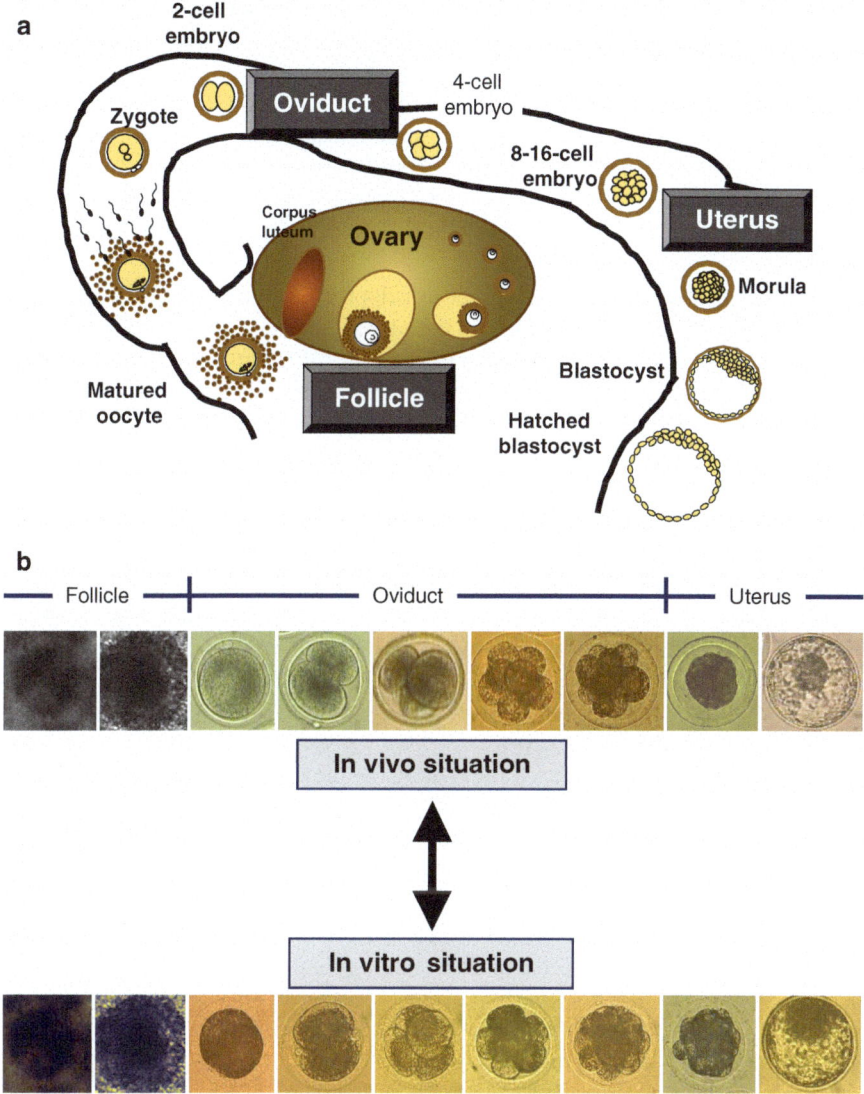

Fig. 12.3 (**a**) Oocyte growth and maturation, fertilization, and early embryonic development (see text). (**b**) Oocyte and embryo development in vivo and in vitro

effects on the donor cow's fertility (Chastant-Maillard et al. 2003) and will result in a higher number of embryos produced per cow per time unit compared to the once a week OPU schedule. Best results are obtained when a 3- and 4-day or 2- and 5-day interval is maintained between OPU sessions (Merton et al. 2003). OPU was initially applied on problem cows that did not respond to superovulation (Kruip et al. 1994; Looney et al. 1994). It can be used to collect COC also from pregnant cows and heifers, including prepubertal heifers (Presicce et al. 1997). The number of follicles can be increased by the use of a hormonal pre-stimulation. The disadvantage of pre-stimulation is that it involves the use of hormones, whereas the lack of requirements for hormones in the case of OPU has been generally considered an advantage compared to MOET (Merton et al. 2003). In contrast to oocytes recovered from abattoir ovaries, the genetic merit and health status of the donor animal including its oocytes are known. The number of oocytes collected from an animal during a single session of OPU depends on a variety of technical and biological factors (Merton et al. 2003). Usually, all follicles between 3 and 8 mm in diameter are aspirated.

OPU has become possible due to the development of ultrasound-guided transvaginal oocyte aspiration in humans and the adoption in the bovine in 1988 (Pieterse et al. 1988). Attempts were undertaken to combine OPU with color Doppler ultrasonography which is a useful, noninvasive technique for evaluating ovarian vascular function, allowing visual observation of the blood flow in the wall of preovulatory follicles (Brannstrom et al. 1998). Blood flow determination of individual preovulatory follicles prior to follicular aspiration for human IVF therapy provides important insight into the intrafollicular environment and may predict the developmental competence of the corresponding oocyte (Coulam et al. 1999; Huey et al. 1999). In cattle, it has been shown that the time interval between the individual OPU sessions had an effect on the quality of oocyte and embryos at the molecular level, whereas differences in the perifollicular blood flow did not (Hanstedt et al. 2010). An increase in blood supply to individual follicles appears to be associated with increased follicular growth rates, while a reduction seems to be closely related to follicular atresia (Acosta et al. 2003; Acosta 2007).

There are various systems used for grading bovine COC by visual assessment of morphological features, including compactness and quantity of surrounding follicular cells and homogeneity of the ooplasm as well as the size of the oocyte. An example is given in Fig. 12.4. Oocytes smaller than 110 μm are transcriptionally active and have a reduced ability to resume meiosis (Fair et al. 1995).

Taken together, OPU can be considered a mature technique and no major improvements should be expected in the technology and its results in the near future.

12.2.2 In Vitro Maturation (IVM)

Cumulus-oocyte complexes (COC) collected from ovaries of slaughtered or euthanized animals or from living animals via OPU require IVM as they are arrested at the GV stage. Maturation involves a series of events that already begin in fetal life with the initiation of meiosis. At birth, the oocytes are arrested at the diplotene stage. After puberty when they are exposed to preovulatory surges of LH and FSH, they proceed with meiosis and are arrested again at the metaphase II, the stage at which they are ovulated (Monniaux et al. 2014). In addition, optimal conditions for cumulus cells surrounding the oocyte need to be considered as there is a complex bi-directional communication between these two cell types (Monniaux 2016; Gilchrist 2011).

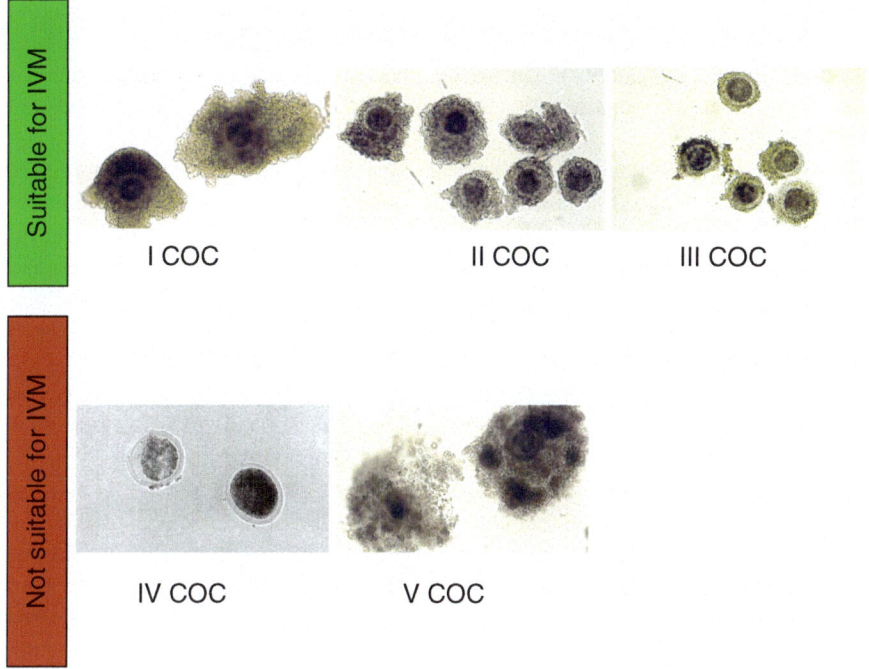

Fig. 12.4 Different categories of COC (I, II, and III COC are suitable for IVM, IV and V COC not; see text)

Maturation is subdivided in nuclear and cytoplasmic maturation. Nuclear maturation refers to the resumption of meiosis, from germinal vesicle breakdown (GVBD), through metaphase I, anaphase, telophase, and rearrest at metaphase II (MII), and can be assessed noninvasively by determining extrusion of the first polar body. Nuclear maturation does not necessarily ensure that cytoplasmic maturation has been completed and should not be used as the sole determinant of oocyte quality. The oocyte must undergo a complex array of cytoplasmic rearrangements that allow the oocyte to support subsequent fertilization and initiate embryonic development. Multiple processes are involved in cytoplasmic maturation, including carbohydrate and lipid metabolism, mitochondrial function and location, reduction of oxygen radicals, accumulation of follistatin, epigenetic programming, communication between cumulus cells and the oocyte, and secretion of oocyte-derived growth factors (Sirard 2016).

Proper maturation of the bovine oocyte from the GV stage to metaphase II takes place within a period of 20–24 h and is a prerequisite for fertilization and preimplantation development. It is possible to achieve blastocyst rates of up to 70% if in vivo matured oocytes are used. In contrast, if oocytes are matured in vitro, blastocyst rates are usually only half that of those matured in vivo. This rather limited success may be attributed to the heterogeneous population of oocytes which are retrieved from follicles of 3–8 mm rather than from preovulatory follicles. In contrast to in vivo ovulated oocytes, these oocytes usually would not make it up to the preovulatory stage and are matured in vitro. Therefore, substantial efforts have been devoted to the establishment of noninvasive and non-perturbing means for selecting the most competent oocytes (Fair 2010; Krisher 2013; Wrenzycki and Stinshoff 2013).

Maturation is initiated immediately following the removal of the immature oocyte from small antral follicles, and such oocytes may have neither the time nor the correct environment to complete the necessary changes required for subsequent successful development (Lonergan and Fair 2016; Krisher 2013; Wrenzycki and Stinshoff 2013).

Much of the success of IVP critically depends on the quality of the starting material, the oocyte. Developmental competence of the oocyte also termed oocyte quality is defined as the oocyte's ability to mature, to be fertilized, and to give rise to normal and healthy offspring (Duranthon and Renard 2001). The success of IVM is also dependent on the composition of the culture media which are usually quite different from in vivo conditions (Roberts et al. 2002; Sutton et al. 2003). A better understanding of intrafollicular conditions is critical for successful oocyte selection and maturation regardless of the species. In this context, profiling bovine follicular fluid revealed important clues about the composition of bovine follicular fluid (Sanchez-Guijo et al. 2016). Therefore, metabolomic analysis of follicular fluid may be a useful tool for characterizing oocyte quality (Sinclair et al. 2008; Bender et al. 2010).

Although substantial progress has been made to improve the efficiency of IVM protocols, there is a lack of consistency in the success rates of conventional in vitro maturation protocols compared to the in vivo situation. Multiple factors likely contribute to the overall poorer quality of in vitro matured oocytes. One of the important factors may be oxidative stress (OS). The generation of prooxidants such as reactive oxygen species (ROS) is an important phenomenon in culture conditions (Khazaei and Aghaz 2017).

The so-called simulated physiological oocyte maturation (SPOM) system had been introduced a few years ago (Albuz et al. 2010). It prevented spontaneous resumption of meiosis after mechanical oocyte retrieval and thereby improved in vitro embryo development. However, due to the fact that these first results were not repeatable, a revised version has been reported (Gilchrist et al. 2015). At the moment, most laboratories practicing IVM of cattle oocytes use a relatively simple oocyte maturation system.

Given that oocyte maturation in vivo takes place in the follicle prior to ovulation, and given the difficulties to mimic the situation in the preovulatory follicle in vitro, it is attractive to mature oocytes within the follicle. To achieve this, a modified ovum pickup equipment has been used to transfer in vitro matured oocytes into the preovulatory follicle of synchronized heifers (follicular recipients), enabling subsequent ovulation, in vivo fertilization (the animals have been artificially inseminated prior to oocyte transfer), and in vivo development. On average, 35% of embryos were recovered in excess after uterine flushing at Day 7. This technique is called intrafollicular oocyte transfer (Kassens et al. 2015). Transfer of frozen-thawed

IFOT-derived blastocysts to synchronized recipients (uterine recipients) resulted in 8 pregnancies out of 19 transfers. In total, seven pregnancies presumed to be IFOT-derived went to term, and microsatellite analysis confirmed that five calves were indeed derived from IFOT. These were the first calves born after IFOT in cattle. The present study established the proof of principle that IFOT is a feasible technique to generate high-quality embryos capable of developing to apparently high-quality blastocysts as well as healthy calves. Blastocysts (Spricigo et al. 2016) and calves (Michael et al. 2017) could also be produced after full in vivo development of immature slaughterhouse derived oocytes transferred to the preovulatory follicle. Therefore, the embryo production by IFOT of immature oocytes represents an alternative for the production of a large number of embryos without requiring hormones and basic laboratory handling only. As an alternative procedure, gamete intrafallopian transfer (GIFT, transfer of in vitro matured COC simultaneously with capacitated spermatozoa) is a viable compromise between in vitro and in vivo conditions and provides an option for reducing impacts of in vitro effects (Wetscher et al. 2005).

A better understanding of the in vivo regulation of follicular development and in particular of final maturation, in concert with oocyte differentiation by intrafollicular factors and intercellular communication, will lead to improved in vitro protocols for final maturation, thus compensating for the lack of preovulatory development of immature oocytes (Wrenzycki and Stinshoff 2013).

12.2.3   In Vitro Fertilization (IVF)

IVF is a complex step whose success requires appropriate oocyte maturation, sperm selection, sperm capacitation, and IVF media. In vivo, fertilization rates are around 90% for heifers and cows (Diskin and Sreenan 1980).

Semen samples contain a heterogeneous population of sperm cells. In vivo, sperm cells are thought to be selected by various mechanisms within the female reproductive tract, with the result that a small number of spermatozoa found near the oocyte are typically those best capable to penetrate the zona pellucida and fertilize the oocyte. However, when using IVF, these natural selection mechanisms are circumvented.

Treatment of bull sperm prior to IVF generally involves the selection of cells with the highest progressive motility. Furthermore, seminal plasma, cryoprotectants, and other factors are removed. One of the most common methods for preparing spermatozoa for IVF is to centrifuge them through a concentration gradient, such as a 45% Percoll mixture layered on a 90% solution.

A variant of colloid centrifugation using only one layer of colloid (in which case there is no gradient) has been developed. Single-layer centrifugation (SLC) through a species-specific colloid has also been shown to be effective in selecting spermatozoa with good motility, normal morphology, and intact chromatin (Thys et al. 2009; Goodla et al. 2014; Morrell et al. 2014; Gloria et al. 2016). An alternative method is the swim-up procedure. The disadvantages of swim-up are that it lasts approximately 45–60 min and allows only 10–20% of the spermatozoa in the sample to be recovered. For colloid centrifugation, only a 25-min preparation time is needed (including the centrifugation), and a recovery rate of >50% is commonly achieved (Thys et al. 2009), which is critically depending on the sperm quality of the original sample.

The changes a sperm cell has to go through before it can successfully fertilize an oocyte are summarized under the term capacitation (Brackett and Oliphant 1975). Capacitation is recognized as a complex series of biochemical and physiological reactions (Breitbart et al. 1995), including the expression of hyperactivated patterns of motility and the acquisition of the capacity to respond to signals originating from the oocyte. Media have been developed to support this process, e.g., TALP medium. As mentioned earlier, the primary capacitation agent in current IVF systems is heparin. The majority of semen samples used for IVF is frozen-thawed. Apparently, fresh semen requires a longer capacitation period than the frozen one. Frozen-thawed semen has been intensively screened prior to cryopreservation. Furthermore, there are numerous studies indicating that the selection of bulls producing sperm cells with a high IVF capacity is an important factor in achieving successful and reproducible IVP results. The marked variability that occurs among bulls in the suitability of semen for IVF may be due to the penetration of the zona and/or ooplasm or processes taking place in the ooplasm. The most common final sperm concentration used in the IVF drop is 1×10^6 sperm/mL, whereas in the cow's oviduct, fertilization is likely to occur in a sperm/oocyte ratio close to 1 to 1 (Hunter 1996).

Once IVM is complete, oocytes are ready to be fertilized. This involves the co-incubation of oocytes with sperm cells. Most laboratories allow for 18–19 h of co-incubation. During this time period, sperm pass through the cumulus cell layers enclosing the oocyte, attach and bind to the zona pellucida, undergo the acrosome reaction, and finally penetrate the zona pellucida. The fertilizing sperm cell will then bind to and fuse with the oolemma, followed by activation of the oocyte and formation of male and female pronuclei. After co-incubation, attached cumulus cells are removed from the zonae. Successful fertilization is characterized by the extrusion of the second polar body and the formation of the female and male pronucleus. In cattle, the pronuclei are not visible due to the amount of lipid vesicles which is in contrast to the picture in for example humans and mice. Among the abnormalities seen after IVF are polyspermy and parthenogenesis (Parrish 2014).

Beside the conventional co-incubation of matured COC and sperm cells, there are four approaches available to assist fertilization when sperm numbers are reduced or when sperm motility is compromised; these approaches are partial zona dissection, zona thinning/zona drilling, subzonal sperm insertion (SUZI), and intracytoplasmic sperm injection (ICSI). ICSI involves the fertilization of MII oocytes by direct injection of a spermatozoon (Goto and Yanagita 1995). However, even haploid parthenogenetic embryos injected with sperm can result in full-term development in mice suggesting that sperm reprogramming sufficient to support full and healthy development can occur not only when M II oocytes are fertilized and in mitotic embryos well after meiotic exit (Suzuki et al. 2016).

ICSI is performed with the aid of a pair of glass pipettes adjusted to an inverted microscope, where the embryologist holds the oocyte with one pipette and injects a selected sperm cell with the second pipette straight into the ooplasm as shown in Fig. 12.5. The clinical use of ICSI was developed many years ago as a solution for some sperm-related male infertility in human-assisted reproduction (Palermo et al. 1992). The use of ICSI in cattle has never attracted much interest and has mainly been used for research purposes, essentially because IVF after heparin capacitation works very efficiently with the majority of bulls.

The success of ICSI in cattle is usually poor (Arias et al. 2014; Sekhavati et al. 2012) and appears to be limited by the low-activating stimulation of the injected spermatozoon (Catt and Rhodes 1995; Malcuit et al. 2006) and asynchronous pronucleus formation (Chen and Seidel 1997). Furthermore, bovine sperm are especially resistant to nuclear decondensation by in vitro matured oocytes, and this deficiency cannot be simply overcome by exogenous activation protocols. Therefore, the inability of a suboptimal ooplasmic environment to induce sperm head decondensation limits the success of ICSI in cattle (Aguila et al. 2017). In addition, the size of bovine sperm cells requires a pipette with a relatively large outer diameter of 10 μm that could be responsible for the damage on the cytoskeleton, thus reducing the developmental potential of the resulting embryos (Galli et al. 2003). Taken together, the efficiency of intracytoplasmic sperm injection (ICSI) in cattle is low compared to other species due in part to inadequate egg activation and sperm nucleus decondensation after injection.

12.2.4 In Vitro Culture (IVC)

IVC of bovine embryos is the last step in IVP and involves approximately 6 days of culture from the presumptive zygote onward. During the early post-fertilization period, several major developmental events occur in the embryo including (1) the first cleavage division, (2) the activation of the embryonic genome, (3) the compaction of the morula, and (4) the formation of the blastocyst.

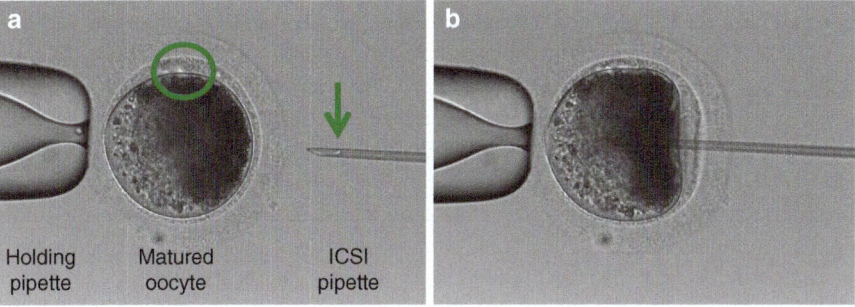

Fig. 12.5 Intracytoplasmic sperm injection (green circle, polar body; green arrow, sperm cell)

Previously embryo requirements were not known, and a temporary in vivo culture in surrogate oviducts of sheep was performed (Lazzari et al. 2010). Oviducts of several other mammalian species including rabbits and mice have also been used to culture bovine embryos in vivo. The isolated mouse oviduct has also been employed as a system to culture bovine embryos successfully (Minami et al. 1988; Rizos et al. 2010). The success of in vitro embryo production techniques demonstrates that it is possible to bypass the oviduct during early development and, to a certain extent, replicate the conditions in vitro. However, overall they do not adequately mimic well enough the complex series of development-specific steps for which the oviduct has evolved a unique and dynamic microenvironment (Kenngott and Sinowatz 2007; Besenfelder et al. 2012).

The most common media for culturing bovine embryos are variations of the original synthetic oviduct fluid (SOF) medium (Tervit et al. 1972). SOF is now part of most routine bovine IVP systems with/without serum. Embryos can be cultured either in only one medium throughout the entire time or in a sequential system in which the medium formulation changes at certain time points in the culture period, i.e., so-called stage-specific media. These sequential media try to mimic the physiological changes that embryos encounter in vivo when they travel down the oviducts into the uterus. Parameters vary from lab to lab, e.g., the volume of medium and the atmosphere in the incubator.

IVC conditions have been improved in the last years, mainly by adjustment of media formulations. However, while more than 30% blastocyst formation could be achieved in most culture systems, it soon became obvious that quantity did not always match quality (Lonergan et al. 2006; Niemann and Wrenzycki 2000; Wrenzycki et al. 2005, 2007) and that serum supplementation was detrimental to embryo/fetal development as one main causal factor of the so-called large offspring syndrome (LOS), characterized by altered embryonic and fetal growth, deviant embryonic and fetal gene expression patterns, and high perinatal losses (Young et al. 1998; Lazzari et al. 2002). A large field study demonstrated that the incidence of LOS was greatly reduced by in vitro culture in cell-free and serum-free SOF media (van Wagtendonk-de Leeuw et al. 2000). These observations highlight the importance of the post-fertilization culture environment for the quality of the resulting blastocysts. However, the existence of diverse embryo culture media and methods makes it very challenging to define the optimal components of embryo culture media. Undefined media supplemented with serum and somatic cells as monolayer have been replaced by semi-defined media culturing embryos in the absence of cells with no serum supplementation. Completely defined media, developed by modifications of the SOF medium with glutamine or citrate and nonessential amino acids (Holm et al. 1999; Keskintepe et al. 1995), and by replacing BSA with polyvinyl alcohol (Keskintepe and Brackett 1996), also support embryo development to the blastocyst stage. Defined conditions not only facilitate the improvement of culture media to provide a better, more consistent environment for the developing embryo but are also critical in disease control (Stringfellow and Givens 2000).

Recently, a culture system for the formation of an in vivo-like oviduct tissue substitute from primary oviduct epithelial cells has been developed. This air-liquid interphase (ALI) culture is fully functional in terms of morphological differentiation (polarization, columnar shape, ciliary activity), generates oviduct fluid surrogates, and enables embryonic development up to the blastocyst stage without addition of embryo culture medium (Chen et al. 2017).

The success of an IVP laboratory may stem not only from improvements of the IVC per se but from the entire IVP system (Baltz 2012; Gardner 2008; Leese 2012). The latter includes incubation conditions, gas phase, culture media, oil overlay, plastic ware, and embryo density and the volume of the medium. In addition, the skills of the staff involved in the entire process are critically important for the success of the system.

The use of noninvasive strategies, such as analysis of follicular fluid and culture media (after culture), also appears to be useful for the search for molecular biomarkers indicative of oocyte competence. The presence of cytokines and growth factors in follicular fluid is crucial for determining oocyte quality (Dumesic et al. 2015). In this context, the metabolic characterization of the culture media, in which IVP embryos are maintained for many hours, may represent an important noninvasive tool to either indicate possible predictive biomarkers of viability or to explain IVP outcome afterward (Munoz et al. 2014).

In general, bovine IVP is at an advanced stage of development. However, an aspect that may change in the future is automation and miniaturization of the IVP process by better mimicking in vivo environment, e.g., using microfluidics (Wheeler et al. 2007) or an encapsulation technology (Blockeel et al. 2009) to obtain IVP embryos of similar quality as the in vivo-derived counterparts. Such systems would facilitate the gradual change of the culture medium to meet the precise requirements of the developing embryo and overcome substantial limitations of conventional culture systems. Nevertheless, commercial production of bovine embryos via IVP is now successful in many laboratories around the world.

12.3 IVP of Embryos in Other Animals

12.3.1 Farm Animals

As mentioned earlier, IVP is best developed in cattle. Nevertheless, there are often species-specific differences, even with cattle (Sartori et al. 2016).

12.3.1.1 Bos indicus/Buffalos

B. indicus animals have greater numbers of retrieved oocytes, due to higher antral follicle counts, resulting in higher percentages of viable oocytes, number of blastocysts, and blastocyst rates when compared with *B. taurus* ones. The better blastocyst rates obtained with *B. indicus* cows and heifers could be attributed to an intrinsically better quality of the oocytes (Viana et al. 2012). OPU/IVP is the only option for embryo production to advance the implementation of embryo-based biotechnologies in buffalo production, although it is still in an early developmental phase (Galli et al. 2014). OPU works better, and it is more cost-effective than superovulation in

Bos indicus females. In this species, OPU has great potential because superovulation (multiple ovulation and embryo transfer) yields poor results compared with cattle (Carvalho et al. 2002). The procedure is similar to the one used in *Bos taurus*. Oocyte recovery, embryo production, and offspring obtained have been described (Galli et al. 2014), but only about 10%–15% of the oocytes recovered develop to transferable embryos. In general, the ovaries of buffalo cows and heifers are small, and, in addition, the follicles tend to be fewer and of small diameter.

12.3.1.2 Horses

In the horse, OPU/IVP is now an established procedure for breeding from infertile and sporting mares throughout the year. It requires ICSI that in the horse, contrary to cattle and buffalo, is very efficient and the only option because conventional IVF does not work (Galli et al. 2014).

To collect in vivo matured oocytes from living mares, the most practical, less invasive, efficient, and repeatable technique used is the ultrasound-guided transvaginal follicular aspiration using a double-lumen needle (Carnevale et al. 2005). Oocytes can be collected from the preovulatory follicle that has reached at least 35 mm in diameter, 24 h after hCG injection with the donor showing signs of uterine edema. The recovery rates are much higher than those for the recovery from immature follicles (see below), because the COC is expanding and is detached from the follicle wall just prior to ovulation. After recovery from the donor mare and sometimes after a few hours of culture to complete maturation, oocytes can either be surgically transferred to the oviduct of inseminated recipients called oocyte transfer (OT) (Carnevale et al. 2005) or can be subjected to intracytoplasmic sperm injection (ICSI). OT has been developed for clinical and research purposes because of frequent failures of conventional IVF. In addition, after ICSI, the fertilized oocytes can be transferred surgically to the oviduct of a synchronized recipient or cultured in vitro up to the blastocyst stage. Nowadays, ICSI and transfer after IVC are the methods of choice.

As only one preovulatory follicle is present at any cycle and only during the breeding season, OPU of immature equine oocytes has emerged as useful alternative. All follicles that are at least 1 cm, also in the nonbreeding season, are flushed. Because of the large size of the follicles and the firm attachment to the follicle wall, it is necessary to use double-lumen needles with separate in- and outflow channels that allow for repeated flushing up to eight to ten times for each follicle (Galli et al. 2007). As mentioned above, due to the firm attachment of the immature oocyte to the follicular wall, follicle flushing is a prerequisite. Mares can be subjected to repeated collections without side effects (Mari et al. 2005).

Immature cumulus-oocyte complexes could also be obtained from slaughterhouse ovaries or after ovariectomy. Because of the tight attachment of the oocyte to the follicular wall, follicle scraping is the method of choice (Hinrichs and Digiorgio 1991) rather than slicing or aspiration as done in cattle. Oocytes and their surrounding cumulus cells are classified by their cumulus morphology as either compact or expanded. Oocytes classified as compact are more likely to have a homogenous cytoplasm which is associated with a lower meiotic competence (Hinrichs and Williams 1997).

The optimum duration of maturation is between 24 and 30 h for oocytes with expanded and between 30 and 36 h for those with compact cumulus. Once matured, there was no difference in developmental competence (rate of blastocyst development) between both types of oocytes (Hinrichs 2005). Evaluation of oocytes after in vitro maturation and preparation for ICSI requires removal of the cumulus; this is more difficult in horses than it is in many other species. The transzonal processes of the equine cumulus are extensive.

Only two foals have been reported from conventional IVF of in vivo matured oocytes (Palmer et al. 1991; Cognie et al. 1992). The first foal, derived from an in vitro matured oocyte fertilized by ICSI, was born in 1996 (Squires 1996). To date, several laboratories have consistently and reproducibly reported the birth of ICSI foals both from in vivo and in vitro matured oocytes. It has been reported that conventional IVF is successful after treatment of sperm with procaine to induce hyperaction (McPartlin et al. 2009). However, recent data showed that procaine induces cytokinesis in equine oocytes associated with elevated levels of parthenogenetic development (Leemans et al. 2015).

Good blastocyst rates (20% to 40% of injected oocytes) are achieved using the Piezo drill (Hinrichs 2005; Galli et al. 2007). No exogenous activation of the oocytes is required. For IVC, a variety of media can be used, and supplementation with 17–19 mM glucose is important (Hinrichs 2005; Herrera et al. 2008). Equine embryos start to develop to the blastocyst stage at Day 7 of culture. In vivo, an acellular capsule forms inside the zona pellucida after entry of the equine embryo into the uterus (Betteridge et al. 1982). The equine capsule is composed of mucin-like glycoproteins produced by the trophectoderm, containing a high proportion of sialic acid (Oriol et al. 1993). Formation of this capsule is not observed in vitro.

12.3.1.3 Pigs

The pig has been a particularly difficult species to obtain high rates of fertilization and subsequent blastocyst development in vitro. Insufficient oocyte cytoplasmic maturation in vitro, high rates of polyspermy, and low embryonic development rates are the major obstacles that still need to be overcome (Gil et al. 2010; Romar et al. 2016).

The developmental potential of porcine IVM oocytes is typically much poorer than that of cattle and sheep IVM oocytes. Pigs are generally slaughtered at 6 or 7 months of age to meet market demands for pork. Therefore, abattoir-derived ovaries are usually collected from prepubertal gilts that have not yet experienced regular estrous cycles. Porcine oocytes are typically matured in vitro for about 40–44 h, much longer than oocytes of other livestock species. Only during the first half of IVM porcine COC are exposed to gonadotropins (Funahashi and Day 1993).

The use of BSA and caffeine as capacitating agents was an important breakthrough and was instrumental for successful in vitro fertilization in the pig. However, the rate of polyspermy has been reported to be over 50% in some laboratories (Mugnier et al. 2009). The degree of polyspermy was found to be closely associated with the number of sperm per oocyte at fertilization (Rath 1992). Unfortunately, simply reducing the number of sperm per oocyte within the insemination droplet has not been useful to overcome the polyspermy problem, because this also results in a decrease of the overall penetration rates. An issue that confounds the polyspermic fertilization in porcine embryo IVP systems is that polyspermic embryos are still able to develop to the blastocyst stage.

Despite these difficulties, IVP blastocyst development rates are in the range from 30% to 50% from monospermically fertilized oocytes in most laboratories (Gil et al. 2010). The addition of particular components such as porcine follicular fluid into the IVM media has been shown to improve the quality of porcine IVM (Algriany et al. 2004). Hormone supplementation at particular stages of maturation and adding insulin-transferrin-selenium to the media have also contributed toward improved maturation rates (Hu et al. 2011). Due to a refinement of techniques, IVM rates now vary from 75% to 85% (Gil et al. 2010). Embryo culture (IVC) has been developed extensively in the pig, and two media compositions are used widely today, including NCSU23 and NCSU-37 (Petters and Wells 1993). The first piglets born from IVP procedures including IVM/IVF/IVC up to the blastocyst stage were reported in 1989 (Mattioli et al. 1989). The rate of live offspring resulting from IVP is relatively low compared to the number of transferred IVP embryos, and, to date, successful production of offspring depends on the transfer of large numbers of blastocysts or earlier embryos, to produce large enough litter sizes. The need to transfer relatively large numbers of embryos to achieve even a comparatively low litter size and the lack of stable non-surgical procedures for embryo transfer remain significant obstacles toward the practical implementation of IVP. However, today, the in vitro procedures used to mature oocytes and culture embryos are integral to the production of transgenic pigs by SCNT (Grupen 2014) or gene editing (Niemann and Petersen 2016).

12.3.1.4 Small Ruminants (Sheep and Goat)

There is less research on assisted reproductive technologies including IVP in small ruminants compared to other livestock species. Results on IVP are still unpredictable and variable which is an important limitation for its commercial application (Paramio and Izquierdo 2016, 2014). This large variation is also seen in in vivo embryo production after superovulation treatment. Oocytes are usually recovered from the follicle of females via laparoscopic ovum pickup (LOPU) which is less traumatic than laparotomy.

The first kid born from IVF of an ovulated oocyte was reported in 1985 and the first lamb in 1986. In 1993 the birth of the first kid from an IVM-IVF oocyte and in 1995 the first kid from IVM-IVF and IVC oocytes was announced (Paramio and Izquierdo 2014). Despite all the improvements to increase blastocyst numbers and blastocyst quality by modifying the different components of IVP, results are still unpredictable and variable with significant differences between laboratories and experiments (Paramio and Izquierdo 2016). The most commonly used medium for IVM is tissue culture medium (TCM199). Oocytes are usually cultured for 24–27 h to achieve maturation. Sperm capacitation is obtained by using heparin and estrus sheep serum (ESS) for fresh and frozen buck and ram semen, respectively. IVF is usually carried out in SOF medium in sheep and in Tyrode's albumin lactate pyruvate medium supplemented with hypotaurine in goats. Sperm and oocytes are co-incubated for 16–24 h. At present, ICSI is not very efficient in small ruminants. Chemical activation of oocytes is an essential part of ICSI protocols in sheep, yielding similar cleavage rates in IVF and sperm-injected oocytes (Shirazi et al. 2009). Goat blastocysts have been obtained by ICSI without further chemical activation of the oocytes (Keskintepe et al. 1997). Currently, the most commonly used IVC medium for culturing small ruminant embryos is SOF with amino acids and the addition of 5%–10% of fetal calf serum or BSA.

It has been suggested that the procedures used in bovine IVP can be applied in small ruminants after minimal modifications (Souza-Fabjan et al. 2014).

12.3.1.5 Camelids

The camelid family includes dromedary and Bactrian camels, llamas, alpacas, vicunas, and guanacos. The first two are Old World camelids, whereas the last four are known as New World camelids or South American camelids.

The development of IVP techniques has been slow in South American camelids (SAC). As for other species, a high number of oocytes can be collected from slaughterhouse ovaries. Oocytes can also be obtained from living animals via laparotomy [llama (Trasorras et al. 2009); alpaca (Ratto et al. 2007)] or ovum pickup [llamas (Brogliatti et al. 2000; Berland et al. 2011)]. There are currently no reports on ultrasound-guided transvaginal follicle aspiration in alpacas (Trasorras et al. 2013). IVM conditions for SAC oocytes are similar to those for ruminants. South American camelid semen has very particular characteristics, such as high structural viscosity (Casaretto et al. 2012). This characteristic feature renders semen handling difficult in the laboratory (Tibary and Vaughan 2006), hindering separation of spermatozoa from the seminal plasma and the isolation of motile from immotile sperm (Giuliano et al. 2010). In addition, spermatozoa from these species show oscillatory movements in the ejaculate rather than progressive motility.

Embryos have been produced in vitro using spermatozoa from the epididymis or from ejaculates. The advantages of using epididymal spermatozoa are that these cells are progressively motile and sample management is easier because of the absence of seminal plasma. The rate of development of IVF produced embryos to the blastocyst stage after 5 days of culture is only 10% in llamas. Although pregnancies by ET using in vivo produced embryos (Trasorras et al. 2010) have been generated, no pregnancy has been achieved until now after transcervical transfer of IVP embryos to the uterus of previously synchronized females.

All research done so far proves that it is possible to produce llama embryos in vitro, but it remains necessary to find an adequate culture medium and conditions to favor embryo development to stages that are compatible with establishing pregnancies (Trasorras et al. 2013).

For Old World camelids, the developmental rate up to the blastocyst stage is 20% in dromedary camels. Only in vivo matured/in vitro fertilized and in vivo matured/fertilized produced embryos continued normal development until term and resulted in the birth of normal and healthy live calves (Tibary et al. 2005).

12.4 Quality Assessment of Preimplantation Embryos

The ultimate test of the quality of an embryo is its ability to produce live and healthy offspring after transfer to a recipient. Morphology and the proportion of embryos developing to the blastocyst stage are important criteria to assess developmental competence. Evaluation of embryo morphology remains the method of choice for selecting viable embryos prior to transfer. It is the most practical and clinically useful approach to assess of embryo viability (Van Soom et al. 2003). A bovine embryo grading system developed previously (Lindner and Wright 1983) is, with minor modifications, still widely applied in this field and published in the IETS Manual. However, embryo morphology alone may not accurate enough to serve as the sole criterion for the prediction of embryo developmental potential in vivo.

Identification of the embryo with the highest potential to implant, establish, and maintain a pregnancy is a primary goal in assisted reproduction techniques. Culture by incubation in a time-lapse imaging system does not harm embryos compared with the standard IVP protocol and results in similar overall embryo development (Holm et al. 2002). Morphokinetic embryo analysis by monitoring the changes in embryo morphology over time is by far the most important noninvasive embryo selection tool today and is routinely used in human IVF laboratories (Pribenszky et al. 2017). However, due to its high costs, it has not entered the commercial veterinary field.

Embryo selection is based on methods that can give a direct or indirect clue regarding the potential of a given embryo to implant. These methodologies are either invasive or noninvasive and can be applied at various stages of development from the oocyte to cleavage-stage embryos and up to the blastocyst stage. In view of the shortcomings of invasive embryo selection, the use of noninvasive selection methods seems to be a better strategy toward identification of viable embryos without the risk of possible impacts due to the investigation itself.

Better noninvasive markers and improved techniques for assessing embryo quality are required. These techniques can provide more information on embryo viability (Rocha et al. 2016). For examples, measurement of oxygen consumption using the nanorespirometer (Lopes et al. 2007) and amino acid profiling (Sturmey et al. 2010) can be employed to predict developmental competence and embryo viability of in vitro produced embryos. Although noninvasive approaches are improving, invasive ones have been extremely helpful in finding candidate genes to determine embryo quality (Wrenzycki et al. 2007; Rizos et al. 2008; Graf et al. 2014).

As epigenetic changes can be induced by environmental factors, understanding how vitro production conditions can interfere with these processes is of critical importance. The embryonic epigenome will continue to be an important area of research, especially to gain a better comprehension how the epigenome influences short- and long-term health. A better understanding of the mechanisms and the role of epigenetics during early embryogenesis will likely improve in vitro protocols and ultimately animal health and productivity. Epigenetic remodeling during preimplantation development is complex and dynamic, including changes in DNA methylation and histone modifications that occur both on a global scale but also differentially at specific loci. Uncovering the bases of these mechanisms will improve our understanding of early development and are promising for improving animal fertility and diagnosing and treating infertility problems (Urrego et al. 2014; Canovas and Ross 2016).

Despite the large number of publications in the field, it is still a long way to better understand and manipulate the mechanisms controlling oocyte maturation and early embryonic development. Overcoming these gaps may allow us to improve ART results. In particular, studies aiming at finding effective biomarkers for oocyte and embryo competence are urgently needed. The field of OMICs seems to be quite promising, especially regarding new findings in transcriptomics, proteomics, and lipidomics in oocytes, CCs, embryos and EVs within the follicular fluid (Krisher et al. 2015; Scott and Treff 2010; Benkhalifa et al. 2015). Future studies on this subject might enable the design of more complex, defined, and efficient culture conditions for oocytes and embryos.

12.5   Challenges and Future Developments

12.5.1 Culture of Post-Hatching Stages

Post-hatching bovine embryonic development in vitro would allow for the establishment of better tools for evaluating developmental potential without the need for transfer to recipient animals. In vitro development of bovine embryos beyond hatching has been reported with the establishment of the hypoblast and contemporary signs of shedding of the polar trophoblast, i.e., Rauber's layer. Moreover, elongation may be achieved upon physical constraints in agar tunnels (Brandao et al. 2004). However, further development of the epiblast is compromised, and embryonic disk formation is lacking (Vejlsted et al. 2006). Further studies are needed to determine the appropriate physical, chemical, and hormonal environment required to pass through this block of development and reach subsequent developmental stages in vitro.

In humans, an in vitro system to culture embryos through implantation stages in the absence of maternal tissues has been established (Shahbazi et al. 2016; Deglincerti et al. 2016). In contrast to cattle where implantation takes place around day 20, the human embryo must implant into the uterus of the mother to survive at the seventh day of development.

12.5.2 Genomic Selection (GS)

In the last decade, significant advances in molecular genetics and bioinformatics have made it possible to establish genomic selection as a new tool to increase the genetic gain in animal breeding (Stock and Reents 2013). The success of genomics in animal breeding begins with the use of whole-genome information (Meuwissen et al. 2001).

A chip was developed already in 2007 that allows to genotype >54,000 single nucleotide polymorphisms (SNP) simultaneously. These markers represented only a small proportion of all discovered SNPs, but they were highly polymorphic in several breeds and evenly spaced over the genome. This chip has immediately been used to genotype existing progeny-tested bulls. With these first reference populations, genomic breeding values (GEBV) were accurate enough to rapidly replace progeny testing. They were made official in 2009 in different countries, allowing the dissemination of semen of young bulls with genomic evaluation. This revolutionized selection (Boichard et al. 2016). In order to decrease genotyping costs, a low-density chip was designed with good imputation accuracy (Boichard et al. 2012), i.e., with excellent prediction of missing markers.

Since the introduction of OPU/IVP, substantial efforts have been made to improve embryo production efficiency. GS provides new tools for improving the efficiency of OPU/IVP programs by selecting donors with high in vitro production results. Heritability for qualitative traits (quality of oocytes, cleavage, and developmental rates) seems to be lower than for quantitative traits (total number of oocytes, number of embryos (Merton et al. 2009)). Different components of reproductive biotechnologies have already been associated with SNPs in cattle and could be added as new traits to improve fertility as well as efficiency of ARTs in donor cows. In particular, SNPs associated with the number of viable oocytes, fertilization, cleavage, and developmental rates have been recently highlighted (Ponsart et al. 2014).

Furthermore, GEBV can already be determined from preimplantation embryo biopsies. Taking embryo biopsies requires highly skilled and trained operators and specific equipment, such as an inverted microscope combined with micromanipulators. Three methods, the microblade biopsy, the aspiration biopsy, and the needle technique, have been described (Cenariu et al. 2012). Usually, five to ten blastomeres are collected as shown in Fig. 12.6. As they do not contain enough DNA for SNP chip analyses, a DNA amplification step (whole-genome amplification, WGA) has to be included. The mean call rate (proportion of called markers) from WGA of genomic DNA stemming from an embryo biopsy should be higher than 85%

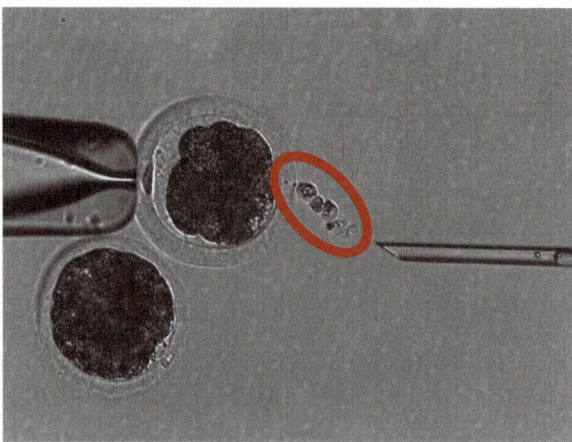

Fig. 12.6 Biopsy taken from a bovine morula (red circle, biopsied blastomeres)

ensuring a low amount of allele drop out (ADO) and replication errors. Another bottleneck for genotyping techniques combined with IVP remains the cryopreservation of IVP biopsied embryos (Ponsart et al. 2014). To overcome this problem, blastocentesis (collection of blastocoel fluid) which is less invasive might be performed as already done in horses (Herrera et al. 2015) and humans (Gianaroli et al. 2014). Furthermore, instead of freezing, biopsied embryos could undergo liquid preservation (Ideta et al. 2013).

Taken together, GS has revolutionized cattle breeding schemes. Strategies to genotype embryos for multiple markers were developed and combined with embryo biopsy.

12.5.3 Exosomes/miRNAs

Recently, a new mechanism of cell communication within the ovarian follicle has been discovered which is performed by extracellular vesicles (EVs). Initially, EVs were described in ovarian follicular fluid of mares (da Silveira et al. 2012). These EVs are lipid bilayer structures secreted by many cell types into the extracellular fluid, serving as a vehicle for membrane and cytosolic proteins, lipids, and RNA (Raposo and Stoorvogel 2013)). Several articles identified miRNAs in bovine (Miles et al. 2012), equine (da Silveira et al. 2012), and human (Santonocito et al. 2014) follicular fluid, suggesting EVs could be a potential mediator of cell-to-cell communication, impacting oocyte and follicle growth (da Silveira et al. 2015). Profiles of miRNAs isolated from EVs present in follicular fluid were described and associated with proper cytoplasmic oocyte maturation; hence, these miRNA profiles can be used to predict oocyte competence (Sohel et al. 2013). The presence of extracellular miRNAs in various biological fluids, including follicular and seminal fluid, their stability, and the relatively easy procedures required to obtain these samples make miRNAs excellent candidates for the use of biomarkers of female and male reproduction and fertility (Pratt and Calcatera 2016; Tesfaye et al. 2016).

The well-orchestrated expression of genes that are derived from the maternal and/or embryonic genome is required for the onset and maintenance of distinct morphological changes during the embryonic development. Optimum regulation of genes or critical gene regulatory events in favor of early embryonic development has been shown to be under control of miRNAs (Hossain et al. 2012). Recently, it has been shown that the addition of extracellular vesicles from oviductal fluid from the isthmus to in vitro culture of bovine embryos in the absence of serum improves development and quality of the embryos (Alminana et al. 2017; Lopera-Vasquez et al. 2017; Pavani et al. 2016).

12.5.4 Microfluidics

During IVP gametes and embryos are exposed to changes in pH, osmolarity, and mechanical stress, which can interfere with successful blastocyst production. Most of the research efforts to reduce these stress factors have been focused on the media rather than finding systems that could reduce the stress introduced by the operator. One possible solution to reduce many of these types of stress on gametes/embryos may be the application of micro- and nanotechnologies, particularly microfluidic technologies. Microfluidic technologies first emerged in the late 1980s and early 1990s (Masuda et al. 1987). The full potential of microfluidic technology has yet to be realized for ARTs in livestock. Major advances in automation and robotics combined with microfluidics have the potential to revolutionize livestock IVP and animal breeding. New technologies in 3D printing may improve our ability to develop physical systems that closely mimic the in vivo environment. Sophisticated pumping and fluid handling methods will allow for the alteration of the fluid milieu that surrounds the gametes and embryos. Subtle changes in media composition delivered at specific time points are achievable using microfluidic devices (Wheeler and Rubessa 2017).

12.5.5 Follicle Culture

For bovine and other domestic species, the development of culture systems capable of supporting the growth of immature follicles to a stage where they could be matured and the oocyte fertilized would ensure a constantly large supply of oocytes for experimental and applied purposes. Development of a successful culture system for preantral follicles with immature oocytes is dependent upon efficient procedures to recover the follicles from the ovary and culture them as well. Basically, there are two ways to culture bovine preantral follicles: (1) enclosed in ovarian tissue fragments (slices or strips), also called "in situ," or (2) using isolated follicles (Araujo et al. 2014). Significant advances have been described, and a major achievement was the in vitro development of secondary follicles up to early antral stages and the production of very low number of embryos (Silva et al. 2016). Although antrum formation was reached, culture of bovine secondary follicles did not lead to the production of meiotically competent oocytes (Gutierrez et al. 2000; McLaughlin and Telfer 2010). A two-step culture system for bovine preantral follicles has been described (McLaughlin and Telfer 2010).

The growth of primordial follicles up to maturation in domestic species is a long process, and a better understanding of the physiological and pharmacological requirements of the various stages of follicle development is inevitable. This technique will be of great value for experimental or diagnostic purposes. Profound similarities in the dynamics of follicle development exist between the menstrual cycle in women and the estrous cycle in cattle and horses (Ginther et al. 2004). In this regard, research using animal models for studying human ovarian function is important to provide a hypothetical basis for further studies in women (Baerwald et al. 2009) as is for the entire IVP procedure (Menezo and Herubel 2002; Santos et al. 2014; Langbeen et al. 2015; Wrenzycki et al. 2007). Further characteristics of early embryo biology are shown in Table 12.3 indicating that bovine and human early development is remarkably similar.

Table 12.3 Comparative aspects of early embryonic development

	Human	Cattle
Oocyte diameter (μm)	150–180	150–180
Time (h) to reach		
2-cell stage	30	36
Blastocyst stage	120	150
Hatching stage	150	200
Stage of EGA[a]	4-cell	8-cell

[a]Embryonic genome activation

12.5.6 Artificial Gametes

Artificial sperm and artificial oocytes generated from pluripotent germline stem cells (GSCs), embryonic stem cells (ESCs), or induced pluripotent stem cells (iPSCs) have resulted in the birth of viable offspring. Artificial sperm and artificial oocytes have also been generated from somatic cells directly, i.e., without documentation of intermediate stages of stem or germ cell development or (epi)genetic status. Albeit embryos showed reduced development, haploidization by transplantation of a somatic cell nucleus into an enucleated donor oocyte has led to fertilized artificial oocytes (Hendriks et al. 2015). Complete restoration of gametogenesis in culture will be important to our understanding of biological events at the cellular and molecular levels that are important for germline development. Understanding these processes will have enormous impact on applications in the biomedical and agricultural communities (Zeng et al. 2015). Fully potent mature oocytes were generated in culture from embryonic stem cells and from induced pluripotent stem cells derived from both embryonic fibroblasts and adult tail tip fibroblasts. Moreover, pluripotent stem cell lines were re-derived from the eggs that were generated in vitro, thereby reconstituting the full female germline cycle in a dish. This culture system will provide a platform for elucidating the molecular mechanisms underlying totipotency and the production of oocytes of other mammalian species in culture (Hikabe et al. 2016).

12.5.7 Mitochondrial Transfer

A promising advancement in human ART involves the replacement of mutant mtDNA in unfertilized oocytes or zygotes by healthy donor mitochondria, thereby allowing women carrying mtDNA mutations to circumvent passage of the condition to their children (Craven et al. 2010). Two microsurgical nuclear transfer procedures termed spindle transfer (ST) and pronuclear transfer (PNT) have been developed. The first approach is conducted at the MII oocyte stage. The spindle is isolated and transplanted into the cytoplasm of a donated unfertilized oocyte that, itself, has been enucleated. The reconstructed oocyte, now free of mutated mtDNA, can be fertilized and subsequently transplanted to the patient (Mitalipov and Wolf 2014). The first baby from this approach has been born in 2016 (Zhang et al. 2017). Human pronuclear transfer has also been reported (Craven et al. 2010; Zhang et al. 2016). Potential concerns with pronuclear and spindle transfer include the impact these procedures might have on the risk of mtDNA carryover during karyoplast transfer, embryogenesis, epigenetics, and genome integrity (Craven et al. 2011). These issues need to be carefully assessed more basic in-depth research (Reznichenko et al. 2016).

12.5.8 Gene Editing

Molecular scissors (MS), including zinc finger nucleases (ZFN), transcription activator-like endonucleases (TALENS), and meganucleases, possess long recognition sites and are thus capable of cutting DNA in a very specific manner. These molecular scissors mediate targeted genetic alterations by enhancing the DNA mutation rate via induction of double-strand breaks at a predetermined genomic site. Compared to conventional homologous recombination-based gene targeting, MS can increase the targeting rate 10,000-fold, and gene disruption via mutagenic DNA repair is stimulated at a similar frequency. The successful application of different MS has been shown in mammals, including humans (Petersen and Niemann 2015). Recently, another novel class of molecular scissors was described that uses RNAs to target a specific genomic site. The CRISPR/Cas9 system is capable of targeting even multiple genomic sites in one shot and thus could be superior to ZFNs or TALEN, especially by its rather simple design. However, careful sequencing of the targeted locus is required for all mutagenesis projects, including the ones by CRISPR/Cas9 (Mianne et al. 2017). Confirmation of the desired on-target mutation and the detection of off-target events is of utmost importance (Zischewski et al. 2017).

The ability to generate gene knockouts is a powerful tool for analysis of gene function and for the generation of animals with novel biotechnological or breeding applications (Fahrenkrug et al. 2010). In livestock species this process traditionally involves the generation of a knockout cell line by utilizing homologous recombination followed by somatic cell nuclear transfer (SCNT). This remains the method of choice for many applications (Kurome et al. 2013); however, application of SCNT strategies requires a high level of technical expertise, a reliable supply of oocytes, and a large recipient herd, features not available in many areas where gene editing might have the greatest impact (Proudfoot et al. 2015). However, in vitro produced embryos can be used for rapid introgression of gene edits into defined populations.

Recently, CRISPR/Cas9-mediated gene editing has been applied to human zygotes (Liang et al. 2015; Kang et al. 2016), and the correction of a pathogenic gene mutation has been reported in 2017 (Ma et al. 2017). Efficiency, accuracy, and safety of the approach suggest that it has the potential to be used for the correction of heritable mutations in human embryos by complementing preimplantation genetic diagnosis. However, much remains to be investigated before clinical applications, including the reproducibility of the technique (Ma et al. 2017).

Conclusion

In vitro production (IVP) of livestock embryos follows a well-developed procedure that is commercially available for most species. Albeit all the improvements in oocyte and embryo culture, at best only 30–35% of immature bovine COC develop to the blastocyst stage. Nevertheless in light of the underlying ovarian mechanisms with the high degree of atresia, this represents a reasonable efficiency. But the in vivo situation cannot yet be mimicked sufficiently well. The quality of the embryos produced is still impaired in comparison with their in vivo-derived counterparts. This suggests that there are still improvements to be made in increasing oocyte and embryo developmental competence. More basic research is needed to unravel the molecular mechanisms, e.g., epigenetic reprogramming during early embryonic development as well as detailed studies on the composition and interactions of culture media. By altering the conditions of oocyte maturation and embryo culture, respectively, to mirror more closely the in vivo conditions, it may be possible to produce not only more blastocysts but, more importantly, blastocysts of better quality.

References

Acosta TJ (2007) Studies of follicular vascularity associated with follicle selection and ovulation in cattle. J Reprod Develop 53(1):39–44. https://doi.org/10.1262/Jrd.18153

Acosta TJ, Hayashi KG, Ohtani M, Miyamoto A (2003) Local changes in blood flow within the preovulatory follicle wall and early corpus luteum in cows. Reproduction 125(5):759–767

Aguila L, Zambrano F, Arias ME, Felmer R (2017) Sperm capacitation pretreatment positively impacts bovine intracytoplasmic sperm injection. Mol Reprod Dev 84(7):649–659. https://doi.org/10.1002/mrd.22834

Albuz FK, Sasseville M, Lane M, Armstrong DT, Thompson JG, Gilchrist RB (2010) Simulated physiological oocyte maturation (SPOM): a novel in vitro maturation system that substantially improves embryo yield and pregnancy outcomes. Hum Reprod 25(12):2999–3011. https://doi.org/10.1093/humrep/deq246

Algriany O, Bevers M, Schoevers E, Colenbrander B, Dieleman S (2004) Follicle size-dependent effects of sow follicular fluid on in vitro cumulus expansion, nuclear maturation and blastocyst formation of sow cumulus oocytes complexes. Theriogenology 62(8):1483–1497. https://doi.org/10.1016/j.theriogenology.2004.02.008

Alminana C, Corbin E, Tsikis G, Alcantara-Neto AS, Labas V, Reynaud K, Galio L, Uzbekov R, Garanina AS, Druart X, Mermillod P (2017) Oviduct extracellular vesicles protein content and their role during oviduct-embryo cross-talk. Reproduction 154(3):153–168. https://doi.org/10.1530/REP-17-0054

Araujo VR, Gastal MO, Figueiredo JR, Gastal EL (2014) In vitro culture of bovine preantral follicles: a review. Reprod Biol Endocrinol 12:78. https://doi.org/10.1186/1477-7827-12-78

Arias ME, Sanchez R, Risopatron J, Perez L, Felmer R (2014) Effect of sperm pretreatment with sodium hydroxide and dithiothreitol on the efficiency of bovine intracytoplasmic sperm injection. Reprod Fertil Dev 26(6):847–854. https://doi.org/10.1071/RD13009

Baerwald AR, Walker RA, Pierson RA (2009) Growth rates of ovarian follicles during natural menstrual cycles, oral contraception cycles, and ovarian stimulation cycles. Fertil Steril 91(2):440–449. https://doi.org/10.1016/j.fertnstert.2007.11.054

Baltz JM (2012) Media composition: salts and osmolality. Methods Mol Biol 912:61–80. https://doi.org/10.1007/978-1-61779-971-6_5

Bender K, Walsh S, Evans ACO, Fair T, Brennan L (2010) Metabolite concentrations in follicular fluid may explain differences in fertility between heifers and lactating cows. Reproduction 139(6):1047–1055. https://doi.org/10.1530/Rep-10-0068

Benkhalifa M, Madkour A, Louanjli N, Bouamoud N, Saadani B, Kaarouch I, Chahine H, Sefrioui O, Merviel P, Copin H (2015) From global proteome profiling to single targeted molecules of follicular fluid and oocyte: contribution to embryo development and IVF outcome. Expert Rev Proteomics 12(4):407–423. https://doi.org/10.1586/14789450.2015.1056782

Berland MA, von Baer A, Ruiz J, Parraguez VH, Morales P, Adams GP, Ratto MH (2011) In vitro fertilization and development of cumulus oocytes complexes collected by ultrasound-guided follicle aspiration in superstimulated llamas. Theriogenology 75(8):1482–1488. https://doi.org/10.1016/j.theriogenology.2010.11.047

Besenfelder U, Havlicek V, Brem G (2012) Role of the oviduct in early embryo development. Reprod Domest Anim 47(Suppl 4):156–163. https://doi.org/10.1111/j.1439-0531.2012.02070.x

Betteridge KJ, Eaglesome MD, Mitchell D, Flood PF, Beriault R (1982) Development of horse embryos up to 22 days after ovulation–observations on fresh specimens. J Anat 135(Aug):191–209

Blockeel C, Mock P, Verheyen G, Bouche N, Le Goff P, Heyman Y, Wrenzycki C, Hoffmann K, Niemann H, Haentjens P, de Los Santos MJ, Fernandez-Sanchez M, Velasco M, Aebischer P, Devroey P, Simon C (2009) An in vivo culture system for human embryos using an encapsulation technology: a pilot study. Hum Reprod 24(4):790–796. https://doi.org/10.1093/humrep/dep005

Boichard D, Chung H, Dassonneville R, David X, Eggen A, Fritz S, Gietzen KJ, Hayes BJ, Lawley CT, Sonstegard TS, Van Tassell CP, VanRaden PM, Viaud-Martinez KA, Wiggans GR, Bovine LDC (2012) Design of a bovine low-density SNP array optimized for imputation. PLoS One 7(3):e34130. https://doi.org/10.1371/journal.pone.0034130

Boichard D, Ducrocq V, Croiseau P, Fritz S (2016) Genomic selection in domestic animals: principles, applications and perspectives. C R Biol 339(7–8):274–277. https://doi.org/10.1016/j.crvi.2016.04.007

Brackett BG, Bousquet D, Boice ML, Donawick WJ, Evans JF, Dressel MA (1982) Normal development following invitro fertilization in the cow. Biol Reprod 27(1):147–158. https://doi.org/10.1095/biolreprod27.1.147

Brackett BG, Oliphant G (1975) Capacitation of rabbit spermatozoa Invitro. Biol Reprod 12(2):260–274. https://doi.org/10.1095/biolreprod12.2.260

Brandao DO, Maddox-Hyttel P, Lovendahl P, Rumpf R, Stringfellow D, Callesen H (2004) Post hatching development: a novel system for extended in vitro culture of bovine embryos. Biol Reprod 71(6):2048–2055. https://doi.org/10.1095/biolreprod.103.025916

Brannstrom M, Zackrisson U, Hagstrom HG, Josefsson B, Hellberg P, Granberg S, Collins WP, Bourne T (1998) Preovulatory changes of blood flow in different regions of the human follicle. Fertil Steril 69(3):435–442

Breitbart H, Shalev Y, Marcus S, Shemesh M (1995) Modulation of prostaglandin synthesis in mammalian sperm acrosome reaction. Hum Reprod 10(8):2079–2084

Brogliatti GM, Palasz AT, Rodriguez-Martinez H, Mapletoft RJ, Adams GP (2000) Transvaginal collection and ultrastructure of llama (Lama glama) oocytes. Theriogenology 54(8):1269–1279. https://doi.org/10.1016/S0093-691x(00)00433-7

Canovas S, Ross PJ (2016) Epigenetics in preimplantation mammalian development. Theriogenology 86(1):69–79. https://doi.org/10.1016/j.theriogenology.2016.04.020

Carnevale EM, da Silva MAC, Panzani D, Stokes JE, Squires EL (2005) Factors affecting the success of oocyte transfer in a clinical program for subfertile mares. Theriogenology 64(3):519–527. https://doi.org/10.1016/j.theriogenology.2005.05.008

Carvalho NA, Baruselli PS, Zicarelli L, Madureira EH, Visintin JA, D'Occhio MJ (2002) Control of ovulation with a GnRH agonist after superstimulation of follicular growth in buffalo: fertilization and embryo recovery. Theriogenology 58(9):1641–1650

Casaretto C, Sarrasague MM, Giuliano S, de Celis ER, Gambarotta M, Carretero I, Miragaya M (2012) Evaluation of Lama glama semen viscosity with a cone-plate rotational viscometer. Andrologia 44:335–341. https://doi.org/10.1111/j.1439-0272.2011.01186.x

Catt JW, Rhodes SL (1995) Comparative intracytoplasmic sperm injection (ICSI) in human and domestic species. Reprod Fertil Dev 7(2):161–166 discussion 167

Cenariu M, Pall E, Cernea C, Groza I (2012) Evaluation of bovine embryo biopsy techniques according to their ability to preserve embryo viability. J Biomed Biotechnol 2012:541384. https://doi.org/10.1155/2012/541384

Chastant-Maillard S, Quinton H, Lauffenburger J, Cordonnier-Lefort N, Richard C, Marchal J, Mormede P, Renard JP (2003) Consequences of transvaginal follicular puncture on Well-being in cows. Reproduction 125(4):555–563

Chen S, Palma-Vera SE, Langhammer M, Galuska SP, Braun BC, Krause E, Lucas-Hahn A, Schoen J (2017) An air-liquid interphase approach for modeling the early embryo-maternal contact zone. Sci Rep 7:42298. https://doi.org/10.1038/srep42298

Chen SH, Seidel GE (1997) Effects of oocyte activation and treatment of spermatozoa on embryonic development following intracytoplasmic sperm injection in cattle. Theriogenology 48(8):1265–1273. https://doi.org/10.1016/S0093-691x(97)00369-5

Cognie Y, Crozet N, Guerin Y, Poulin N, Bezard J, Duchamp G, Magistrini M, Palmer E (1992) Invitro fertilization in ovine, caprine and equine species. Ann Zootech 41(3–4):353–359. https://doi.org/10.1051/animres:19920316

Colleau JJ (1991) Using embryo sexing within closed mixed multiple ovulation and embryo transfer schemes for selection on dairy-cattle. J Dairy Sci 74(11):3973–3984

Coulam CB, Goodman C, Rinehart JS (1999) Colour Doppler indices of follicular blood flow as predictors of pregnancy after in-vitro fertilization and embryo transfer. Hum Reprod 14(8):1979–1982

Craven L, Elson JL, Irving L, Tuppen HA, Lister LM, Greggains GD, Byerley S, Murdoch AP, Herbert M, Turnbull D (2011) Mitochondrial DNA disease: new options for prevention. Hum Mol Genet 20(R2):R168–R174. https://doi.org/10.1093/hmg/ddr373

Craven L, Tuppen HA, Greggains GD, Harbottle SJ, Murphy JL, Cree LM, Murdoch AP, Chinnery PF, Taylor RW, Lightowlers RN, Herbert M, Turnbull DM (2010) Pronuclear transfer in human embryos to prevent transmission of mitochondrial DNA disease. Nature 465(7294):82–85. https://doi.org/10.1038/nature08958

da Silveira JC, de Andrade GM, Nogueira MF, Meirelles FV, Perecin F (2015) Involvement of miRNAs and cell-secreted vesicles in mammalian ovarian Antral follicle development. Reprod Sci 22(12):1474–1483. https://doi.org/10.1177/1933719115574344

Deglincerti A, Croft GF, Pietila LN, Zernicka-Goetz M, Siggia ED, Brivanlou AH (2016) Self-organization of the in vitro attached human embryo. Nature 533(7602):251–254. https://doi.org/10.1038/nature17948

Diskin MG, Sreenan JM (1980) Fertilization and embryonic mortality-rates in beef heifers after artificial-insemination. J Reprod Fertil 59(2):463–468

Dumesic DA, Meldrum DR, Katz-Jaffe MG, Krisher RL, Schoolcraft WB (2015) Oocyte environment: follicular fluid and cumulus cells are critical for oocyte health. Fertil Steril 103(2):303–316. https://doi.org/10.1016/j.fertnstert.2014.11.015

Duranthon V, Renard JP (2001) The developmental competence of mammalian oocytes: a convenient but biologically fuzzy concept. Theriogenology 55(6):1277–1289

Fahrenkrug SC, Blake A, Carlson DF, Doran T, Van Eenennaam A, Faber D, Galli C, Gao Q, Hackett PB, Li N, Maga EA, Muir WM, Murray JD, Shi D, Stotish R, Sullivan E, Taylor JF, Walton M, Wheeler M, Whitelaw B, Glenn BP (2010) Precision genetics for complex objectives in animal agriculture. J Anim Sci 88(7):2530–2539. https://doi.org/10.2527/jas.2010-2847

Fair T (2010) Mammalian oocyte development: checkpoints for competence. Reprod Fert Develop 22(1):13–20. https://doi.org/10.1071/RD09216

Fair T, Hyttel P, Greve T (1995) Bovine oocyte diameter in relation to maturational competence and transcriptional activity. Mol Reprod Dev 42(4):437–442. https://doi.org/10.1002/mrd.1080420410

Fukuda Y, Ichikawa M, Naito K, Toyoda Y (1990) Birth of normal calves resulting from bovine oocytes matured, fertilized, and cultured with cumulus cells-invitro up to the blastocyst stage. Biol Reprod 42(1):114–119. https://doi.org/10.1095/biolreprod42.1.114

Funahashi H, Day BN (1993) Effects of the duration of exposure to hormone supplements on cytoplasmic maturation of pig oocytes invitro. J Reprod Fertil 98(1):179–185

Galli C, Colleoni S, Duchi R, Lagutina I, Lazzari G (2007) Developmental competence of equine oocytes and embryos obtained by in vitro procedures ranging from in vitro maturation and ICSI to embryo culture, cryopreservation and somatic cell nuclear transfer. Anim Reprod Sci 98(1–2):39–55. https://doi.org/10.1016/j.anireprosci.2006.10.011

Galli C, Duchi R, Colleoni S, Lagutina I, Lazzari G (2014) Ovum pick up, intracytoplasmic sperm injection and somatic cell nuclear transfer in cattle, buffalo and horses: from the research laboratory to clinical practice. Theriogenology 81(1):138–151. https://doi.org/10.1016/j.theriogenology.2013.09.008

Galli C, Vassiliev I, Lagutina I, Galli A, Lazzari G (2003) Bovine embryo development following ICSI: effect of activation, sperm capacitation and pre-treatment with dithiothreitol. Theriogenology 60(8):1467–1480

Gardner DK (2008) Dissection of culture media for embryos: the most important and less important components and characteristics. Reprod Fertil Dev 20(1):9–18

Gianaroli L, Magli MC, Pomante A, Crivello AM, Cafueri G, Valerio M, Ferraretti AP (2014) Blastocentesis: a source of DNA for preimplantation genetic testing. Results from a pilot study. Fertil Steril 102(6):1692–1699 e1696. https://doi.org/10.1016/j.fertnstert.2014.08.021

Gil MA, Cuello C, Parrilla I, Vazquez JM, Roca J, Martinez EA (2010) Advances in swine in vitro embryo production technologies. Reprod Domest Anim 45:40–48. https://doi.org/10.1111/j.1439-0531.2010.01623.x

Gilchrist RB (2011) Recent insights into oocyte-follicle cell interactions provide opportunities for the development of new approaches to in vitro maturation. Reprod Fert Develop 23(1):23–31

Gilchrist RB, Zeng HT, Wang X, Richani D, Smitz J, Thompson JG (2015) Reevaluation and evolution of the simulated physiological oocyte maturation system. Theriogenology 84(4):656–657. https://doi.org/10.1016/j.theriogenology.2015.03.032

Ginther OJ, Gastal EL, Gastal MO, Bergfelt DR, Baerwald AR, Pierson RA (2004) Comparative study of the dynamics of follicular waves in mares and women. Biol Reprod 71(4):1195–1201. https://doi.org/10.1095/biolreprod.104.031054

Giuliano S, Carretero M, Gambarotta M, Neild D, Trasorras V, Pinto M, Miragaya M (2010) Improvement of llama (Lama glama) seminal characteristics using collagenase. Anim Reprod Sci 118(1):98–102. https://doi.org/10.1016/j.anireprosci.2009.06.005

Gloria A, Carluccio A, Wegher L, Robbe D, Befacchia G, Contri A (2016) Single and double layer centrifugation improve the quality of cryopreserved bovine sperm from poor quality ejaculates. J Anim Sci Biotech 7:30. https://doi.org/10.1186/s40104-016-0088-6

Goodla L, Morrell JM, Yusnizar Y, Stalhammar H, Johannisson A (2014) Quality of bull spermatozoa after preparation by single-layer centrifugation. J Dairy Sci 97(4):2204–2212. https://doi.org/10.3168/jds.2013-7607

Goto K, Yanagita K (1995) Normality of calves obtained by intracytoplasmic sperm injection. Hum Reprod 10(6):1554

Graf A, Krebs S, Heininen-Brown M, Zakhartchenko V, Blum H, Wolf E (2014) Genome activation in bovine embryos: review of the literature and new insights from RNA sequencing experiments. Anim Reprod Sci 149(1–2):46–58. https://doi.org/10.1016/j.anireprosci.2014.05.016

Grupen CG (2014) The evolution of porcine embryo in vitro production. Theriogenology 81(1):24–37. https://doi.org/10.1016/j.theriogenology.2013.09.022

Gutierrez CG, Ralph JH, Telfer EE, Wilmut I, Webb R (2000) Growth and antrum formation of bovine preantral follicles in long-term culture in vitro. Biol Reprod 62(5):1322–1328

Hanstedt A, Wilkening S, Bruning K, Honnens A, Wrenzycki C (2010) Effect of perifollicular blood flow on the quality of oocytes collected during repeated opu sessions. Reprod Fert Develop 22(1):223–223

Hendriks S, Dancet EA, van Pelt AM, Hamer G, Repping S (2015) Artificial gametes: a systematic review of biological progress towards clinical application. Hum Reprod Update 21(3):285–296. https://doi.org/10.1093/humupd/dmv001

Herrera C, Morikawa MI, Castex CB, Pinto MR, Ortega N, Fanti T, Garaguso R, Franco MJ, Castanares M, Castaneira C, Losinno L, Miragaya MH, Mutto AA (2015) Blastocele fluid from in vitro- and in vivo-produced equine embryos contains nuclear DNA. Theriogenology 83(3):415–420. https://doi.org/10.1016/j.theriogenology.2014.10.006

Herrera C, Revora M, Vivani L, Miragaya MH, Quintans C, Pasqualini RS, Losinno L (2008) In vitro production of equine embryos from young and old mares by intracytoplasmic sperm injection. Reprod Fert Develop 20(1):145. https://doi.org/10.1071/Rdv20n1ab129

Hikabe O, Hamazaki N, Nagamatsu G, Obata Y, Hirao Y, Hamada N, Shimamoto S, Imamura T, Nakashima K, Saitou M, Hayashi K (2016) Reconstitution in vitro of the entire cycle of the mouse female germ line. Nature 539(7628):299–303. https://doi.org/10.1038/nature20104

Hinrichs K (2005) Update on equine ICSI and cloning. Theriogenology 64(3):535–541. https://doi.org/10.1016/j.theriogenology.2005.05.010

Hinrichs K, Digiorgio LM (1991) Embryonic-development after intra-follicular transfer of horse oocytes. J Reprod Fertil 44:369–374

Hinrichs K, Williams KA (1997) Relationships among oocyte-cumulus morphology, follicular atresia, initial chromatin configuration, and oocyte meiotic competence in the horse. Biol Reprod 57(2):377–384. https://doi.org/10.1095/biolreprod57.2.377

Holm P, Booth PJ, Callesen H (2002) Kinetics of early in vitro development of bovine in vivo- and in vitro-derived zygotes produced and/or cultured in chemically defined or serum-containing media. Reproduction 123(4):553–565

Holm P, Booth PJ, Schmidt MH, Greve T, Callesen H (1999) High bovine blastocyst development in a static in vitro production system using SOFaa medium supplemented with sodium citrate and myo-inositol with or without serum-proteins. Theriogenology 52(4):683–700. https://doi.org/10.1016/S0093-691x(99)00162-4

Hossain MM, Salilew-Wondim D, Schellander K, Tesfaye D (2012) The role of microRNAs in mammalian oocytes and embryos. Anim Reprod Sci 134(1–2):36–44. https://doi.org/10.1016/j.anireprosci.2012.08.009

Hu J, Ma X, Bao JC, Li W, Cheng D, Gao Z, Lei A, Yang C, Wang H (2011) Insulin-transferrin-selenium (ITS) improves maturation of porcine oocytes in vitro. Zygote 19(3):191–197. https://doi.org/10.1017/S0967199410000663

Huey S, Abuhamad A, Barroso G, Hsu MI, Kolm P, Mayer J, Oehninger S (1999) Perifollicular blood flow Doppler indices, but not follicular pO2, pCO2, or pH, predict oocyte developmental competence in in vitro fertilization. Fertil Steril 72(1):707–712

Hunter RHF (1996) Ovarian control of very low sperm/egg ratios at the commencement of mammalian fertilisation to avoid polyspermy. Mol Reprod Dev 44(3):417–422

Ideta A, Aoyagi Y, Tsuchiya K, Kamijima T, Nishimiya Y, Tsuda S (2013) A simple medium enables bovine embryos to be held for seven days at 4 degrees C. Sci Rep 3:1173. https://doi.org/10.1038/srep01173

Kang X, He W, Huang Y, Yu Q, Chen Y, Gao X, Sun X, Fan Y (2016) Introducing precise genetic modifications into human 3PN embryos by CRISPR/Cas-mediated genome editing. J Assist Reprod Genet 33(5):581–588. https://doi.org/10.1007/s10815-016-0710-8

Kassens A, Held E, Salilew-Wondim D, Sieme H, Wrenzycki C, Tesfaye D, Schellander K, Hoelker M (2015) Intrafollicular oocyte transfer (IFOT) of abattoir-derived and in vitro-matured oocytes results in viable blastocysts and birth of healthy calves. Biol Reprod 92(6):150. https://doi.org/10.1095/biolreprod.114.124883

Kenngott RA, Sinowatz F (2007) Prenatal development of the bovine oviduct. Anat Histol Embryol 36(4):272–283. https://doi.org/10.1111/j.1439-0264.2006.00762.x

Keskintepe L, Brackett BG (1996) In vitro developmental competence of in vitro-matured bovine oocytes fertilized and cultured in completely defined media. Biol Reprod 55(2):333–339

Keskintepe L, Burnley CL, Brackett BG (1995) Production of viable bovine blastocysts in defined in-vitro conditions. Biol Reprod 52(6):1410–1417. https://doi.org/10.1095/biolreprod52.6.1410

Keskintepe L, Morton PC, Smith SE, Tucker MJ, Simplicio AA, Brackett BG (1997) Caprine blastocyst formation following intracytoplasmic sperm injection and defined culture. Zygote 5(3):261–265

Khazaei M, Aghaz F (2017) Reactive oxygen species generation and use of antioxidants during in vitro maturation of oocytes. Int J Fertil Steril 11(2):63–70. https://doi.org/10.22074/ijfs.2017.4995

Krisher RL (2013) In vivo and in vitro environmental effects on mammalian oocyte quality. Annu Rev Anim Biosci 1:393–417. https://doi.org/10.1146/annurev-animal-031412-103647

Krisher RL, Schoolcraft WB, Katz-Jaffe MG (2015) Omics as a window to view embryo viability. Fertil Steril 103(2):333–341. https://doi.org/10.1016/j.fertnstert.2014.12.116

Kruip TA, Boni R, Wurth YA, Roelofsen MW, Pieterse MC (1994) Potential use of ovum pick-up for embryo production and breeding in cattle. Theriogenology 42(4):675–684

Kruip TAM, Pieterse MC, Vanbeneden TH, Vos PLAM, Wurth YA, Taverne MAM (1991) A new method for bovine embryo production–a potential alternative to superovulation. Vet Rec 128(9):208–210

Kurome M, Geistlinger L, Kessler B, Zakhartchenko V, Klymiuk N, Wuensch A, Richter A, Baehr A, Kraehe K, Burkhardt K, Flisikowski K, Flisikowska T, Merkl C, Landmann M, Durkovic M, Tschukes A, Kraner S, Schindelhauer D, Petri T, Kind A, Nagashima H, Schnieke A, Zimmer R, Wolf E (2013) Factors influencing the efficiency of generating genetically engineered pigs by nuclear transfer: multi-factorial analysis of a large data set. BMC Biotechnol 13:43. https://doi.org/10.1186/1472-6750-13-43

Langbeen A, De Porte HF, Bartholomeus E, Leroy JL, Bols PE (2015) Bovine in vitro reproduction models can contribute to the development of (female) fertility preservation strategies. Theriogenology 84(4):477–489. https://doi.org/10.1016/j.theriogenology.2015.04.009

Lazzari G, Colleoni S, Lagutina I, Crotti G, Turini P, Tessaro I, Brunetti D, Duchi R, Galli C (2010) Short-term and long-term effects of embryo culture in the surrogate sheep oviduct versus in vitro culture for different domestic species. Theriogenology 73(6):748–757. https://doi.org/10.1016/j.theriogenology.2009.08.001

Lazzari G, Wrenzycki C, Herrmann D, Duchi R, Kruip T, Niemann H, Galli C (2002) Cellular and molecular deviations in bovine in vitro-produced embryos are related to the large offspring syndrome. Biol Reprod 67(3):767–775

Leemans B, Gadella BM, Stout TAE, Heras S, Smits K, Ferrer-Buitrago M, Claes E, Heindryckx B, De Vos WH, Nelis H, Hoogewijs M, Van Soom A (2015) Procaine induces cytokinesis in horse oocytes via a pH-dependent mechanism. Biol Reprod 93(1):23. https://doi.org/10.1095/biolreprod.114.127423

Leese HJ (2012) Metabolism of the preimplantation embryo: 40 years on. Reproduction 143(4):417–427. https://doi.org/10.1530/REP-11-0484

Liang P, Xu Y, Zhang X, Ding C, Huang R, Zhang Z, Lv J, Xie X, Chen Y, Li Y, Sun Y, Bai Y, Songyang Z, Ma W, Zhou C, Huang J (2015) CRISPR/Cas9-mediated gene editing in human tripronuclear zygotes. Protein Cell 6(5):363–372. https://doi.org/10.1007/s13238-015-0153-5

Lindner GM, Wright RW Jr (1983) Bovine embryo morphology and evaluation. Theriogenology 20(4):407–416

Lonergan P, Fair T (2016) Maturation of oocytes in vitro. Annu Rev Anim Biosci 4(4):255–268. https://doi.org/10.1146/annurev-animal-022114-110822

Lonergan P, Fair T, Corcoran D, Evans AC (2006) Effect of culture environment on gene expression and developmental characteristics in IVF-derived embryos. Theriogenology 65(1):137–152. https://doi.org/10.1016/j.theriogenology.2005.09.028

Lonergan P, Fair T, Forde N, Rizos D (2016) Embryo development in dairy cattle. Theriogenology 86(1):270–277. https://doi.org/10.1016/j.theriogenology.2016.04.040

Looney CR, Lindsey BR, Gonseth CL, Johnson DL (1994) Commercial aspects of oocyte retrieval and in-vitro fertilization (Ivf) for embryo production in problem cows. Theriogenology 41(1):67–72. https://doi.org/10.1016/S0093-691x(05)80050-0

Lopera-Vasquez R, Hamdi M, Maillo V, Gutierrez-Adan A, Bermejo-Alvarez P, Ramirez MA, Yanez-Mo M, Rizos D (2017) Effect of bovine oviductal extracellular vesicles on embryo development and quality in vitro. Reproduction 153(4):461–470. https://doi.org/10.1530/REP-16-0384

Lopes AS, Wrenzycki C, Ramsing NB, Herrmann D, Niemann H, Lovendahl P, Greve T, Callesen H (2007) Respiration rates correlate with mRNA expression of G6PD and GLUT1 genes in individual bovine in vitro-produced blastocysts. Theriogenology 68(2):223–236. https://doi.org/10.1016/j.theriogenology.2007.04.055

Ma H, Marti-Gutierrez N, Park SW, Wu J, Lee Y, Suzuki K, Koski A, Ji D, Hayama T, Ahmed R, Darby H, Van Dyken C, Li Y, Kang E, Park AR, Kim D, Kim ST, Gong J, Gu Y, Xu X, Battaglia D, Krieg SA, Lee DM, Wu DH, Wolf DP, Heitner SB, Belmonte JCI, Amato P, Kim JS, Kaul S, Mitalipov S (2017) Correction of a pathogenic gene mutation in human embryos. Nature 548(7668):413–419. https://doi.org/10.1038/nature23305

Malcuit C, Maserati M, Takahashi Y, Page R, Fissore RA (2006) Intracytoplasmic sperm injection in the bovine induces abnormal [Ca2+]i responses and oocyte activation. Reprod Fertil Dev 18(1–2):39–51

Mari G, Barbara M, Eleonora I, Stefano B (2005) Fertility in the mare after repeated transvaginal ultrasound-guided aspirations. Anim Reprod Sci 88(3–4):299–308. https://doi.org/10.1016/j.anireprosci.2005.01.002

Masuda M, Kuriki H, Komiyama Y, Nishikado H, Egawa H, Murata K (1987) Measurement of membrane fluidity of polymorphonuclear leukocytes by flow cytometry. J Immunol Methods 96(2):225–231

Mattioli M, Bacci ML, Galeati G, Seren E (1989) Developmental competence of pig oocytes matured and fertilized invitro. Theriogenology 31(6):1201–1207. https://doi.org/10.1016/0093-691x(89)90089-7

McLaughlin M, Telfer EE (2010) Oocyte development in bovine primordial follicles is promoted by activin and FSH within a two-step serum-free culture system. Reproduction 139(6):971–978. https://doi.org/10.1530/REP-10-0025

McPartlin LA, Suarez SS, Czaya CA, Hinrichs K, Bedford-Guaus SJ (2009) Hyperactivation of stallion sperm is required for successful in vitro fertilization of equine oocytes. Biol Reprod 81(1):199–206. https://doi.org/10.1095/biolreprod.108.074880

Menezo YJ, Herubel F (2002) Mouse and bovine models for human IVF. Reprod Biomed Online 4(2):170–175

Merton JS, Ask B, Onkundi DC, Mullaart E, Colenbrander B, Nielen M (2009) Genetic parameters for oocyte number and embryo production within a bovine ovum pick-up-in vitro production embryo-production program. Theriogenology 72(7):885–893. https://doi.org/10.1016/j.theriogenology.2009.06.003

Merton JS, de Roos AP, Mullaart E, de Ruigh L, Kaal L, Vos PL, Dieleman SJ (2003) Factors affecting oocyte quality and quantity in commercial application of embryo technologies in the cattle breeding industry. Theriogenology 59(2):651–674

Meuwissen TH, Hayes BJ, Goddard ME (2001) Prediction of total genetic value using genome-wide dense marker maps. Genetics 157(4):1819–1829

Mianne J, Codner GF, Caulder A, Fell R, Hutchison M, King R, Stewart ME, Wells S, Teboul L (2017) Analysing the outcome of CRISPR-aided genome editing in embryos: screening, genotyping and quality control. Methods 121-122:68–76. https://doi.org/10.1016/j.ymeth.2017.03.016

Michael H, Ana K, Dessie SW, Harald S, Christine W, Dawit T, Christiane N, Karl S, Eva HH (2017) Birth of healthy calves after intra-follicular transfer (IFOT) of slaughterhouse derived immature bovine oocytes. Theriogenology 97:41–49. https://doi.org/10.1016/j.theriogenology.2017.04.009

Miles JR, McDaneld TG, Wiedmann RT, Cushman RA, Echternkamp SE, Vallet JL, Smith TP (2012) MicroRNA expression profile in bovine cumulus-oocyte complexes: possible role of let-7 and miR-106a in the development of bovine oocytes. Anim Reprod Sci 130(1–2):16–26. https://doi.org/10.1016/j.anireprosci.2011.12.021

Minami N, Bavister BD, Iritani A (1988) Development of hamster two-cell embryos in the isolated mouse oviduct in organ culture system. Gamete Res 19(3):235–240. https://doi.org/10.1002/mrd.1120190303

Mitalipov S, Wolf DP (2014) Clinical and ethical implications of mitochondrial gene transfer. Trends Endocrinol Metab 25(1):5–7. https://doi.org/10.1016/j.tem.2013.09.001

Monniaux D (2016) Driving folliculogenesis by the oocyte-somatic cell dialog: lessons from genetic models. Theriogenology 86(1):41–53. https://doi.org/10.1016/j.theriogenology.2015.04.017

Monniaux D, Clement F, Dalbes-Tran R, Estienne A, Fabre S, Mansanet C, Monget P (2014) The ovarian reserve of primordial follicles and the dynamic reserve of antral growing follicles: what is the link? Biol Reprod 90(4):85. https://doi.org/10.1095/biolreprod.113.117077

Morrell JM, Richter J, Martinsson G, Stuhtmann G, Hoogewijs M, Roels K, Dalin AM (2014) Pregnancy rates after artificial insemination with cooled stallion spermatozoa either with or without single layer centrifugation. Theriogenology 82(8):1102–1105. https://doi.org/10.1016/j.theriogenology.2014.07.028

Mugnier S, Dell'Aquila ME, Pelaez J, Douet C, Ambruosi B, De Santis T, Lacalandra GM, Lebos C, Sizaret PY, Delaleu B, Monget P, Mermillod P, Magistrini M, Meyers SA, Goudet G (2009) New insights into the mechanisms of fertilization: comparison of the fertilization steps, composition, and structure of the Zona Pellucida between horses and pigs. Biol Reprod 81(5):856–870. https://doi.org/10.1095/biolreprod.109.077651

Munoz M, Uyar A, Correia E, Diez C, Fernandez-Gonzalez A, Caamano JN, Trigal B, Carrocera S, Seli E, Gomez E (2014) Non-invasive assessment of embryonic sex in cattle by metabolic fingerprinting of in vitro culture medium. Metabolomics 10(3):443–451. https://doi.org/10.1007/s11306-013-0587-9

Niemann H, Petersen B (2016) The production of multi-transgenic pigs: update and perspectives for xenotransplantation. Transgenic Res 25(3):361–374. https://doi.org/10.1007/s11248-016-9934-8

Niemann H, Wrenzycki C (2000) Alterations of expression of developmentally important genes in preimplantation bovine embryos by in vitro culture conditions: implications for subsequent development. Theriogenology 53(1):21–34

Oriol JG, Sharom FJ, Betteridge KJ (1993) Developmentally-regulated changes in the glycoproteins of the equine embryonic capsule. J Reprod Fertil 99(2):653–664

Palermo G, Joris H, Devroey P, Van Steirteghem AC (1992) Pregnancies after intracytoplasmic injection of single spermatozoon into an oocyte. Lancet 340(8810):17–18

Palmer E, Bezard J, Magistrini M, Duchamp G (1991) Invitro fertilization in the horse–a retrospective study. J Reprod Fertil 44:375–384

Paramio MT, Izquierdo D (2014) Current status of in vitro embryo production in sheep and goats. Reprod Domest Anim 49:37–48. https://doi.org/10.1111/rda.12334

Paramio MT, Izquierdo D (2016) Recent advances in in vitro embryo production in small ruminants. Theriogenology 86(1):152–159. https://doi.org/10.1016/j.theriogenology.2016.04.027

Parrish JJ (2014) Bovine in vitro fertilization: in vitro oocyte maturation and sperm capacitation with heparin. Theriogenology 81(1):67–73. https://doi.org/10.1016/j.theriogenology.2013.08.005

Parrish JJ, Susko-Parrish JL, Leibfried-Rutledge ML, Critser ES, Eyestone WH, First NL (1986) Bovine in vitro fertilization with frozen-thawed semen. Theriogenology 25(4):591–600

Pavani KC, Alminana C, Wydooghe E, Catteeuw M, Ramirez MA, Mermillod P, Rizos D, Van Soom A (2016) Emerging role of extracellular vesicles in communication of preimplantation embryos in vitro. Reprod Fertil Dev 29(1):66–83. https://doi.org/10.1071/RD16318

Petersen B, Niemann H (2015) Molecular scissors and their application in genetically modified farm animals. Transgenic Res 24(3):381–396. https://doi.org/10.1007/s11248-015-9862-z

Petters RM, Wells KD (1993) Culture of pig embryos. J Reprod Fertil 48:61–73

Pieterse MC, Kappen KA, Kruip TAM, Taverne MAM (1988) Aspiration of bovine oocytes during trans-vaginal ultrasound scanning of the ovaries. Theriogenology 30(4):751–762. https://doi.org/10.1016/0093-691x(88)90310-X

Pieterse MC, Vos PLAM, Kruip TAM, Wurth YA, Vanbeneden TH, Willemse AH, Taverne MAM (1991) Transvaginal ultrasound guided follicular aspiration of bovine oocytes. Theriogenology 35(4):857–861. https://doi.org/10.1016/0093-691x(91)90426-E

Ponsart C, Le Bourhis D, Knijn H, Fritz S, Guyader-Joly C, Otter T, Lacaze S, Charreaux F, Schibler L, Dupassieux D, Mullaart E (2014) Reproductive technologies and genomic selection in dairy cattle. Reprod Fert Develop 26(1):12–21. https://doi.org/10.1071/RD13328

Pratt SL, Calcatera SM (2016) Expression of microRNA in male reproductive tissues and their role in male fertility. Reprod Fertil Dev 29(1):24–31. https://doi.org/10.1071/RD16293

Presicce GA, Jiang S, Simkin M, Zhang L, Looney CR, Godke RA, Yang XZ (1997) Age and hormonal dependence of acquisition of oocyte competence for embryogenesis in prepubertal calves. Biol Reprod 56(2):386–392. https://doi.org/10.1095/biolreprod56.2.386

Pribenszky C, Nilselid AM, Montag M (2017) Time-lapse culture with morphokinetic embryo selection improves pregnancy and live birth chances and reduces early pregnancy loss: a meta-analysis. Reprod Biomed Online 35(5):511–520. https://doi.org/10.1016/j.rbmo.2017.06.022

Proudfoot C, Carlson DF, Huddart R, Long CR, Pryor JH, King TJ, Lillico SG, Mileham AJ, McLaren DG, Whitelaw CB, Fahrenkrug SC (2015) Genome edited sheep and cattle. Transgenic Res 24(1):147–153. https://doi.org/10.1007/s11248-014-9832-x

Raposo G, Stoorvogel W (2013) Extracellular vesicles: exosomes, microvesicles, and friends. J Cell Biol 200(4):373–383. https://doi.org/10.1083/jcb.201211138

Rath D (1992) Experiments to improve invitro fertilization techniques for invivo-matured porcine oocytes. Theriogenology 37(4):885–896. https://doi.org/10.1016/0093-691x(92)90050-2

Ratto M, Gomez C, Berland M, Adams GP (2007) Effect of ovarian superstimulation on COC collection and maturation in alpacas. Anim Reprod Sci 97(3–4):246–256. https://doi.org/10.1016/j.anireprosci.2006.02.002

Reznichenko AS, Huyser C, Pepper MS (2016) Mitochondrial transfer: implications for assisted reproductive technologies. Appl Transl Genom 11:40–47. https://doi.org/10.1016/j.atg.2016.10.001

Rizos D, Clemente M, Bermejo-Alvarez P, de La Fuente J, Lonergan P, Gutierrez-Adan A (2008) Consequences of in vitro culture conditions on embryo development and quality. Reprod Domest Anim 43(Suppl 4):44–50. https://doi.org/10.1111/j.1439-0531.2008.01230.x

Rizos D, Ramirez MA, Pintado B, Lonergan P, Gutierrez-Adan A (2010) Culture of bovine embryos in intermediate host oviducts with emphasis on the isolated mouse oviduct. Theriogenology 73(6):777–785. https://doi.org/10.1016/j.theriogenology.2009.10.001

Roberts R, Franks S, Hardy K (2002) Culture environment modulates maturation and metabolism of human oocytes. Hum Reprod 17(11):2950–2956. https://doi.org/10.1093/humrep/17.11.2950

Rocha JC, Passalia F, Matos FD, Maserati MP Jr, Alves MF, Almeida TG, Cardoso BI, Basso AC, Nogueira MF (2016) Methods for assessing the quality of mammalian embryos: how far we are from the gold standard? JBRA Assist Reprod 20(3):150–158. https://doi.org/10.5935/1518-0557.20160033

Romar R, Funahashi H, Coy P (2016) In vitro fertilization in pigs: new molecules and protocols to consider in the forthcoming years. Theriogenology 85(1):125–134. https://doi.org/10.1016/j.theriogenology.2015.07.017

Sanchez-Guijo A, Blaschka C, Hartmann MF, Wrenzycki C, Wudy SA (2016) Profiling of bile acids in bovine follicular fluid by fused-core-LC-MS/MS. J Steroid Biochem 162:117–125. https://doi.org/10.1016/j.jsbmb.2016.02.020

Santonocito M, Vento M, Guglielmino MR, Battaglia R, Wahlgren J, Ragusa M, Barbagallo D, Borzi P, Rizzari S, Maugeri M, Scollo P, Tatone C, Valadi H, Purrello M, Di Pietro C (2014) Molecular characterization of exosomes and their microRNA cargo in human follicular fluid: bioinformatic analysis reveals that exosomal microRNAs control pathways involved in follicular maturation. Fertil Steril 102(6):1751–1761 e1751. https://doi.org/10.1016/j.fertnstert.2014.08.005

Santos RR, Schoevers EJ, Roelen BA (2014) Usefulness of bovine and porcine IVM/IVF models for reproductive toxicology. Reprod Biol Endocrinol 12:117. https://doi.org/10.1186/1477-7827-12-117

Sartori R, Monteiro PLJ, Wiltbank MC (2016) Endocrine and metabolic differences between Bos taurus and Bos indicus cows and implications for reproductive management. Anim Reprod 13(3):168–181. https://doi.org/10.21451/1984-3143-AR868

Scott RT Jr, Treff NR (2010) Assessing the reproductive competence of individual embryos: a proposal for the validation of new "-omics" technologies. Fertil Steril 94(3):791–794. https://doi.org/10.1016/j.fertnstert.2010.03.041

Sekhavati MH, Shadanloo F, Hosseini MS, Tahmoorespur M, Nasiri MR, Hajian M, Nasr-Esfahani MH (2012) Improved bovine ICSI outcomes by sperm selected after combined heparin-glutathione treatment. Cell Reprogram 14(4):295–304. https://doi.org/10.1089/cell.2012.0014

Shahbazi MN, Jedrusik A, Vuoristo S, Recher G, Hupalowska A, Bolton V, Fogarty NNM, Campbell A, Devito L, Ilic D, Khalaf Y, Niakan KK, Fishel S, Zernicka-Goetz M (2016) Self-organization of the human embryo in the absence of maternal tissues. Nat Cell Biol 18(6):700–708. https://doi.org/10.1038/ncb3347

Shirazi A, Ostad-Hosseini S, Ahmadi E, Heidari B, Shams-Esfandabadi N (2009) In vitro developmental competence of ICSI-derived activated ovine embryos. Theriogenology 71(2):342–348. https://doi.org/10.1016/j.theriogenology.2008.07.027

Silva JR, van den Hurk R, Figueiredo JR (2016) Ovarian follicle development in vitro and oocyte competence: advances and challenges for farm animals. Domest Anim Endocrinol 55:123–135. https://doi.org/10.1016/j.domaniend.2015.12.006

da Silveira JC, Veeramachaneni DN, Winger QA, Carnevale EM, Bouma GJ (2012) Cell-secreted vesicles in equine ovarian follicular fluid contain miRNAs and proteins: a possible new form of cell communication within the ovarian follicle. Biol Reprod 86(3):71. https://doi.org/10.1095/biolreprod.111.093252

Sinclair KD, Lunn LA, Kwong WY, Wonnacott K, Linforth RST, Craigon J (2008) Amino acid and fatty acid composition of follicular fluid as predictors of in-vitro embryo development. Reprod Biomed Online 16(6):859–868

Sirard MA (2016) Somatic environment and germinal differentiation in antral follicle: the effect of FSH withdrawal and basal LH on oocyte competence acquisition in cattle. Theriogenology 86(1):54–61. https://doi.org/10.1016/j.theriogenology.2016.04.018

Sohel MM, Hoelker M, Noferesti SS, Salilew-Wondim D, Tholen E, Looft C, Rings F, Uddin MJ, Spencer TE, Schellander K, Tesfaye D (2013) Exosomal and non-exosomal transport of extracellular microRNAs in follicular fluid: implications for bovine oocyte developmental competence. PLoS One 8(11):e78505. https://doi.org/10.1371/journal.pone.0078505

Souza-Fabjan JMG, Locatelli Y, Duffard N, Corbin E, Touze JL, Perreau C, Beckers JF, Freitas VJF, Mermillod P (2014) In vitro embryo production in goats: slaughterhouse and laparoscopic ovum pick up-derived oocytes have different kinetics and requirements regarding maturation media. Theriogenology 81(8):1021–1031. https://doi.org/10.1016/j.theriogenology.2014.01.023

Spricigo JF, Sena Netto SB, Muterlle CV, Rodrigues Sde A, Leme LO, Guimaraes AL, Caixeta FM, Franco MM, Pivato I, Dode MA (2016) Intrafollicular transfer of fresh and vitrified immature bovine oocytes. Theriogenology 86(8):2054–2062. https://doi.org/10.1016/j.theriogenology.2016.07.003

Squires EL (1996) Maturation and fertilization of equine oocytes. Vet Clin N Am-Equine 12(1):31

Stock KF, Reents R (2013) Genomic selection: status in different species and challenges for breeding. Reprod Domest Anim 48(Suppl 1):2–10. https://doi.org/10.1111/rda.12201

Stringfellow DA, Givens MD (2000) Infectious agents in bovine embryo production: hazards and solutions. Theriogenology 53(1):85–94

Sturmey RG, Bermejo-Alvarez P, Gutierrez-Adan A, Rizos D, Leese HJ, Lonergan P (2010) Amino acid metabolism of bovine blastocysts: a biomarker of sex and viability. Mol Reprod Dev 77(3):285–296. https://doi.org/10.1002/mrd.21145

Sutton ML, Gilchrist RB, Thompson JG (2003) Effects of in-vivo and in-vitro environments on the metabolism of the cumulus-oocyte complex and its influence on oocyte developmental capacity. Hum Reprod Update 9(1):35–48. https://doi.org/10.1093/humupd/dmg009

Suzuki T, Asami M, Hoffmann M, Lu X, Guzvic M, Klein CA, Perry AC (2016) Mice produced by mitotic reprogramming of sperm injected into haploid parthenogenotes. Nat Commun 7:12676. https://doi.org/10.1038/ncomms12676

Tervit HR, Whittingham DG, Rowson LE (1972) Successful culture in vitro of sheep and cattle ova. J Reprod Fertil 30(3):493–497

Tesfaye D, Salilew-Wondim D, Gebremedhn S, Sohel MM, Pandey HO, Hoelker M, Schellander K (2016) Potential role of microRNAs in mammalian female fertility. Reprod Fertil Dev 29(1):8–23. https://doi.org/10.1071/RD16266

Thys M, Vandaele L, Morrell JM, Mestach J, Van Soom A, Hoogewijs M, Rodriguez-Martinez H (2009) In vitro fertilizing capacity of frozen-thawed bull spermatozoa selected by single-layer (Glycidoxypropyltrimethoxysilane) Silane-coated silica colloidal centrifugation. Reprod Domest Anim 44(3):390–394. https://doi.org/10.1111/j.1439-0531.2008.01081.x

Tibary A, Anouassi A, Khatir H (2005) Update on reproductive biotechnologies in small ruminants and camelids. Theriogenology 64(3):618–638. https://doi.org/10.1016/j.theriogenology.2005.05.016

Tibary A, Vaughan J (2006) Reproductive physiology and infertility in male south American camelids: a review and clinical observations. Small Ruminant Res 61(2–3):283–298. https://doi.org/10.1016/j.smallrumres.2005.07.018

Trasorras VL, Chaves MG, Miragaya MH, Pinto M, Rutter B, Flores M, Aguero A (2009) Effect of eCG superstimulation and buserelin on cumulus-oocyte complexes recovery and maturation in llamas (Lama glama). Reprod Domest Anim 44(3):359–364. https://doi.org/10.1111/j.1439-0531.2007.00972.x

Trasorras V, Chaves MG, Neild D, Gambarotta M, Aba M, Aguero A (2010) Embryo transfer technique: factors affecting the viability of the corpus luteum in llamas. Anim Reprod Sci 121(3–4):279–285. https://doi.org/10.1016/j.anireprosci.2010.06.004

Trasorras V, Giuliano S, Miragaya M (2013) In vitro production of embryos in south American camelids. Anim Reprod Sci 136(3):187–193. https://doi.org/10.1016/j.anireprosci.2012.10.009

Urrego R, Rodriguez-Osorio N, Niemann H (2014) Epigenetic disorders and altered gene expression after use of assisted reproductive technologies in domestic cattle. Epigenetics 9(6):803–815. https://doi.org/10.4161/epi.28711

Van Soom A, Mateusen B, Leroy J, De Kruif A (2003) Assessment of mammalian embryo quality: what can we learn from embryo morphology? Reprod Biomed Online 7(6):664–670

van Wagtendonk-de Leeuw AM, Mullaart E, de Roos APW, Merton JS, den Daas JHG, Kemp B, de Ruigh L (2000) Effects of different reproduction techniques: AI, MOET or IVP, on health and welfare of bovine offspring. Theriogenology 53(2):575–597. https://doi.org/10.1016/S0093-691x(99)00259-9

Vejlsted M, Du Y, Vajta G, Maddox-Hyttel P (2006) Post-hatching development of the porcine and bovine embryo—defining criteria for expected development in vivo and in vitro. Theriogenology 65(1):153–165. https://doi.org/10.1016/j.theriogenology.2005.09.021

Viana JHM, Siqueira LGB, Palhao MP, Camargo LSA (2012) Features and perspectives of the Brazilian in vitro embryo industry. Anim Reprod 9(1):12–18

Wetscher F, Havlicek V, Huber T, Gilles M, Tesfaye D, Griese J, Wimmers K, Schellander K, Muller M, Brem G, Besenfelder U (2005) Intrafallopian transfer of gametes and early stage embryos for in vivo culture in cattle. Theriogenology 64(1):30–40. https://doi.org/10.1016/j.theriogenology.2004.11.018

Wheeler MB, Rubessa M (2017) Integration of microfluidics in animal in vitro embryo production. Mol Hum Reprod 23(4):248–256. https://doi.org/10.1093/molehr/gaw048

Wheeler MB, Walters EM, Beebe DJ (2007) Toward culture of single gametes: the development of microfluidic platforms for assisted reproduction. Theriogenology 68:S178–S189. https://doi.org/10.1016/i.theriogenology.2007.04.042

Wrenzycki C, Herrmann D, Lucas-Hahn A, Korsawe K, Lemme E, Niemann H (2005) Messenger RNA expression patterns in bovine embryos derived from in vitro procedures and their implications for development. Reprod Fertil Dev 17(1–2):23–35

Wrenzycki C, Herrmann D, Niemann H (2007) Messenger RNA in oocytes and embryos in relation to embryo viability. Theriogenology 68:S77–S83. https://doi.org/10.1016/j.theriogenology.2007.04.028

Wrenzycki C, Stinshoff H (2013) Maturation environment and impact on subsequent developmental competence of bovine oocytes. Reprod Domest Anim 48:38–43. https://doi.org/10.1111/rda.12204

Young LE, Sinclair KD, Wilmut I (1998) Large offspring syndrome in cattle and sheep. Rev Reprod 3(3):155–163

Zeng F, Huang F, Guo J, Hu X, Liu C, Wang H (2015) Emerging methods to generate artificial germ cells from stem cells. Biol Reprod 92(4):89. https://doi.org/10.1095/biolreprod.114.124800

Zhang J, Liu H, Luo S, Lu Z, Chavez-Badiola A, Liu Z, Yang M, Merhi Z, Silber SJ, Munne S, Konstantinidis M, Wells D, Tan JJ, Huang T (2017) Live birth derived from oocyte spindle transfer to prevent mitochondrial disease. Reprod Biomed Online 34(4):361–368. https://doi.org/10.1016/j.rbmo.2017.01.013

Zhang J, Zhuang G, Zeng Y, Grifo J, Acosta C, Shu Y, Liu H (2016) Pregnancy derived from human zygote pronuclear transfer in a patient who had arrested embryos after IVF. Reprod Biomed Online 33(4):529–533. https://doi.org/10.1016/j.rbmo.2016.07.008

Zischewski J, Fischer R, Bortesi L (2017) Detection of on-target and off-target mutations generated by CRISPR/Cas9 and other sequence-specific nucleases. Biotechnol Adv 35(1):95–104. https://doi.org/10.1016/j.biotechadv.2016.12.003

The manufacturer's authorised representative in the EU is Springer
Nature Customer Service Centre GmbH, Europaplatz 3, 69115 Heidelberg,
Germany. If you have any concerns regarding our products, please
contact ProductSafety@springernature.com

Printed and bound by CPI Group (UK) Ltd, Croydon, CR0 4YY
29/04/2026
02099550-0001